Digital Geometry
in Image Processing

IIT Kharagpur Research Monograph Series

Published Titles:

Digital Geometry in Image Processing, *Jayanta Mukhopadhyay, Partha Pratim Das, Samiran Chattopadhyay, Partha Bhowmick, and Biswa Nath Chatterji*

Mathematical Techniques for Wave Interaction with Flexible Structures, *Trilochan Sahoo*

Microfluidics and Microscale Transport Processes, *edited by Suman Chakraborty*

Modeling of Responsive Supply Chain, *M.K. Tiwari, B. Mahanty, S. P. Sarmah, and M. Jenamani*

Micellar Enhanced Ultrafiltration: Fundamentals & Applications, *Sirshendu De and Sourav Mondal*

IIT KHARAGPUR RESEARCH MONOGRAPH SERIES

Digital Geometry
in Image Processing

Jayanta Mukhopadhyay
Partha Bhowmick
Partha Pratim Das
Samiran Chattopadhyay
Biswa Nath Chatterji

CRC Press
Taylor & Francis Group
Boca Raton London New York

CRC Press is an imprint of the
Taylor & Francis Group, an **informa** business

CRC Press
Taylor & Francis Group
6000 Broken Sound Parkway NW, Suite 300
Boca Raton, FL 33487-2742

First issued in paperback 2019

ISBN-13: 978-1-4665-0567-4 (hbk)
ISBN-13: 978-0-367-38021-2 (pbk)

Library of Congress Cataloging-in-Publication Data

Mukhopadhyaya, Jayanta.
    Digital geometry in image processing / authors, Jayanta Mukhopadhyay [and six others].
        pages cm -- (Iit kharagpur research monograph series)
        Includes bibliographical references and index.
        ISBN 978-1-4665-0567-4 (hardback)
        1. Image processing--Digital techniques--Mathematics. 2. Geometry--Data processing. I. Title.

    TA1637.M8348 2013
    006.601'516--dc23   2012050939

Contents

3 Digitization of Straight Lines and Planes

Series Preface

About the Series

IIT Kharagpur had been a forerunner in research publications and this monograph series is a natural culmination. Empowered with vast experience of over sixty years, the faculty now gets together with their glorious alumni to present bibles of information under the *IIT Kharagpur Research Monograph Series*.

Initiated during the Diamond Jubilee Year of the Institute, the Series aims at collating research and developments in various branches of science and engineering in a coherent manner. The Series which will be an ongoing endeavour is expected to be a source reference to fundamental research as well as to provide directions to young researchers. The presentations are in a format that these can serve as stand alone texts or reference books.

> *The specific objective of this research monograph series is to encourage the eminent faculty and coveted alumni to spread and share knowledge and information to the global community for the betterment of mankind*

The Institute

Indian Institute of Technology Kharagpur is one of the pioneering Technological Institutes in India and it is the first of its kind to be established immediately after the independence of India. It was founded on August 18, 1951, at Hijli, Kharagpur, West Bengal, India. The IIT Kharagpur has the largest campus of all IITs, with an area of 2,100 acres. At present, it has 34 departments, centers and schools and about 10,000 undergraduate, postgraduate and research students with faculty strength of nearly 600; the number of faculty is expected to double within approximately five years. The faculty and the alumni of IIT Kharagpur are having wide global exposures with the advances in science and engineering. The experience and the contributions of the faculty, students and the alumni are expected to gain exposure through this monograph series.

More on IIT Kharagpur is available at www.iitkgp.ac.in

Preface

In a digital image, each element has an integral coordinate position and has a finite set of points in its neighborhood. On the other hand, a point in a Euclidean space has an infinite number of neighboring points. This indicates that the geometry in the digital image space is non-Euclidean. One must study this geometry and its approximation to the Euclidean world, to correlate the measurements and shape of geometric objects with our common notion of Euclidean geometry. The geometry in the digital space is called digital geometry. However, one should note that this geometry is not unique. In different ways, digital geometry may be defined depending upon the neighborhood definition of a point in the space or the distance functions used for the same purpose. In this book, we discuss different digital geometries in multi-dimensional integral coordinate spaces, and also study some of their interesting properties, including their metric and topological properties, shapes of circles (for 2-D space) and spheres (for 3-D space), proximity to Euclidean norms, and a number of theoretic representations of different geometric objects such as straight lines, circles, etc. We will demonstrate how these concepts and properties are useful in different techniques for image processing and analysis. In particular, their applications in object representation and shape analysis are extensively covered in different chapters of this book.

At IIT Kharagpur, there has always been a strong research group working in this area for about the last three decades. Though there exist some good texts and reference books covering the topics of image processing and digital geometry in general, it is felt by this group of authors that there is a need to connect these two topics highlighting the important results of digital geometry which are used in image analysis and processing. This book is the outcome of that endeavor. The project was also initiated in the diamond jubilee year of this prestigious institute as a part of publishing a series of monographs highlighting the continuing research and development activities in different areas. However, while writing this book we treated all the topics comprehensively elucidating the important and significant results of other researchers. Nevertheless in all the topics, as there is a significant research contribution by the authors themselves, it adds more depth and clarity in their presentation. We hope the book will be useful to the researchers in these areas, and also in teaching advanced topics in image processing and digital geometry to graduate and research students.

There are seven chapters in this book. In Chapter 1, fundamentals related to digital topology are discussed. It includes an introduction to the concept of digital grid and connectivity in a grid. Different measures and operations in digital topology are also discussed.

Chapter 2 introduces fundamental concepts of neighborhoods and paths in a metric space. It also elucidates different types of digital metrics, which are formed from various definitions of the digital neighborhood. The distances covered in this book include distances from the definitions of different types of topological adjacency (m-Neighbors), varying costs among the neighbors (t-cost and weighted distances), path-dependent neighborhoods (generalized octagonal distances), and combinations of distance functions (weighted t-cost distance function). Properties of hyperspheres in respective metric spaces are also discussed in this chapter, and finally, proximity of these distance functions to respective Euclidean metrics is analyzed.

Chapter 3 covers digitization of straight lines and planes. It introduces different digitization schemes and number theoretic characterization of segments of digital straight lines and planes from a set of discrete points in respective dimensions. An iterative technique for estimating parameters of these geometric segments is illustrated in this chapter, which is later generalized for estimation of parametric curves in Chapter 5. Number theoretic characterization is also used for estimating lengths and areas of straight-line and planar segments, respectively.

The concept of digital straightness is further elaborated in Chapter 4. In this case, the degree of approximation is also quantified and used for polygonal approximation of digital contours.

Chapter 5 considers the topic of parametric curve estimation and reconstruction. It generalizes the technique of iterative refinement of estimates discussed in Chapter 3. Using number theoretic representation of digital conics, it also provides measures for estimating their lengths.

Medial Axis Transform (MAT), which represents an object by a set of overlapping maximal disks contained in the object, is discussed in Chapter 6. Various techniques for computing MAT using digital distance functions are discussed in this chapter. This is followed by illustration of applications of MAT for processing 2-D and 3-D binary images. They include thinning of binary patterns, geometric transformation, computation of boundary normals of 2-D objects, computation of cross-sections of 3-D objects, etc.

The last chapter (Chapter 7) of this book discusses digital geometric approaches for modeling surfaces. In 3-D digital space, a surface consists of voxels having integer coordinates. Most of the algorithms, on the contrary, work in the real/Euclidean space from which the mapping to digital space is nontrivial. This chapter discusses a few problems on voxelation of surface of revolution, its significance, the existing procedures, and a digital-geometric approach to construct the voxel set. It also presents a technique for computing isothetic polyhedral cover of a digital point set in a multi-resolution grid.

As the book is written by several authors, it was a stupendous task to manage the uniformity of presentation and cross-referencing of different concepts. We believe there are improvements that could be made in this text. However, due to the time constraint, we had to settle for this version to meet the deadline and other commitments. We would greatly appreciate receiving any comments from readers concerning any errors encountered in this book.

While working in this area, the authors worked with various distinguished researchers, and the outcome of this book would not have been possible without their support and contribution. The authors gratefully acknowledge their contribution. In particular, Jayanta Mukhopadhyay, P. P. Das, and B. N. Chatterji acknowledge the contribution made by Prof. Ashwatha Kumar of M.S. Ramaiah Institute of Technology, Bangalore, for his involvement in the development of MAT-based image processing algorithms. We also acknowledge Prof. P. P. Chakrabarti, IIT Kharagpur for his significant contribution in the development of theory of digital distances in the initial years of our research engagement. The algorithm of finding straight edges in a gray-scale image (Chapter 4) has been contributed by Mr. Sanjoy Pratihar. The theories and techniques discussed in Section 7.1 of Chapter 7 are based on collaborative research by P. Bhowmick with Ms. Nilanjana Karmakar, Dr. Arindam Biswas, and Prof. Bhargab B. Bhattacharya. We sincerely acknowledge them all for their contributions, a part of which is discussed in that chapter. The authors also acknowledge Prof. P. K. Biswas and Prof. P. K. Dutta of IIT, Kharagpur for their support and encouragement in the planning and execution of this project. We also thank Prof. D. Acharya, Director, IIT Kharagpur, who mooted the idea of publishing this type of book as a part of the Diamond Jubilee celebration of the Institute. We deeply appreciate the support and encouragement provided by Ms. Aastha Sharma and Ms. Laurie Schlags of the CRC Press toward its completion.

Jayanta Mukhopadhyay,
P. P. Das,
Samiran Chattopadhyay,
Partha Bhowmick, and B. N. Chatterji

15 July, 2012

List of Figures

List of Tables

Symbol Description

Sets and Spaces

$\mathbb{R}$ — The set of all real numbers.

$\mathbb{R}^+$ — The set of positive real numbers.

$\mathbb{Z}$ — The set of all integers.

$\mathbb{N}$ — The set of all non-negative integers.

$\mathbb{P}$ — The set of positive integers.

$\mathbb{Q}$ — The set of rational numbers.

$\mathbb{Z}_N$ — $\{0, 1, 2, \ldots, N - 1\}$

$\mathbb{C}$ — The set of all complex numbers.

$L^2(\mathbb{R})$ — The space of all square integrable functions.

$L^2(\mathbb{Z})$ — The space of all square integrable functions over the integer grid.

$[a, b]$ — $\{x | x \in \mathbb{R} \text{ and } a \le x \le b\}$

$(a, b]$ — Left open, right closed interval, i.e. $\{x | a < x \text{ and } x \le b\}$.

$\underline{u}$ (or $\mathbf{u}$) — $\underline{u}$ (or $\mathbf{u}$) is a point in $\mathbb{Z}^n$ or $\mathbb{R}^n$ and also denoted as $\underline{u} = (u(1), u(2), \ldots, u(n))$ (or $\underline{u} = (u_1, u_2, \ldots, u_n)$).

$|\underline{u}|$ — Norm of $\underline{u}$.

$\underline{0}$ — $(0, 0, \ldots, 0)$. The dimension n is implicit.

$\underline{k}$ — $(k, k, \ldots, k)$. The dimension n is implicit.

$\vec{x}$ — $\vec{x}$ is a vector in $\mathbb{R}^n$ and also denoted as $\vec{x} = (x(1), x(2), \ldots, x(n))$.

$|\vec{x}|$ — The norm of $\vec{x}$.

Functions, Sequences, Operators, and Symbols

$\langle h, g \rangle$ — The inner product of two functions $h(x)$ and $g(x)$.

$\vec{h} . \vec{g}$ — The dot product of two vectors $\vec{h}$ and $\vec{g}$.

$\vec{h} \times \vec{g}$ — The cross product of two vectors $\vec{h}$ and $\vec{g}$.

$a + jb$ — A complex number with j as $\sqrt{-1}$.

x^* — The complex conjugate of $x \in \mathbb{C}$.

$|x|$ — The magnitude of $x \in \mathbb{C}$.

$\angle x$ — The phase of $x \in \mathbb{C}$.

$< x >_N$ — $x \bmod N$ for $x \in \mathbb{Z}$.

$|x|$ — The absolute value of $x \in \mathbb{R}$.

$\lceil x \rceil$ — Ceiling of x, i.e., the smallest integer greater than or equal to $x \in \mathbb{R}$.

$\lfloor x \rfloor$ — Floor of x i.e., the greatest integer smaller than or equal to $x \in \mathbb{R}$.

$sign(x)$ — -1, 0, and 1 depending on the sign of $x \in \mathbb{R}$.

$round(x)$ — The nearest integer approximation of $x \in \mathbb{R}$.

$\overline{w(n)}$ — The conjugate reflection of $w(n) \in \mathbb{C}$.

$||\vec{x}||$ — The magnitude of the vector $\vec{x}$.

$|u_j|$ — jth maximum magnitude of components of $\underline{u}$.—

$\begin{pmatrix} P \\ Q \end{pmatrix}$ — P choose Q.

$\sum_{i=a}^{b} x_i$ — $x_a + x_{a+1} + \ldots + x_b$, $a, b \in \mathbb{Z}$ and $b > a$. Alternate notations are $\sum_{a \le i \le b} x_i$, or $\sum_{i \in [a, b]} x_i$.

$\prod_{i=a}^{b} x_i$ — $x_a x_{a+1} \ldots x_b$, $a, b \in \mathbb{Z}$ and $b > a$. Alternate notations are $\prod_{a \le i \le b} x_i$, or $\prod_{i \in [a, b]} x_i$.

Matrices and Operators

X^T — The transpose of matrix X.

X^H — The Hermitian transpose of matrix X.

X^{-1} — The inverse of matrix X.

Chapter 1

Digital Topology: Fundamentals

The conventional geometry of our surrounding space is Euclidean. Strictly speaking, it is the geometry in a three-dimensional Euclidean space. If we restrict the geometry in a plane (e.g. the floor of a building) it turns out to be two dimensional Euclidean space. Again if a person is conservative enough to take into account of the curvature of earth, the 2-D planar floor is not strictly the Euclidean one. One may approximate it more accurately to the Riemannian space, which consists of the points lying on a spherical surface and the distances between two points are computed by the length of the arc defined by the circle with center and radius that are the same as those of the sphere.

A digital image is usually defined in two dimensions. Each element or pixel of the image has integral coordinate positions, contrary to the respective 2-D Euclidean space, which is continuous and has infinite points in the neighborhood of any point in the space. This indicates that the geometry in the image space is a non-Euclidean one. One must study this geometry and its approximation to the Euclidean world. In general, the digital image space is referred as the digital space. In this space, the points are represented by integral coordinate positions (for example, by row and column numbers). Hence,

1

the geometry in this space is called digital geometry. However, we should note that this geometry is not unique. In different ways, digital geometry may be defined depending upon the neighborhood definition of a point in the space. Digital images from different imaging technologies such as CT Scan, MRI, PET, etc. are also available in 3-D integral coordinate space. Hence, it is also necessary to understand the geometry in such a 3-D space. This book addresses several aspects of digital geometry, such as different metric spaces and shapes of hyperspheres in those spaces, analysis of discrete curves, straight lines and surfaces, Medial Axis Transform (MAT), etc. It also discusses how these concepts are used in developing image processing algorithms. In this chapter, we provide a brief introduction to digital topology and related concepts. This will help a beginner to understand the subject matter in subsequent chapters.

In digital topology, we study the topological properties in a digital space. In particular, we study these properties with respect to relationships among the pixels, and voxels of binary images in 2-D and 3-D, respectively. Several image processing operations, such as connected component labeling, contour following, thinning, etc., are defined using the concepts of digital topology. Various models, such as graph theoretic modeling [177], abstract cellular complex [121], etc., are proposed in studying the topology in a discrete space. In this text, we restrict our discussion to the topology modeled by a graphical representation over a rectangular lattice (or grid) in Euclidean space.

1.1 Tessellation of a Continuous Space

A *tessellation* is an aggregate of cells that covers the continuous space without overlapping. The dimension of each unit cell is the same as that of the space. A tessellation provides the digitization of space, so that each cell is represented by a unique tuple of integers. Usually, a cell in a tessellation is restricted to be a convex set, and there is a homogeneity in the occurrences of cells of identical shapes. For example, in a 2-D plane, every vertex of a *homogeneous tessellation* shares r regular polygons of sides $a_1, a_2, \ldots, a_r$. Such a tessellation is denoted by *the crystallographic set* $\{a_1, a_2, \ldots, a_r\}$. There are only 11 different homogeneous tessellations of the plane, called *Archimedean Tilings*. In only three, all the cells are uniform. We refer to them as *regular tessellations*. They are shown in Fig. 1.1, namely triangular (Fig. 1.1(a)), rectangular (Fig. 1.1(b)), and hexagonal tessellations (Fig. 1.1(b)).

Similarly, in 3-D we define a regular tessellation, in which the space is partitioned by a regular polyhedron. However, in contrast to 2-D, there is only one type of regular tessellation in 3-D, which is composed of cubic cells, a three-dimensional form of rectangular tessellation. In higher dimensions too, rectangular tessellations are regular. In fact, except in 4-D, in all other dimensions rectangular tessellation is the only form of regular tessellation. In

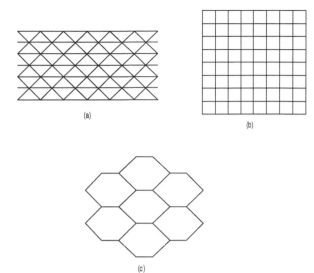

FIGURE 1.1: Regular tessellations in 2-D: (a) triangular, (b) rectangular, and (c) hexagonal.

this book, we restrict out discussion to the geometry of digital space, which is obtained through rectangular tessellation in the corresponding continuous space.

1.2 Digital Grid

A *digital grid* is a finite subset of integral coordinate space, so that it can be represented by a rectangular array of respective dimension. Let us represent a digital grid in n-dimension as G^n. For example, a set of points $G^2 = \{(i,j) \in \mathbb{Z}^2 | i = 0, 1, 2, \ldots, M - 1; j = 0, 1, 2, \ldots, N - 1\}$ is a digital grid in 2-D ($\mathbb{Z}^2$) of size $M \times N$. A point in a 2-D digital grid is called a *pixel*. Similarly, a 3-D digital grid element is called a *voxel*. We define the border of G^n as the set of points whose coordinate values of one of the dimensions are either 0 (the least value) or the maximum. The border of G^2 in the previous example is $B(G^2) = \{(i,j) | (i = 0) \text{ or } (i = M - 1) \text{ or } (j = 0) \text{ or } (j = N - 1)\}$.

Two pixels are said to be *8 -adjacent*, if they are distinct, and any one of their coordinates (or both) differs (or differ) by 1. An 8-adjacent pixel of $p \in G^2$ is called its *8-neighbor*. The coordinates of a 4-adjacent pixel of p differ by 1 only in one of their dimensions and the corresponding pixel is named its *4-neighbor*. In 3-D we have similar definitions of 6, 18, and 26 adjacency. A voxel is 6 adjacent of v, if its coordinate differs from v by 1 at only one of the dimensions. It is 18 -adjacent when it is 6 adjacent or the ordinal values in two of its coordinate dimensions differ by 1, and it is 26 adjacent, when it is 18 adjacent or all the coordinate values differ by 1 from v. They are also called 6, 18, and 26 *neighbors* of v, respectively. Let us denote the set of neighbors of a point p in a digital grid as either $N(p)$ in general or $N_k(p)$, where $k \in \{4, 8\}$ in 2-D and $k \in \{6, 18, 26\}$ in 3-D, to denote a specific type of neighbor.

A binary image in n-D is defined as a function I such that $I : G^n \rightarrow \{0, 1\}$. Let us call the grid points with the value 1 as the *object* or *foreground* *points*, and points with the value 0 the *background points*. In particular in 2-D we denote the image $P : G^2 \rightarrow \{0, 1\}$, partitioned into its foreground and background pixels. In 3-D we represent it as $V : G^3 \rightarrow \{0, 1\}$. Let us assume that the border of G^n always contains background (0) points.

Using the adjacency relationship, we define a few properties of a set of grid points from [177], and [118] as follows: A *path* between two points a and b in a digital grid is k-*adjacent* if there exists a sequence of points $p_0(= a), p_1, p_2, \ldots, p_l(= b)$, $l \geq 1$, such that every two consecutive points p_i, p_{i+1}, $0 \leq i \leq l - 1$ are k-adjacent. Two sets of points S_1 and S_2 are k-*connected*, if there exists a pair of points $p \in S_1$, and $q \in S_2$, such that p and q are k-adjacent. A set of grid points S is k-connected, if it cannot be partitioned into two subsets that are not k-adjacent to each other. A k-component of a set of grid points S is a non-empty k-connected subset of S that is not k-adjacent to any other point in S.

1.2.1   Jordan's Theorem on Closed Curves

Let us discuss how Jordan's theorem for a *simple closed curve* in the Euclidean plane gets extended in a digital grid. The theorem states that [118], *the simple closed curve separates the remainder of the plane into two connected components, one of which is called inside and the other is denoted as the outside of the curve.* It implies that removal of any point from the curve would make the remainder of the plane connected.

Let us define a curve in a digital grid using the notion of adjacency relationship between a pair of neighboring points. A path in S is called a *simple arc* if all but two of its points (its *end points*) have exactly two neighbors in S, while those two end points have exactly one, and a *simple closed curve* is a path in S when all its points have exactly two neighbors in S. In a 2-D digital grid too, a simple closed curve (with foreground pixels) partitions the remaining background into two connected components. Hence, Jordan's theorem is equally applicable here. However, the definitions of connectivity differ for the background and foreground pixels in this case. This we elaborate with the help of an example as shown in Fig. 1.2.

Consider a background pixel (white pixel) p enclosed by four foreground pixels, which form a curve if they are 8-adjacent. However, the number of 8-connected components of background is also 1, which includes the point p

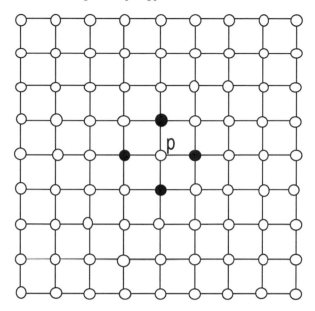

FIGURE 1.2: Partitioning by an 8-connected curve in a digital grid.

as well. This violates the notion of partitioning as discussed in Jordan's curve theorem for a 2-D plane. On the other hand, if we use the definition of 4-adjacency, though there are two 4-connected components in the background, the foreground (of black pixels) consists of 4 isolated points, which do not form a curve. To resolve this paradox of connectivity, Rosenfeld [177] proposed to use two different types of connectivity for foreground and background, i.e., either 4-connectivity for foreground and 8-connectivity for background, or the reverse. With this notion of connectivity, the grid topology is distinguished.

1.3 Grid Topology

Following the notations of [118], we denote a grid topology for 2-D and 3-D binary images by a quadruple (U, m, n, S), where U could be either G^2 or G^3; $(m, n) \in \{(4, 8), (8, 4)\}$ if U is G^2, else for G^3 it is an element of $\{(6, 18), (18, 6), (6, 26), (26, 6)\}$; and S is the set of foreground points in U. The type of adjacency used for defining neighbors in S is denoted by m, and n is the corresponding type for the background points, i.e., for $\bar{S} = U - S$. The quadruple grid topology is also referred to here as a *digital picture*. We also denote the border of U as $B(U)$, and the connected components of background

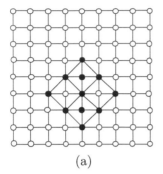   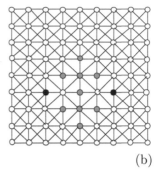

          (a)   (b)

FIGURE 1.3: Topological configurations on the same set of foreground pixels with (a) (8,4), and (b) (4,8). (See color insert.)

points containing the border, the *infinite background component*. So, any finite connected background component is called a *hole* in 2-D (or a *cavity* in 3-D).

For the same set of points, how the topological configuration varies on account of different pairs of adjacency types is shown in Figs. 1.3 (a) and (b), respectively. In the configuration with $(8,4)$ connectivity, there is only one component of foreground pixels containing two holes (refer to Fig. 1.3 (a)), whereas with $(4,8)$ connectivity there are three components, the largest of them is shown with orange color, and the remaining two with black. There is no hole in the latter (refer to Fig. 1.3 (b)).

1.3.1   Component Labeling

In various applications, the distinct components in an image must be determines. This computation is called *component labeling*. A component could be obtained from a point p by accumulating the set of points *connected* to it. This may be achieved by recursively visiting a neighboring point and labeling the visited point with the component number. The recursion terminates if there exists no unvisited neighboring point. The algorithm primarily employs the strategy of the depth first search (DFS) in a graph for obtaining its connected component from a vertex. Though the algorithm runs in linear time complexity, it requires a large stack size, and because the computing environment may not permit stack building beyond a certain limit, it is not suitable to handle large images.

There exists an efficient computation, which is performed by scanning the image twice (for 2-D images), namely, the forward scanning (from left to right and top to bottom) and the reverse scanning (from right to left and bottom to top). In each scan, a separate mask identifying the adjacent points in the grid is used. In Figs. 1.4 and 1.5, masks for forward and reverse scans for labeling connected components in grids $(4,8)$, and $(8,4)$ are shown. In these figures,

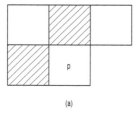

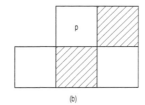

FIGURE 1.4: Masks for 4-connected component labeling: (a) during forward scan, and (b) during reverse scan.

the neighboring points, which are to be checked for labeling are shown with shaded boxes around the point p. Using these masks the labels of the visited neighbors for a foreground point, if any, are checked and decisions are taken as follows:

1. If none of the visited neighbors is a foreground point, and if the point is unlabeled, a unique label (a non-zero positive integer value) is assigned to p.

2. The distinct labels of its visited neighbors (including the point itself) are declared as equivalent, and the minimum of them is assigned to the point. In a subsequent scan, all equivalent labels are renumbered with the minimum of the set.

At the end of the two scans, a candidate label is chosen from each equivalent set of labels, and each component of the image gets a unique identification label. A similar algorithm exists in 3-D.

1.3.2 Containment

Let us define the property of containment between two sets of points in a digital picture. Let X and Y be two sets of points in (U, m, n, S), such that X is connected. In that case, X surrounds Y, if each point in Y is contained in a finite component of $U - X$. The relation is described as X *surrounds* Y. It is an asymmetric and transitive relation, and it induces a partial order on the connected subsets of U. In a finite digital picture, the *infinite background component* surrounds the set of all foreground points. Consider an example as shown in Fig. 1.6. The black pixels belong to the foreground, whereas the white pixels belong to the background. Let us consider a set of pixels (colored orange in the figure). This set is surrounded by the connected set of foreground pixels. We should note that in this example the corresponding pair of adjacency types is given by $(8, 4)$.

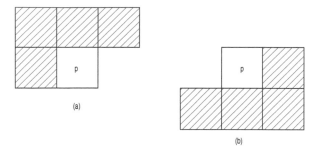

FIGURE 1.5: Masks for 8-connected component labeling: (a) during forward scan, and (b) during reverse scan.

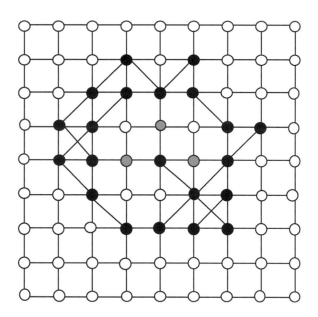

FIGURE 1.6: A connected component of foreground pixels surrounds orange pixels in background in this $(8, 4)$ digital grid. (See color insert.)

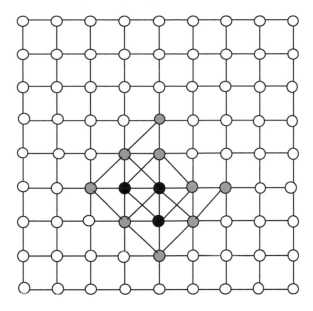

FIGURE 1.7: The border (orange pixels) and interior (black pixels) of a connected component of foreground pixels in an $(8, 4)$ digital grid. (See color insert.)

By using the 'surround' relationship, we can provide a more precise definition of a hole or cavity. In a 2-D digital picture, a finite background component that is adjacent to and surrounded by a foreground component X is called a *hole* in X. For a three-dimensional digital picture, such a background component is called a *cavity*. In Fig. 1.6, the finite background component containing those orange pixels is a *hole* in the set of foreground pixels forming the connected component.

1.3.3   Boundary and Interior

In a digital picture, a foreground point is said to be *isolated*, if it has no foreground neighbor. For example, in Fig. 1.3(b), two black foreground points are isolated. We should note that in that figure the orange pixels are also part of the foreground pixels and form a connected component using 4-adjacency. We call a foreground point a *border point*, if it is *n-adjacent* to one or more background points in a digital grid of (m, n) type of adjacency; otherwise, it is an *interior point*. The *boundary* or *border* of a connected component of foreground pixels X is formed by the set of all border points in X, and similarly, its interior is formed by all interior points belonging to it. In Fig. 1.7, the border and interior of a connected foreground component are shown by orange and black pixels, respectively.

1.3.3.1   Contour Tracing

The contour of a connected component is defined as the connected sequence of border points, so that the starting point and end point are neighbors and no point is revisited in the sequence. In this case, a contour is *a simple closed curve* in the digital grid, as defined previously. Such a contour is called a *simple contour*. However, a contour may not always be simple; there may be a border point having more than two neighbors. Given a simple contour and an object point on it, we discuss an algorithm [162] in an $(8,4)$ digital grid to trace the sequence. The task is also called *contour following*. Let p be a border point and q be its neighbor. We take this pair (p,q) as the starting configuration of the contour and move forward to obtain succeeding pairs in it. To get the next border point in the contour, we perform a clock-wise search among the 8-neighbors starting from q. In Fig. 1.8 (a), the sequence of points to be searched is shown as $r_0(=q), r_1, \ldots, r_7$. Suppose that r_i the next immediate object point occurs at. In that case, r_i becomes the next point in the sequence and r_{i-1} is the corresponding neighboring background point of r_i. We would repeat the same process with the pair of (r_i, r_{i-1}) and it continues till we reach at p again. A typical example of a contour-trace is shown in Fig. 1.8 (b). In the figure, the path is shown by red arrows. The staring pair of points is taken at (p_0, q_0) and it continues till we get $p_9 = p_0$. It may be noted that a background border point $(q_3 = q_4)$ occurs twice in the trace. The contour may also be traced counter-clockwise if the order of search for a neighboring border point (refer to Fig. 1.8(a)) is made counter-clockwise.

There could be multiple contours in a picture. In that case, we perform contour tracing after obtaining each unflagged border point. Once a contour is obtained, all the points in it are flagged. However, the above algorithm fails for non-simple contours. Interested readers may refer to work reported in [196, 171] for the details about techniques for tracing complex contours.

There are various applications of contour tracing. Contours can be further approximated by simple polygons (with many fever vertices). Polygons are useful in representing shapes and other information. Using the sequence of points, it is possible to compute various geometric features such as normals, curvatures, etc., at any point of the contour.

1.3.3.2   Chain Code

A contour or an arc in a 2-D digital grid can be encoded by a sequence of discrete orientations. The neighboring points [86] may be considered connected by line segments in a digital arc, which is on a fixed grid of possible orientations. There are 8 or 4 different orientations depending on the neighborhood definitions as shown in Figs. 1.9 (a) and 1.10 (a). Freeman [86] used this directional information in encoding the sequence of points (paths and contours). The respective code is called a *chain code*. Here, the starting point is explicitly stored and the subsequent points of the curve are represented by encoding successive displacements from the preceding points. Contour trac-

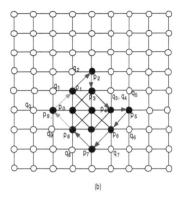

(a)

(b)

FIGURE 1.8: (a) The order of searching a foreground pixel in the neighborhood of a border pixel at p with a background neighbor at q in an $(8,4)$ digital grid. The order follows clockwise movement starting from q. (b) The sequence of pairs of border pixels (p_i, q_i), where p_i belongs to the foreground, and q_i belongs to background, respectively, for the point set as shown in Fig. 1.7. (See color insert.)

ing or path following is required to obtain the chain code of a sequence of border points. The advantage of this scheme is in its linear representation. Using chain codes various local and global properties of digital curves can be derived. In Chapter 3 and 4 we discuss its use in characterization of a digital straight line segment and estimation of its length. The drawback of representing a curve using chain code is that it is sometimes quite long and sensitive to small disturbances. It is also difficult to perform set operations like union and intersection using the coded data.

1.3.3.3 Neighborhood Plane Sct (NPS)

Like 2-D grids, in 3-D also, discrete orientations of primitive planar patches are used to encode the local surface features. In [148] in a $(26,6)$ 3-D digital grid 9 neighborhood planes around a point p are identified, whose normals are in the directions of one of its 18-neighbors. These planes are called *digital neighborhood planes* (DNP) of p. They are illustrated in Fig. 1.11.

The neighboring condition of p determines the plane in which p lies. Let P_i be the set of points assigned to the i-th neighboring plane as described in Figure 1.11. In a digital picture, $P = (G^3, 18, 6, O)$, the *neighborhood plane set (NPS)* of a point p belonging to an object O, is defined as:

$$[p]_k = \{i \mid |N_{18}(p) \cap O \cap P_i| > k)\}, \ k \geq 3.$$

Here, $|S|$ defines the cardinality of set S. The value of k usually lies between 3 and 5. It may be noted that in the above, $N(p)$ denotes the set of neighboring points of p. The NPS of a 3-D point refers to the set of DNPs that have a

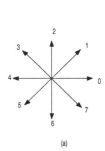

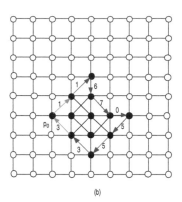

(a)   (b)

FIGURE 1.9: (a) Codes of discrete orientations in an $(8,4)$ 2-D grid, and (b) chain code of a contour starting from $p_0 \equiv 116705533$. (See color insert.)

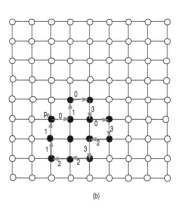

(a)   (b)

FIGURE 1.10: (a) Codes of discrete orientations in a $(4,8)$ 2-D grid, and (b) chain code of a contour starting from $p_0 \equiv 010303232211$. (See color insert.)

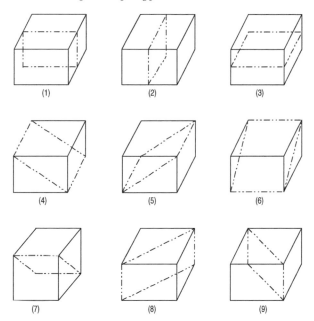

FIGURE 1.11: Digital neighborhood planes in an $(18, 6)$ 3-D grid.

sufficient number of points around p in the image. We can simply denote the *NPS* of p as $[p]$ considering k remains constant in an application. This definition is also extended [148] for a set of points $P = \{p_1, p_2, ..., p_n\}$ as follows:

$$[P] = [p_1] \cap [p_2] \cap ..... \cap [p_n]$$

The following two properties of the NPS are easily obtained from the above definition. These are used in segmenting and characterizing 3-D surfaces [148, 149, 19].

Lemma 1.1. *if $M = P \cup Q$, then $[M] = [P] \cap [Q]$.* □

Lemma 1.2. *If $Q \subset P$, then $[P] \subseteq [Q]$.* □

In [18] four more additional DNPs are identified, each corresponding to the normal defined between the point and its diagonal neighbor in the $3 \times 3 \times 3$ mask.

1.4   Topology Preserving Operations

Following the notations of a digital picture, we define here addition and deletion operations [118] on the foreground point set. A *deletion* of a foreground point of a picture implies turning the point into its background. Similarly, an *addition* implies the reverse operation of bringing a background point into the foreground. For a set of points $X \subset S$ in a picture $P = (U, m, n, S)$, the deletion of X transforms it into a picture $P' = (U, m, n, S - X)$. A deletion (or addition) operation is topology preserving if it keeps the number of connected components of foreground points and the number of holes in 2-D (or cavities and tunnels in 3-D) the same. In [197], the criteria for a topology a preserving deletion operation in 2-D is provided as follows:

Let $P = (G^2, m, n, S)$ be a two-dimensional digital picture. Then a subset X of S can be deleted safely to get a picture $P' = (G^2, m, n, S - X)$ preserving the topology of P if and only if:

1. each foreground component of S contains exactly one foreground component of P, and

2. each background component of P contains exactly one background component of X.

1.4.1   Skeletonization

If in 2-D a topologically equivalent picture P consists of only simple arcs and closed curves, we call it a skeleton of P. Similar notion is extended to 3-D with simple surfaces, simple closed surfaces, simple arcs, and simple closed curves to define a skeleton of a 3-D picture. Let us consider the case of 2-D skeletonization first. We define a point *simple*, if its deletion still preserves the topology of the picture. A simple point can be characterized in various ways. The following theorem [155, 175] states a characterization of simple points in 2-D.

Theorem 1.1. *Let p be a non-isolated border point in a digital picture. Let S be its foreground pixel set, and let $S = S - \{p\}$. Then the following are equivalent:*

1. p is a simple point.

2. p is adjacent to just one component of $N(p) \bigcap S$.

3. p is adjacent to just one component of $N(p) - S$.

$\square$

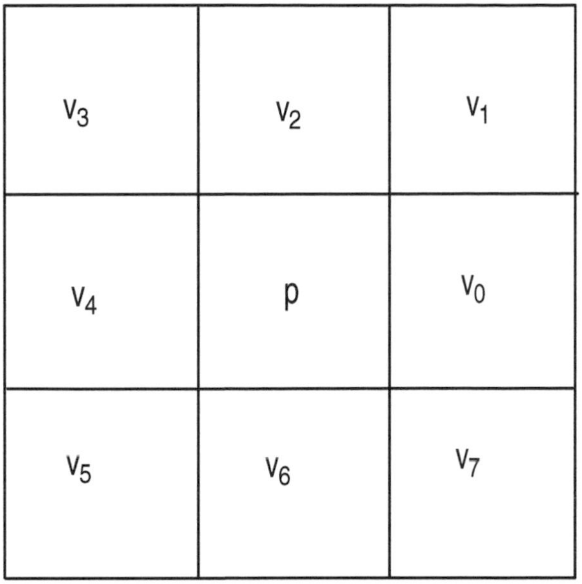

v_3	v_2	v_1
v_4	p	v_0
v_5	v_6	v_7

FIGURE 1.12: The Boolean neighbors of p.

Using the above characterization, it is possible to delete simple border points iteratively till no more deletion is possible. In addition, care should be taken so that a foreground component does not get eroded. This would result in a skeleton of the pattern. Let us consider the technique proposed in [156]. Let us denote the neighbors of a point p with a set of Boolean variables $v_0, v_2, \ldots, v_7$ as shown in Fig. 1.12, such that a background pixel assumes the value *false*, otherwise it is *true* for a foreground pixel. The *safe point thinning algorithm* (SPTA) [156] works in an $(8, 4)$ digital grid, and characterizes a border pixel into any of the four types, namely *left, right, bottom,* and *top*. For example, a foreground pixel p is a *left-border pixel*, if $v_4 \in N_8(p)$ is *false*. Similarly other types of border pixels are also defined, when any of v_0, v_6, and v_2 is *false*. Naccache and Shinghal showed four neighboring conditions for which a point should not be deleted as it would either destroy the connectivity of the pattern, or erode the pattern completely. These conditions are shown in Fig. 1.13. In the figure, the foreground pixels are shown with dark circles, background pixels with empty circles, and the dotted circles denote 'don't care' states. A left border pixel is *simple* or *safe* for deletion when the following Boolean expression (refer Eq. (1.1)) is false.

$$C_4 = v_0.(v_1 + v_2 + v_6 + v_7).(v_2 + \bar{v}_3).(v_6 + \bar{v}_5) \tag{1.1}$$

In the above equation, '.' and '+' denote 'Boolean AND' and 'Boolean OR' operations, respectively. Similarly, Boolean conditions for other types of border points could be derived to check whether they are simple. Using

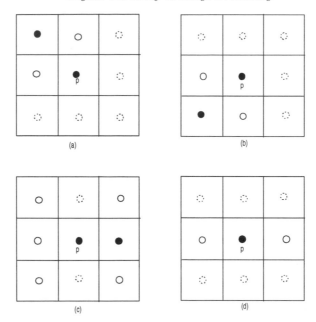

FIGURE 1.13: Neighboring conditions of unsafe deletion of p.

these checks, border points are deleted or saved in the resulting pattern in successive iteration. If at any iteration, there is no such deletion, the process stops and the resulting pattern becomes the thinned pattern or skeleton of foreground points. For removing any bias of deletion of a particular types of border points, Naccache and Shinghal adopted a strategy of dividing an iteration into two scans, one for left and right border points, and the other for top and bottom border points. While scanning these points, if they fail to satisfy respective Boolean conditions, they are marked (or flagged) for deletion. The deletion takes place for all the points at the final stage of the iteration. A brief description of the algorithm is given below.

A typical example of a thinned pattern obtained after applying the thinning algorithm SPTA is shown in Fig. 1.14.

1.4.1.1   Extended SPTA (ESPTA) for 3-D Images

An extension of the safe point thinning algorithm (SPTA) for a 3-D picture $P = (G^3, 26, 6, S)$ has been reported in [146, 147]. In 3-D, let us denote the Boolean variables in the $3 \times 3 \times 3$ neighborhood of a point p as shown in Fig. 1.16.

The concepts used in 2-D are extended in 3-D. Here, the border points are classified into *six* categories. In addition to the previous four types, the other two are referred as *front* and *back* edge points. Hence, in this case, every

---

Algorithm 1: Safe Point Thinning Algorithm for 2-D Objects

Algorithm SPTA

Input: A 2-D digital picture $P = (G^2, 8, 4, S)$.

Output: The skeleton of P.

1. Scan left and right border points. Check the simple point conditions (an example shown in Eq. (1.1)) for each point. If the condition fails, the point is flagged and considered to be deleted for subsequent check on deletion condition.

2. Repeat the above step for top and bottom border points.

3. Delete all the flagged points. If there is no deletion, exit from the computation, else go to step 1.

End SPTA

(a)

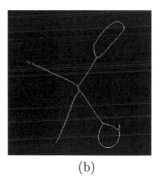

(b)

FIGURE 1.14: Thinned pattern obtained by SPTA (a) original image (scissors), and (b) thinned pattern.

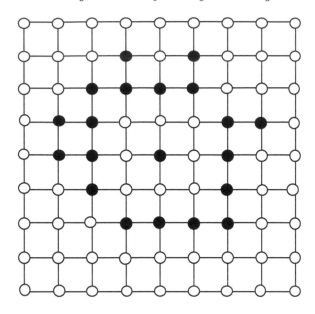

FIGURE 1.15: A 2-D point set in a grid.

iteration consists of three scans for deleting border points. One additional scan is required to check the safe deletion condition for *Front* and *Back* border points. However, in the 3-D neighborhood, for each type of edge point the safe point condition should be checked in four different neighboring planes. For example, for the left- edge point the planes containing both the v_4 and v_0 points are considered. They are shown in Fig. 1.17. In each of these planes, the safe point condition should be *true* for the safe removal of the point. The condition is stated below:

$$C_4 = C_4^{(a)}.C_4^{(b)}.C_4^{(c)}.C_4^{(d)}$$

where,

$$C_4^{(a)} = v_0.(v_1 + v_2 + v_6 + v_7).(v_2 + \bar{v_3}).(v_6 + \bar{v_5}) \qquad (1.2)$$

$$C_4^{(b)} = v_0.(u_0 + u_8 + w_0 + w_8).(u_8 + \bar{u_4}).(w_8 + \bar{w_4}) \qquad (1.3)$$

$$C_4^{(c)} = v_0.(u_1 + u_2 + w_7 + w_6).(u_2 + \bar{u_3}).(w_6 + \bar{w_5}) \qquad (1.4)$$

$$C_4^{(d)} = v_0.(w_1 + w_2 + u_7 + u_6).(w_2 + \bar{w_3}).(u_6 + \bar{u_5}) \qquad (1.5)$$

The skeleton obtained by the above algorithm preserves the 26-connectivity of the pattern. A typical result is shown in Fig. 1.18.

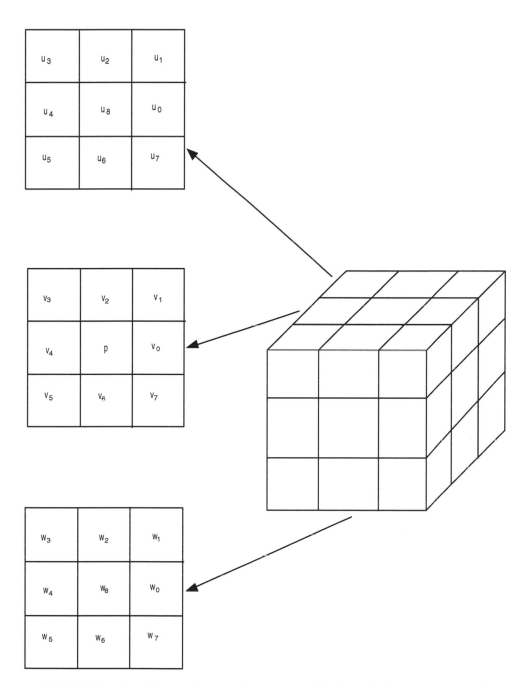

FIGURE 1.16: 3-D neighboring Boolean variables in a $3 \times 3 \times 3$ mask around a point p.

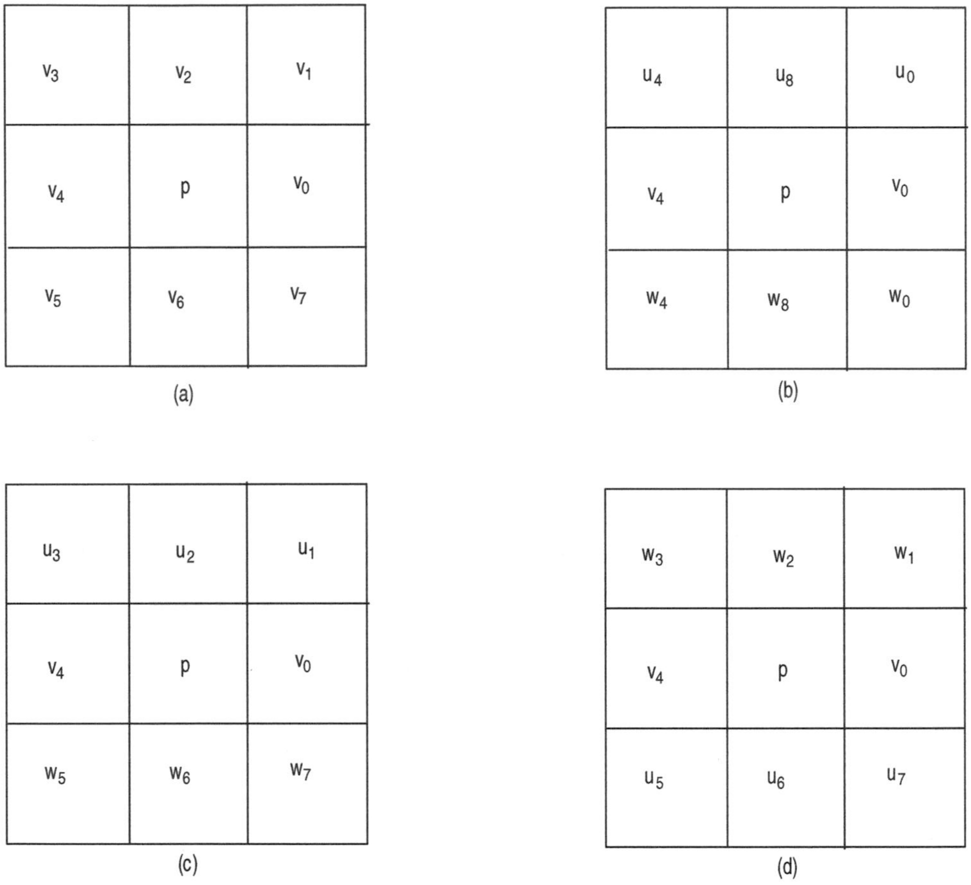

Digital Geometry in Image Processing

FIGURE 1.17: Four planes for which the safe point deletion check is to be carried out for a left edge border point in 3-D. The points are referred with respect to Fig. 1.16.

(a) (b)

FIGURE 1.18: Thinned pattern obtained by ESPTA (a) original 3-D object, and (b) thinned pattern.

1.4.2 Adjacency Tree

Let $\mathcal{C} = \{C_0, C_1, \ldots C_k\}$ be the set of k disjoint connected components of foreground and background of a 2-D digital picture $P = (G^2, m, n, S)$. It may be noted that foreground components are m-connected and the background components are n-connected. We further denote $\mathcal{F} \subset \mathcal{C}$ as the set of foreground components and $\mathcal{B} \subset \mathcal{C}$ as the set of background components. Two components C_i and C_j form an edge, if there exist $p \in C_i$ and $q \in C_j$ such that they are adjacent. In this case, if C_i is a foreground component, C_j must be a background component, and the reverse is also true. Using this adjacency information, the picture is described in the form of a graph [36], where vertices are mapped to the foreground and background components, and edges between two vertices are formed if they are adjacent. It is shown in [36], that the graph is connected and acyclic, thus forming a tree. This tree is called an *adjacency tree*. Let us consider an example of how an adjacency tree could be formed given a 2-D picture. In Fig. 1.15 a set of foreground object points (black in color) is shown. If the grid is taken of type $(8,4)$ connectivity, its set of connected components for foreground and background is shown in Fig. 1.19(a). In this case, C_1 and C_2 are the foreground components, and B_1 and B_2 are background components. The respective adjacency tree formed by nodes corresponding to components and edges between two adjacent components is shown in Fig. 1.19(b). However, if the pattern of Fig. 1.15 is taken in a $(4,8)$ grid, the set of connected components and their adjacency tree differ significantly (see Fig. 1.20). From these figures, we may observe the following for an adjacency tree, which is also in general true. These properties are discussed in [36].

1. The background component containing the border of the picture forms

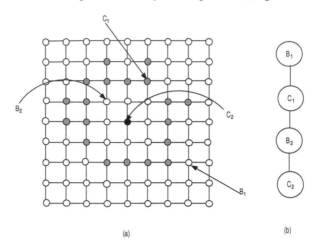

(a)

(b)

FIGURE 1.19: (a) Connected components of foreground points (differently colored) and background points (white) of the 2-D point set in an (8,4) grid, and (b) corresponding adjacency tree. (See color insert.)

the root of the tree. This component surrounds every other component of the picture.

2. One of the vertices of an edge would be a foreground component and the other would be a background component.

3. If a node X lies in a path from Y to the root of the tree, X surrounds Y.

1.5 The Euler Characteristics

In continuous 2-D space, a polyhedral set is defined as a subset of a plane that is the union of all the points, closed line segments, and closed triangles. In 3-D continuous space, in addition to these primitives, we also include all the closed tetrahedra in a 3-D polyhedral set. A topologically invariant feature of a polyhedral set is its Euler characteristics, which is a number and defined as follows.

Definition 1.1. *Let the Euler number of a polyhedral set S be denoted as $\chi(S)$, then:*

1. $\chi(\Phi) = 0$, where Φ is the null set.

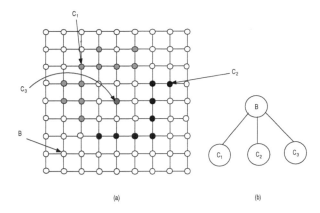

(a)   (b)

FIGURE 1.20: (a) Connected components of foreground points (differently colored) and background points (white) of the 2-D point set in a (4,8) grid, and (b) corresponding adjacency tree. (See color insert.)

2. $\chi(S) = 1$, *if S is convex and not null.*

3. *For any two polyhedral sets S_1 and S_2, the following property holds:*

$$\chi(S_1 \bigcup S_2) = \chi(S_1) + \chi(S_2) - \chi(S_1 \bigcap S_2).$$

Following this definition, one can compute the Euler number of an arbitrary triangulation of S in 3-D as follows:

$$\chi(S) = n_p - n_e + n_f - n_v \tag{1.6}$$

where n_p, n_e, n_f, and n_v denote numbers of points, edges, triangles, and tetrahedra, respectively. In 2-D the above number can be found to be equivalent to the difference between the the number of connected components and holes for a polyhedral set. In 3-D it is equal to the sum of the number of connected components and cavities minus the number of tunnels present in it. For example, the Euler number of a hollow cube is 2, as it has one connected component and a cavity. But the Euler number of a hollow prism is zero, as it has a tunnel.

In a digital grid, Euler characteristics are defined by transforming a set of digital points into an equivalent analog form. For a digital picture $P = (G^2, m, n, S)$, its transformed topologically equivalent analog form $T(P)$ is obtained by taking the union of all the points in S, all the edges formed by a pair of neighbors in S, and all the unit squares or unit right-angled triangles depending upon the type of grid connectivity. For a $(4, 8)$ grid, it takes all unit squares formed by points in S so that its sides are of unit length. On the other hand, in $(8, 4)$ grid, it takes all the unit right an angled triangles, so

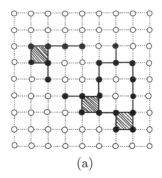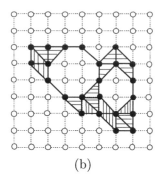

(a)   (b)

FIGURE 1.21: Transformation of a point set (denoted by dark points) in a digital grid to an analog form (a) $(4, 8)$, and (b) $(8, 4)$. In this example, the set of points is the same for both the grids.

that each of its smaller sides is of length 1, and the hypotenuse is of length $\sqrt{2}$. Typical examples of such conversion are shown in Fig. 1.21 (a) and (b). For the same point set, the Euler number may differ. In this example in $(4, 8)$ grid, it is 1 (2 components and 1 hole). In an $(8, 4)$ grid, the Euler number is -1 (1 component and 2 holes).

Various methods are advanced to compute the Euler number using the local features of a picture. In [94], it has been shown that the Euler number could be computed by considering the 2×2 subsets of a digital grid. Such a unit is referred as a *quad cell*. There could be six classes of quad structure out of sixteen arrangements of foreground and background pixels. Four of these classes are formed by the presence of a specific number of foreground points, unless they are linearly inseparable from background points. There are two types of inseparable configurations, each forming diagonals in a quad structure with two background and two foreground points. Typical configurations of these six classes are shown in Fig. 1.22.

Let Q_i, $i = 1, 2, 3$, and 4 denote the classes corresponding to equivalent configurations of Fig. 1.22 (a)-(d), and Q_d denote a configuration of Fig. 1.22 (e) or (f). In that case, the Euler number E_8 in an $(8, 4)$ grid is given by the following expression:

$$E_8 = \frac{1}{4} \left(n(Q_1) - n(Q_3) - 2n(Q_D) \right) \tag{1.7}$$

where $n(.)$ denotes the number of quads in that class in a digital grid. Similarly, the Euler number E_4 in a $(4, 8)$ grid is expressed as follows:

$$E_4 = \frac{1}{4} \left(n(Q_1) - n(Q_3) + 2n(Q_D) \right) \tag{1.8}$$

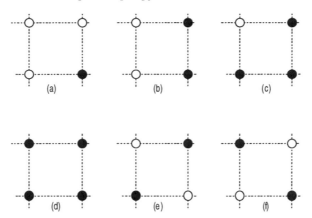

FIGURE 1.22: Representative configuration of object points in each group. The groups (a) to (c) have four members, each could be obtained by rotating the configuration. The empty configuration is ignored here.

1.6 Summary

The digital topology is based on the notion of a finite neighborhood around a point in that space. It has been found that graph theoretic modeling is convenient in defining different topological features in a rectangular lattice space. They include the definition of adjacency, neighbors of a point, connected components, connected paths, borders, interiors, skeletons, etc. This chapter summarized all these important concepts and introduced some of the nontrivial techniques of computing those features. However, in the neighborhood definition of a topological space, it is interesting to explore whether it is guided by any algebraic form of distance function, and whether the function is a metric. In the next chapter, we review several such metrics in the digital spaces, and reveal their relationships with the digital topology considered in this chapter.

Exercises

1. Suppose you have a digital picture $P = (G^3, 6, 26, S)$. What is the type of connectivity among the foreground points? What is the type of connectivity of background points? Discuss why a digital picture in the form of $(G^3, 6, 6, S)$ is not topologically well defined.

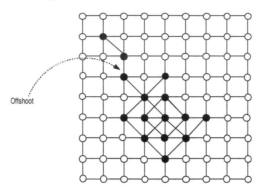

FIGURE 1.23: A non-simple contour with an offshoot.

2. How do you define the adjacency relationship in a hexagonal grid? How many types of adjacency relationships could be there? Discuss the advantage and disadvantage of using hexagonal grids for image representation with respect to the conventional rectangular grid.

3. Suppose there are offshoots in a contour that make it non-simple. A typical example is shown in Fig. 1.23. In the absence of those offshoots, the contour becomes simple. Discuss an image processing operation to make it a simple contour.

4. Write the safe point deletion conditions for right, top, and bottom edge points in 2-D following the same notations given in Section 1.4.1.

5. Identify the digital neighborhood planes, where simple point check for a front edge point in 3-D is to be carried out in the algorithm ESPTA (refer to Section 1.4.1.1).

6. What is the Euler number of a hollow cylinder that contains another hollow cylinder inside? Discuss why in 2-D a skeleton preserves both the adjacency tree and Euler number of the original pattern.

7. Modify the ESPTA to apply in a $(18, 6)$ grid.

8. Find the unit normal vectors of digital neighborhood planes as discussed in Section 1.3.3.3.

9. Prove both the lemmas related to NPS given in the Section 1.3.3.3.

10. Using the same notations of a set of points in the $3 \times 3 \times 3$ neighborhood of a point in a 3-D digital grid (refer to Fig. 1.16), enumerate the set of points for each DNP as shown in Fig. 1.11.

Chapter 2

Distance Functions in Digital Geometry

Image processing in more than two dimensions has attracted a lot of interest recently. Three-dimensional image processing has several applications, such as computed tomography. The inclusion of time has increased the three spatial dimensions to four in studies involving moving objects. Various applications involving gray-scale pictures and objects with several features pertaining to a single dimension require representation in higher dimensions.

Studies on the topological properties of quantized spaces are required in image processing applications. This has led to a non-Euclidean geometry known as digital geometry (or digital topology). A few results in 2-D and 3-D digital geometry were established early [182, 177, 178] during the evolution of this area and the need to handle information in higher dimensions necessitated explorations in arbitrary dimensions (n-D). Many such studies as undertaken in the past couple of decades are summarized in [118, 114].

Since the measurement of distance between elements is essential in many applications, a digital analogue of a non-Euclidean distance in n-D forms an important and interesting part of digital geometry. Such distances find use in segmentation, merging, thinning, clustering, and the like.

In this chapter, we introduce the generalized notions of neighborhoods, paths and distances in n-D digital geometry to present a number of classes of digital distances based on them. We present their shortest path algorithms, prove the conditions for metricity, study the structures of their hyperspheres and estimate the errors in geometric (shapes of hyperspheres with respect to the Euclidean hypersphere) and direct (differences in measures from Euclidean norm) terms to elucidate various good approximations of the Euclidean norm for digital geometry. We also highlight a number of interesting special cases in 2-D and 3-D.

2.1 Mathematical Definitions and Notation

We start with a few basic definitions that are frequently used.

Definition 2.1. All-positive monotone hyperoctant Σ_n *is defined as follows:*

$$\Sigma_n = \{\mathbf{x} : \mathbf{x} \in \mathbb{Z}^n \text{ and } \forall i, 1 \leq i < n, x_i \geq x_{i+1} \geq 0\}$$

That is, $\mathbf{x}$ *belongs to the all-positive hyperoctant and its components are sorted in non-increasing order. By definition,* $\mathbf{0} \in \Sigma_n$. □

Definition 2.2. $\phi(\cdot)$ *is defined as the* 2^n **Symmetry Function** *of an n-D point. That is, given* $\mathbf{x} \in \mathbb{Z}^n$, *and* $x_i \geq 0, \forall i$, $\phi(\mathbf{x})$ *gives the set of points in* $\mathbb{Z}^n$ *obtained by reflections and permutations of* $\mathbf{x}$. *For example, if* $\mathbf{x} = (4, 1, 1)$, *we get the set of symmetric points as* $\phi(\mathbf{x}) = \{(\pm 4, \pm 1, \pm 1), (\pm 1, \pm 4, \pm 1), (\pm 1, \pm 1, \pm 4)\}$. □

Definition 2.3. Binary representation *of a number* $x = \sum_{i=1}^{n}[b(i).2^{n-i}]$ *is denoted by* $[b(1), b(2), \ldots, b(n)]_2$, *where* $b(i) \in \{0, 1\}, \forall i, 1 \leq i \leq n, n \geq 1$. *In particular, we write* x *as* $[1^r 0^{n-r}]_2$ *when*

$$
\begin{aligned}
b(i) &= 1, & 1 \leq i \leq r \\
&= 0, & r < i \leq n, 1 \leq r \leq n;
\end{aligned}
$$

□

Definition 2.4. $f_i : \mathbb{Z}^n \to \mathbb{P}, 1 \leq i \leq n$ *is the* i^{th} **maximum absolute component function** *for a vector* $\mathbf{u}, \forall \mathbf{u} \in \mathbb{Z}^n$. *That is, if* $k_1, k_2, \ldots, k_n$ *are n distinct indices,* $1 \leq k_j \leq n$, $1 \leq j \leq n$, *then* $|u(k_1)| \geq |u(k_2)| \geq |u(k_3)| > \cdots \geq |u(k_n)|$ *and* $f_i(\mathbf{u}) = |u(k_i)|, 1 \leq i \leq n$. *By definition,* $f_0(\mathbf{u}) = 0$. *Clearly,*

$$
\begin{aligned}
f_1(\mathbf{u}) &= \max_{i=1}^{n}\{|u_i|\} & \text{\textit{Largest component}} \\
f_2(\mathbf{u}) &= \max_{1 \leq i < j \leq n}\{\min(|u_i|, |u_j|)\} & \text{\textit{Second largest component}} \\
f_n(\mathbf{u}) &= \min_{i=1}^{n}\{|u_i|\} & \text{\textit{Smallest component}}
\end{aligned}
$$

□

Definition 2.5. $h_i : \mathbb{Z}^n \to \mathbb{P}, 1 \leq i \leq n$ *is the* **sum of** i **maximum absolute components function** *for a vector* $\mathbf{u}, \forall \mathbf{u} \in \mathbb{Z}^n$. *That is,* $\forall \mathbf{u} \in \mathbb{Z}^n$, $h_i(\mathbf{u}) = \sum_{j=1}^{i} f_j(\mathbf{u})$. *Clearly,* $h_0(\mathbf{u}) = 0$ *and* $h_n(\mathbf{u}) = \sum_{i=1}^{n} |u(i)|$. □

2.1.1 Properties of Integer Functions

Next we state a few properties of floor ($\lfloor . \rfloor$) and ceiling ($\lceil . \rceil$) integer functions without proof. These are frequently used in this chapter.

Property 2.1. $\forall x \in \mathbb{R}$, $x \leq \lceil x \rceil < x + 1$ *and* $x \geq \lfloor x \rfloor > x - 1$

Property 2.2. *For* $p \in \mathbb{N}$, $p = \lceil x \rceil$ *iff* $p \geq x > p - 1$

Property 2.3. $\forall x \in \mathbb{R}$ *and* $\forall a \in \mathbb{Z}$, $\lceil x + a \rceil = \lceil x \rceil + a$ *and* $\lfloor x + a \rfloor = \lfloor x \rfloor + a$

Property 2.4. $\forall x, y \in \mathbb{R}$ *and* $x \geq y$, $\lceil x \rceil \geq \lceil y \rceil$ *and* $\lfloor x \rfloor \geq \lfloor y \rfloor$

Property 2.5. $\forall a, b, m \in \mathbb{N}$, $\lceil a/m \rceil + \lceil b/m \rceil \geq \lceil (a+b)/m \rceil$

Property 2.6. $\forall a, b, m \in \mathbb{N}$ *and* $a \geq b$, $\lceil a/m \rceil \geq \lceil b/m \rceil$

Property 2.7. $\forall a, r, s \in \mathbb{N}$ *and* $r \leq s$, $\lceil a/r \rceil \geq \lceil a/s \rceil$

Property 2.8. $\forall a, b, m \in \mathbb{N}$, $b \geq a$ *and* $m = \lceil b/a \rceil$, $a \geq \lceil b/m \rceil$

Property 2.9. $\forall a, b \in \mathbb{P}$ *and* $b \neq 0$,

$$\left\lceil \frac{a}{b} \right\rceil = \left\lfloor \frac{a+b-1}{b} \right\rfloor .$$

Property 2.10. $\forall r, p, t \in \mathbb{P}$, $p \neq 0$ *and* $0 \leq t \leq p$,

$$\sum_{j=0}^{t-1} \left\lceil \frac{r-j}{p} \right\rceil + \sum_{j=0}^{p-t-1} \left\lfloor \frac{r+j}{p} \right\rfloor = r.$$

2.2 Neighborhoods, Paths, and Distances

Unlike Euclidean geometry, where neighborhoods are based on the continuum limit, shortest paths are *unique* and *straight*, and distances are *Euclidean*, digital spaces, due to their very nature of discreteness, offer a wide variety of neighborhood and path structures and associated shortest paths and distances. The discrete points are conceived to be *vertices* of an underlying graph where *neighborhoods* define varying forms of *adjacency*, *costs* are attached to the constraints of adjacency, paths are based on the continuity of adjacency, shortest paths are non-unique, and distances are interesting sets of metrics that offer a wealth of structural and mathematical properties to explore.

In this section, we introduce the notions of neighborhoods (adjacency), paths and distances in a generic framework of an underlying graph before exploring their specific properties in the subsequent sections.

2.2.1 Neighborhoods

Definition 2.6. *The n-**D digital space or grid** is the infinite set $\mathbb{Z}^n$. The elements of this set are referred to interchangeably as* Points *or* Vectors *or* hypervoxels *or* vertices (of the underlying graph). □

Definition 2.7. *The **neighborhood** of a point $\mathbf{u} \in \mathbb{Z}^n$ is a set of points $Neb(\mathbf{u})$ from $\mathbb{Z}^n$ that are* adjacent *to $\mathbf{u}$ in some sense.* □

We associate a non-negative (finite or infinite) cost $\delta : \mathbb{Z}^n \times \mathbb{Z}^n \to \mathbb{R}^+ \cup \{0\}$ between $\mathbf{u}$ and its neighbor $\mathbf{v}$ so that $\delta(\mathbf{u}, \mathbf{v}) = c$ where $\mathbf{v} \in Neb(\mathbf{u})$. Note that though the cost may be real-valued in general, it is usually integral in most cases.

Example 2.1. *In 2-D, $\mathbf{u} = (2,3)$ has a neighborhood $Neb(2,3) = \{(3,3),(1,3),(2,2),(2,4)\}$ with all 4 costs being 1. This is called the 4-neighborhood [182].* □

The above notion naturally induces a weighted graph over $\mathbb{Z}^n$ where we define shortest paths and distances. However, it is impractical to enumerate the neighborhood of every vertex (point) in an infinite graph, and we need a compact repeatable structure for the neighborhood at every point for meaningfully building up a geometry. This is done through the introduction of neighborhood sets.

Definition 2.8. *A **neighborhood set** N is a (finite) set of (difference) vectors from $\mathbb{Z}^n$ such that $\forall \mathbf{u} \in \mathbb{Z}^n$, $Neb(\mathbf{u}) = \{\mathbf{v} : \exists \mathbf{w} \in N, \mathbf{v} = \mathbf{u} \pm \mathbf{w}\}$. With N, we associate a cost function $\delta : N \to \mathbb{P}$, where $\delta(\mathbf{w})$ is the incremental distance or arc cost between neighbors separated by $\mathbf{w}$. Hence, $\forall \mathbf{v} \in Neb(\mathbf{u})$, $\delta(\mathbf{u}, \mathbf{v}) = \delta(\mathbf{u} - \mathbf{v})$.* □

Note that since the neighborhood sets apply on difference vectors, the neighborhoods induced by them are **translation invariant**, that is, the choice of origin has no effect on the overall geometry, specifically the neighborhood structures and paths that are defined using them. We shall often use this fact in the ensuing analysis.

We often denote a neighborhood set as $N(\cdot)$ to indicate the existence of one or more parameters on which the set may depend. Various choices of neighborhood sets and associated cost functions, therefore, induce different graph structures with different notions of paths and distances.

2.2.1.1 Characterizations of Neighborhood Sets

While any choice of a neighborhood set is possible, certain structures in them often make the distance geometry interesting and useful. Hence, usually neighborhood sets are characterized by the following factors [71]. $\forall \mathbf{w} \in N(\cdot) \subset \mathbb{Z}^n$:

1. *Proximity:* Any two neighbors are proximal and share a common hyperplane. That is, $\max_{i=1}^{n} |w_i| \leq 1$.

2. *Separating Dimension:* The dimension m of the separating hyperplane is bounded by a constant r such that $0 \leq r \leq m < n$. For example, in 2-D, 4-neighbors have $r = 1$, $m = 1$ and consequently only line separation is allowed. 8-neighbors, on the other hand, have $r = 0$, $m = 0, 1$, and both point and line separations are allowed. That is, $n - m = \sum_{i=1}^{n} |w_i| \leq n - r$.

3. *Separating Cost:* The cost between neighbors is integral. That is, $\delta(\mathbf{w}) \in \mathbb{P}$. Often the cost is taken to be unity.

4. *Isotropy and Symmetry:* The neighborhood is isotropic in all (discrete) directions. That is, all permutations and/or reflections of $\mathbf{w}$, $\phi(\mathbf{w}) \in N(\cdot)$.

5. *Uniformity:* The neighborhood relation is identical at all points along a path and at all points of the space $\mathbb{Z}^n$.

   In addition, *translation invariance* follows directly from the difference vector definition of neighborhood sets.

Though most distances in digital geometry follow the above characterization, there are many exceptions where one or more of the above properties are violated such as:

1. Knight's distance [67] does not obey proximity (see Section 2.3.4),

2. t-cost distances [58] use non-unity costs (see Section 2.3.2),

3. hyperoctagonal distances [59] use path-dependent neighborhoods, albeit cyclically, and thus violate uniformity (see Section 2.4.1).

Common neighborhood definitions in n-D that generalize the notions of well-known 2-D and 3-D distances are presented in Table 2.1.

Example 2.2. *In 2-D, City Block (4-neighbors) and Chessboard (8-neighbors) distances are defined by neighborhood sets* $\{(\pm 1, 0), (0, \pm 1)\}$ *and* $\{(\pm 1, 0), (0, \pm 1), (\pm 1, \pm 1)\}$, *respectively, with costs associated with every adjacency being 1.* □

Using simple combinatorial reasoning we count the number of m-neighbors of a point in n-D as follows [60]:

Lemma 2.1. $\forall \mathbf{x} \in \mathbb{Z}^n$ *and* $\forall m, 1 \leq m \leq n$, *the number of m-neighbors and $O(m)$-adjacent neighbors of $\mathbf{x}$ are given by* $2^m \binom{n}{m}$ *and* $\sum_{i=1}^{m} 2^i \binom{n}{i}$, *respectively.* □

Note that $O(m)$-adjacent neighbors do not count the central point $\mathbf{x}$. Should this point be included in the count, the above summation should start from 0 instead of 1.

Example 2.3. *We illustrate well-known neighborhoods in low dimensions.*

TABLE 2.1: Common neighborhood definitions in n-D [60, 58]. Neighborhood conditions and costs are shown for neighbors $\mathbf{u}, \mathbf{v} \in \mathbb{Z}^n$.

Neighborhood Set $[N(\cdot)]$	Condition $[\mathbf{w} = \mathbf{u} - \mathbf{v}]$	Cost $[\delta(\mathbf{w})]$	Remarks
m-neighbors $1 \le m \le n$	$\max_{i=1}^{n} |w_i| \le 1$ $\sum_{i=1}^{n} |w_i| = m$	1	$\mathbf{u}$ and $\mathbf{v}$ are separated by m-D hyperplane
$O(m)$-neighbors $1 \le m \le n$	$\max_{i=1}^{n} |w_i| \le 1$ $\sum_{i=1}^{n} |w_i| \le m$	1	r-neighbors for $1 \le r \le m$
t-cost-neighbors $1 \le t \le n$	$\max_{i=1}^{n} |w_i| \le 1$	$min(t, \sum_{i=1}^{n} |w(i)|)$	Cost bounded by t and dimension of separating hyperplane

2-D: *For any* $\mathbf{x} \in \mathbb{Z}^2$, *there are* 4 *$O(1)$-adjacent neighbors that share an edge with* $\mathbf{x}$ *and* 8 *$O(2)$-adjacent neighbors that share an edge or a corner with* $\mathbf{x}$. *These are known as* **4-neighbors** *and* **8-neighbors**, *respectively. We illustrate these neighborhoods in Fig. 2.1.*

$N(Knight) = \{(\pm 1, \pm 2), (\pm 2, \pm 1)\}$ is defined as a move of a Knight on a chessboard. It is an example of a non-proximal neighborhood set.

3-D: *For any* $\mathbf{x} \in \mathbb{Z}^3$, *there are* 6 *$O(1)$-adjacent neighbors that share a face with* $\mathbf{x}$, 18 *$O(2)$-adjacent neighbors that share a face or an edge with* $\mathbf{x}$, *and* 26 *$O(3)$-adjacent neighbors that share a face or an edge or a corner with* $\mathbf{x}$. *These are known as* **6-neighbors**, **18-neighbors**, *and* **26-neighbors**, *respectively. We illustrate these neighborhoods in Fig. 2.2.*

$\square$

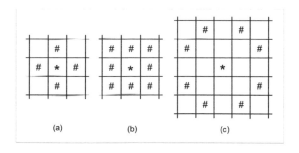

(a) (b) (c)

FIGURE 2.1: Neighborhoods of a point (marked with "*") in 2-D. (a) 4-Neighbors, (b) 8-Neighbors, and (c) Knight Neighbors.

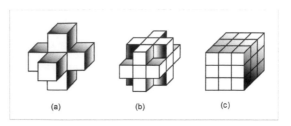

Reprinted from Sadhana 18(1993), P. P. Das and B. N. Chatterji, *Digital Distance Geometry: A Survey*, 159–187,

Copyright (1993), with permission from Indian Academy of Sciences.

FIGURE 2.2: Neighborhoods of a point in 3-D.

2.2.2   Digital Paths

Having introduced the underlying graph for the digital space with the notions of adjacency through neighborhood sets, we are now ready to span paths between pairs of points in n-D.

Definition 2.9. *Given a neighborhood set $N(\cdot)$, a **digital path** $\Pi(\mathbf{u}, \mathbf{v}; N(\cdot))$ between $\mathbf{u}, \mathbf{v} \in \mathbb{Z}^n$, is defined as a sequence of points in $\mathbb{Z}^n$ where all pairs of consecutive points are neighbors. That is,*

$$\Pi(\mathbf{u}, \mathbf{v}; N(\cdot)) : \{\mathbf{u} = \mathbf{x}_0, \mathbf{x}_1, \mathbf{x}_2, ..., \mathbf{x}_i, \mathbf{x}_{i+1}, ..., \mathbf{x}_{M-1}, \mathbf{x}_M = \mathbf{v}\}$$

such that $\forall i, 0 \leq i < M$, $\mathbf{x}_i, \mathbf{x}_{i+1} \in \mathbb{Z}^n$ and $\mathbf{x}_{i+1} \in N(\mathbf{x}_i)$.   □

Definition 2.10. *The **length of a digital path** denoted by $|\Pi(\mathbf{u}, \mathbf{v}; N(\cdot))|$, is defined as $\sum_{i=0}^{M-1} \delta(\mathbf{x}_{i+1} - \mathbf{x}_i)$. Usually there are many paths from $\mathbf{u}$ to $\mathbf{v}$ and the path with the smallest length is denoted as $\Pi^*(\mathbf{u}, \mathbf{v}; N(\cdot))$. It is called the **minimal path** or **shortest path**.*

If the neighborhood costs are all unity, then the length of the minimal path is given by $|\Pi^*(\mathbf{u}, \mathbf{v}; N(\cdot))| = M$. It is the number of points we need to touch after starting from $\mathbf{u}$ to reach $\mathbf{v}$.   □

If the context of the neighborhood set is clear, we may simply refer to a path by $\Pi(\mathbf{u}, \mathbf{v})$. If $\mathbf{u} = \mathbf{0}$, the path is marked as $\Pi(\mathbf{v}; N(\cdot))$ or $\Pi(\mathbf{v})$.

Interestingly, unlike Euclidean geometry, the shortest path between two points in the digital grid is often not unique. That is, there are multiple paths having the same minimum length. Hence, we usually talk about *a* shortest path rather than *the* shortest path.

Example 2.4. *We explore $O(m)$-neighbor (Table 2.1) paths in low dimensions [60]. Neighborhood set $N(\cdot)$ is represented simply by m.*
    2-D: *Consider $\mathbf{u} = (2, 3)$ and $\mathbf{v} = (5, 8)$.*

$$\Pi_1(\mathbf{u}, \mathbf{v}; 1) \quad = \quad \{(2,3),(3,3),(4,3),(5,4),(6,5),(6,6),$$
$$(7,7),(8,8),(7,8),(6,8),(5,8)\}$$
$$\Pi_2(\mathbf{u}, \mathbf{v}; 1) \quad = \quad \{(2,3),(3,3),(4,3),(5,4),(6,5),(6,6),$$
$$= \quad (7,7),(6,8),(5,8)\}$$
$$\Pi_3(\mathbf{u}, \mathbf{v}; 1) \quad = \quad \{(2,3),(3,3),(3,4),(3,5),(4,5),(5,5),$$
$$= \quad (5,6),(5,7),(5,8)\}$$

Note: $|\Pi_1| = 10$ & $|\Pi_2| = |\Pi_3| = 8$

$$\Pi_4(\mathbf{u}, \mathbf{v}; 2) \quad = \quad \{(2,3),(3,4),(4,5),(5,6),(6,7),(7,5),$$
$$= \quad (6,6),(5,7),(5,8)\}$$
$$\Pi_5(\mathbf{u}, \mathbf{v}; 2) \quad = \quad \{(2,3),(3,4),(3,5),(4,6),(5,7),(5,8)\}$$
$$\Pi_6(\mathbf{u}, \mathbf{v}; 2) \quad = \quad \{(2,3),(2,4),(2,5),(3,6),(4,7),(5,8)\}$$

Note: $|\Pi_4| = 8$ *and* $|\Pi_5| = |\Pi_6| = 5$

3-D: *Consider* $\mathbf{u} = (3,5,6)$ *and* $\mathbf{v} = (2,7,4)$.

$$\Pi_1(\mathbf{u}, \mathbf{v}; 1) \quad = \quad \{(3,5,6),(4,5,6),(4,6,6),(4,6,7),(4,5,7),(3,5,7),$$
$$(3,6,7),(3,6,6),(2,6,6),(2,6,5),(2,7,5),(2,7,4)\}$$
$$\Pi_2(\mathbf{u}, \mathbf{v}; 1) \quad = \quad \{(3,5,6),(3,5,6),(3,6,6),(3,7,6),(3,7,5),(2,7,4)\}$$
$$\Pi_3(\mathbf{u}, \mathbf{v}; 1) \quad = \quad \{(3,5,6),(2,5,6),(2,6,6),(2,6,5),(2,6,4),(2,7,4)\}$$

Note: $|\Pi_1| = 11$ *and* $|\Pi_2| = |\Pi_3| = 5$

$$\Pi_4(\mathbf{u}, \mathbf{v}; 2) \quad = \quad \{(3,5,6),(3,4,6),(2,4,5),(2,5,6),(2,6,5),(2,7,4)\}$$
$$\Pi_5(\mathbf{u}, \mathbf{v}; 2) \quad = \quad \{(3,5,6),(2,6,6),(2,7,5),(2,7,4)\}$$
$$\Pi_6(\mathbf{u}, \mathbf{v}; 2) \quad = \quad \{(3,5,6),(3,6,5),(3,7,5),(2,7,4)\}$$

Note: $|\Pi_4| = 5$ *and* $|\Pi_5| = |\Pi_6| = 3$

$$\Pi_7(\mathbf{u}, \mathbf{v}; 3) \quad = \quad \{(3,5,6),(4,5,7),(4,6,6),(3,7,5),(2,7,4)\}$$
$$\Pi_8(\mathbf{u}, \mathbf{v}; 3) \quad = \quad \{(3,5,6),(2,6,5),(2,7,4)\}$$
$$\Pi_9(\mathbf{u}, \mathbf{v}; 3) \quad = \quad \{(3,5,6),(3,6,5),(2,7,4)\}$$

Note: $|\Pi_7| = 4$ *and* $|\Pi_8| = |\Pi_9| = 2$

$\square$

Different paths in 2-D and 3-D are illustrated in Figs. 2.3 and 2.4, respectively.

As we observe in the above example, there are several (often, infinitely many) $O(m)$-paths between two points and finitely many of them may be shortest.

For example, in Example 2.4, in 2-D Π_2^* and Π_3^* are $O(1)$ shortest paths while Π_5^* and Π_6^* are $O(2)$ shortest paths. Π_1 and Π_4 are other paths. Similar observations are made in 3-D.

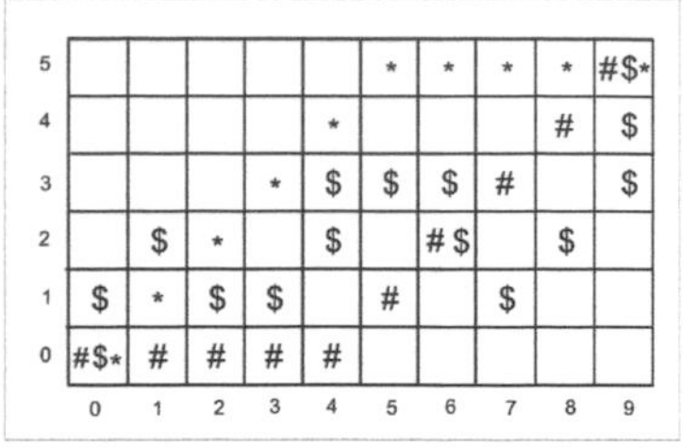

	0	1	2	3	4	5	6	7	8	9
5						*	*	*	*	#$*
4					*				#	$
3				*	$	$	$	#		$
2		$	*		$		#$		$	
1	$	*	$	$		#		$		
0	#$*	#	#	#	#					

Reprinted from Sadhana 18(1993), P. P. Das and B. N. Chatterji, *Digital Distance Geometry: A Survey*, 159–187,

Copyright (1993), with permission from Indian Academy of Sciences.

FIGURE 2.3: $O(2)$ or 8-paths between two points $\mathbf{u} = \mathbf{0}$ and $\mathbf{v} = (9,5)$ in 2-D. The paths Π_1 (marked by '*') and Π_2 (marked by '#') are both minimal while the path Π (marked by '$') is not minimal. Note that $|\Pi_1^*|=|\Pi_2^*|=9$ and $|\Pi|=14$.

2.2.2.1   Shortest Path Algorithm

For a neighborhood set $N(\cdot)$, there may be infinitely many paths in $\mathbb{Z}^n$ between $\mathbf{u}$ and $\mathbf{v}$. Here, we formulate a shortest path tracing algorithm to trace one of the many shortest paths between the two points. Once the path has been traced, we analyze the algorithm to find the length of the shortest path.

Without loss of generality, we assume that $\mathbf{u} = \mathbf{0}$ and $\mathbf{v} = \mathbf{x}$, where $\mathbf{x} \in \Sigma_n$. That is, $\mathbf{x}$ belongs to the all-positive hyperoctant and its components are sorted in non-increasing order. $\mathbf{x}$ is obtained from $\mathbf{v}$ by first taking the absolute value of the components of $\mathbf{v}$ and then sorting them in non-increasing order.

By our assumption of *translation invariance*, the choice of origin is immaterial. Hence, $\mathbf{u} = \mathbf{0}$ offers no restriction. In addition, an $N(\cdot)$ usually is *isotropic and symmetric*. Therefore the choice of $\mathbf{v} = \mathbf{x}$ is justified. Combining the two assumptions, it is clear that for any arbitrary choice of $\mathbf{u}$ and $\mathbf{v}$ in any hyperoctant, there is a corresponding $\mathbf{x}$ satisfying the condition above such that any shortest path from $\mathbf{u}$ to $\mathbf{v}$ bears a point-by-point correspondence with a shortest path from $\mathbf{0}$ to $\mathbf{x}$.

We now present an algorithm **TRACE** to trace such a shortest path. Though there may be a number shortest paths between $\mathbf{0}$ and $\mathbf{x}$, **TRACE** traces any one of them.

Note on the choice of the next point: Usually, there are many choices for the next point on the path **TRACE**$(\mathbf{x}, n, N(\cdot))$. Depending on $N(\cdot)$, we need to make an appropriate choice to ensure that the algorithm approaches the origin $(\mathbf{0})$ in every step and terminates in as many steps as there are points on the shortest path. For a specific neighborhood set, this choice is guided by the fact that shortest paths can be concatenated. That is, at any stage of

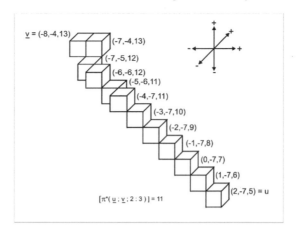

Reprinted from Sadhana 18(1993), P. P. Das and B. N. Chatterji, *Digital Distance Geometry: A Survey*, 159–187,

Copyright (1993), with permission from Indian Academy of Sciences.

FIGURE 2.4: A minimal $O(2)$ or 18-path between two points (2,-7,5) and (-8,-4,13) in 3-D.

the algorithm, $\mathbf{z} \in \Sigma_n \cap Neb(\mathbf{y}; N(\cdot))$, is chosen if $\Pi^*(\mathbf{y}, \mathbf{0})$ is a concatenation of $\Pi^*(\mathbf{y}, \mathbf{z})$ and $\Pi^*(\mathbf{z}, \mathbf{0})$ (evidently, $|\Pi^*(\mathbf{y}, \mathbf{0})| = |\Pi^*(\mathbf{y}, \mathbf{z})| + |\Pi^*(\mathbf{z}, \mathbf{0})|$). In Algorithm 3 we illustrate such a choice for $O(m)$-neighborhood set for d_m^n distance.

2.2.3 Distances and Metrics

In $\mathbb{Z}^n$, Euclidean distance E_n is defined as follows:

Definition 2.11. *In n-D,* $E_n : \mathbb{Z}^n \times \mathbb{Z}^n \to \mathbb{R}^+ \cup \{0\}$ *is the* **Euclidean distance** *where* $\forall \mathbf{u}, \mathbf{v} \in \mathbb{Z}^n$, $E_n(\mathbf{u}, \mathbf{v}) = \sqrt{\sum_{i=1}^n (u(i) - v(i))^2}$ □

From Euclidean geometry, we know that the length of the shortest path (that is, a straight line joining two points) is given by the Euclidean distance. A similar result holds for various neighborhood-defined paths in digital n-D geometry. Once we define a neighborhood, the resulting shortest path between two points has a length that is represented by a nice closed form distance function that gives the length of the shortest path in terms of the coordinates of the two points and the parameters of the neighborhood set.

A distance function usually has this property of representing the length of the shortest path, provided it is a *metric* in the following sense.

Definition 2.12. *A distance function* $d : \mathbb{Z}^n \times \mathbb{Z}^n \to \mathbb{R}^+ \cup \{0\}$ *is called a* **metric** *if* $\forall \mathbf{u}, \mathbf{v}, \mathbf{w} \in \mathbb{Z}^n$,

    1. d is a total: $d(\mathbf{u}, \mathbf{v})$ *is defined and finite.*

Algorithm 2: TRACE(x, n, $N(\cdot)$): Trace an $N(\cdot)$ path from **0** to **x**

Require: x $\in \Sigma_n$ and $N(\cdot)$ is isotropic and symmetric.

  y $\leftarrow$ **x**
  length $\leftarrow 0$
  print x
  while y $\neq$ **0 do**
    // Next Point
    Choose **z** $\in \Sigma_n \cap Neb(\mathbf{y}; N(\cdot))$ such that **z** is *closer* to **0** than **y**
    // Update Length
    length $\leftarrow$ *length* $+ \delta(\mathbf{z} - \mathbf{y})$
    // Update Point
    y $\leftarrow$ **z**
    print y
  end while
  print *length*

---

2. d is a positive: $d(\mathbf{u}, \mathbf{v}) \geq 0$.

3. d is a definite: $d(\mathbf{u}, \mathbf{v}) = 0$ *iff* $\mathbf{u} = \mathbf{v}$.

4. d is symmetric: $d(\mathbf{u}, \mathbf{v}) = d(\mathbf{v}, \mathbf{u})$.

5. d is triangular: $d(\mathbf{u}, \mathbf{v}) + d(\mathbf{v}, \mathbf{w}) \geq d(\mathbf{u}, \mathbf{w})$.

Note that the condition of being total *is important because* $\Pi(\mathbf{u}, \mathbf{v})$ *may not exist* $\forall \mathbf{u}, \mathbf{v} \in \mathbb{Z}^n$, *and hence* $d(\mathbf{u}, \mathbf{v})$ *may not be defined. Examples include Bishop's Neighborhood* $N(Bishop) = \{\pm 1, \pm 1\}$ *or Super-Knight's Neighborhood (see Section 2.3.4, [72]).*

If d violates one or more of (1) through (4), it is usually not of any interest and it is called a **non-metric***. If d violates only (5), it is called a* **semi-metric***.* $\square$

Though we primarily deal with the digital n-D space $\mathbb{Z}^n$, we at times generalize the results for the real n-D space $\mathbb{R}^n$ as well. The distance function then is defined as $d : \mathbb{R}^n \times \mathbb{R}^n \to \mathbb{R}^+ \cup \{0\}$ and the Euclidean distance as $E_n : \mathbb{R}^n \times \mathbb{R}^n \to \mathbb{R}^+ \cup \{0\}$. The notion of the metric property holds in a similar manner over $\mathbb{R}^n$.

2.2.3.1   Distance as a Norm

The neighborhoods, paths, shortest paths, and associated distance functions in this chapter are all *translation invariant*. That is, the choice of the origin does not affect these notions in any way. $\forall \mathbf{u}_1, \mathbf{v}_1, \mathbf{u}_2, \mathbf{v}_2 \in \mathbb{Z}^n$ with $|\mathbf{u}_1 - \mathbf{v}_1| = |\mathbf{u}_2 - \mathbf{v}_2|$, every path $\Pi(\mathbf{u}_1, \mathbf{v}_1)$ corresponds, point-by-point, with a path $\Pi(\mathbf{u}_2, \mathbf{v}_2)$ where the same neighborhood difference vector is used at every step of the two paths. Hence, the sets of paths between two points are topologically identical as long as the absolute difference vectors between them are same. It then follows that for any distance function d,

$$d(\mathbf{u}_1, \mathbf{v}_1) = d(\mathbf{u}_2, \mathbf{v}_2), \quad if \; |\mathbf{u}_1 - \mathbf{v}_1| = |\mathbf{u}_2 - \mathbf{v}_2|.$$

That is, the difference vector between any two points decides the distance between them and the same is not dependent on the end point vectors. We simplify the distance functions to its *norm* as follows:

Definition 2.13. $\forall \mathbf{u}, \mathbf{v} \in \mathbb{Z}^n$, norm $d_N : \mathbb{Z}^n \to \mathbb{R}^+ \cup \{0\}$ *of a distance function* $d : \mathbb{Z}^n \times \mathbb{Z}^n \to \mathbb{R}^+ \cup \{0\}$ *is defined as:*

$$d_N(\mathbf{x}) = d(\mathbf{u}, \mathbf{v}) = d(\mathbf{0}, \mathbf{u} - \mathbf{v}), \quad where \; \mathbf{x} = \mathbf{u} - \mathbf{v}$$

If d is symmetric, $d(\mathbf{u}, \mathbf{v}) = d(\mathbf{v}, \mathbf{u})$ *and* $d_N(\mathbf{x}) = d_N(-\mathbf{x}) = d_N(|\mathbf{x}|)$. $\quad\square$

We use the distance function and its norm interchangeably in this chapter and represent them by the same symbol whenever the context is clear.

2.2.4 Metricity Preserving Transforms

Throughout this chapter, we explore various neighborhoods and study the metric property of their corresponding distance functions. In the process, we often need to prove the metric property of distance functions and derive necessary and sufficient conditions for their metricity. While these proofs can be obtained independently in every case, we present a few generic results here that help shorten many of the arguments.

We start with the observation that most distance functions presented here are transformations or compositions of a number of simpler functions for which the metric property is either already known or is easy to establish. So the metricity of a composite distance is trivial if the transformation function satisfies a set of properties. Mathematically, we pose the following question:

Let $\sigma : \mathbb{R}^+ \cup \{0\} \to \mathbb{R}^+ \cup \{0\}$ be a transformation function and $d : A \times A \to \mathbb{R}^+ \cup \{0\}$ be a metric (where A is $\mathbb{Z}^n$ or $\mathbb{R}^n$ or some other set). Is their composition $D = \sigma \circ d$ a metric on A or not? Naturally,

$$D : A \times A \to \mathbb{R}^+ \quad and$$
$$D(x, y) = \sigma(d(x, y)) \quad \forall x, y \in A.$$

Examples of such transformations include:

Normalization of the range of d with two finite constant C_1 and C_2 such that $C_1 \le D(x, y) \le C_2, \forall x, y \in A$, or

Approximation of d to the quantized space where the range of d is restricted to the set of non-negative integers $\mathbb{P}$.

Example 2.5. *Table 2.2 shows examples of transformations for the Euclidean distance* $E_2 : \mathbb{R}^2 \times \mathbb{R}^2 \to \mathbb{R}^+ \cup \{0\}$ *in 2-D.* $\quad\square$

TABLE 2.2: Illustrations for metric behaviour of transformed Euclidean distance in 2-D.

Transformed Function	Remarks
$\lfloor e \rceil$, round(e), e^2	Violates triangularity
$(1 - exp(-e(x, y)))$	Is normalized ($C_1 = 0$ and $C_2 = 1$) and a metric
$\lceil e \rceil$	Integer approximated and a metric

In general the issue of the metricity of D is settled iff both σ and d are known. Often this involves a lot of algebraic manipulations that are easily avoided if a certain property of the transformation σ is known. Theorem 2.1 from [61] provides a characterization for the class of transforms σ which produce a metric D for *any* given metric d over *any* A.

Definition 2.14. σ *is a* **Metricity Preserving Transform or MPT** *if for all metrics d over any set A, $D = \sigma \circ d$ is a metric on A.* □

It may be noted, however, that if σ is not an MPT, we cannot say anything regarding the metricity of D unless d is known. Depending on the choice of d, D may or may not be a metric. It merely ensures that there exists at least one metric d on some A (for example, E_2 on $\mathbb{R}^2$) for which D is not a metric.

Theorem 2.1. $\sigma : \mathbb{R}^+ \cup \{0\} \to \mathbb{R}^+ \cup \{0\}$ *is an MPT iff σ satisfies the following conditions:*

1. σ *is total*

2. $\sigma(x) = 0$ *iff $x = 0$*

3. $\sigma(x) + \sigma(y) \geq \max_{z=|x-y|}^{x+y} \sigma(z), \forall x, y \in \mathbb{R}^+ \cup \{0\}$

Proof. We prove the sufficiency first. Given a σ satisfying (1)-(3), a set A, and a metric d on A, consider the metric properties of $D = \sigma \circ d$.

Positive-definiteness: Clearly $D : A \times A \to \mathbb{R}^+ \cup \{0\}$ and hence $D(x, y) \geq 0, \forall x, y \in A$. Again $D(x, y) = \sigma(d(x, y)) = 0$ and if $D(x, y) = 0$, then, $\sigma(d(x, y)) = 0$. By (2) then $d(x, y) = 0$ and $x = y$. Hence, D is positive-definite.

Symmetry: $D(x, y) = \sigma(d(x, y)) = \sigma(d(y, x)) = D(y, x), \forall x, y \in A$.

Triangularity: Since d is a metric on A, from triangularity of d, we have $|d(x, y) - d(y, z)| \leq d(x, z) \leq d(x, y) + d(y, z), \forall x, y, z \in A$. So

$$D(x, y) = \sigma(d(x, y)) \leq \max_{u=|d(x,y)-d(y,z)|}^{d(x,y)+d(y,z)} \{\sigma(u)\}.$$

Now,

$$D(x,y) + D(y,z) = \sigma(d(x,y)) + \sigma(d(y,z))$$

$$\geq \max_{u=|d(x,y)-d(y,z)|}^{d(x,y)+d(y,z)} \{\sigma(u)\} \; [From \; Condition(3)]$$

$$\geq D(x,z), \; \forall x,y \in A.$$

Hence D is a metric on A.

Next we prove the necessity by contradiction. Let σ be an MPT that does not satisfy one or more of the conditions (1), (2), and (3). Immediately the violation of conditions (1) and (2) are meaningless and impossible for an MPT. So, we assume that $\exists x, y \in \mathbb{R}$ such that $\exists z, |x - y| \leq z \leq x + y$ for which $\sigma(x) + \sigma(y) < \sigma(z)$. Let $A = \mathbb{R}^2$, $d = E_2$, and $D = D_{E_2} = \sigma \circ E_2$. Now it is always possible to find $u, v, w \in \mathbb{R}^2$ such that $E_2(u,v) = x$, $E_2(v,w) = y$ and $E_2(u,w) = z$, so that $\sigma(E_2(u,v)) + \sigma(E_2(v,w)) < \sigma(E_2(u,w))$ [see Fig. 2.2.4]. Or, $D_{E_2}(u,v) + D_{E_2}(v,w) < D_{E_2}(u,w)$. Hence, D_{E_2} is not triangular. But D_{E_2} should be triangular since σ is metricity preserving, which yields the contradiction. $\qquad\square$

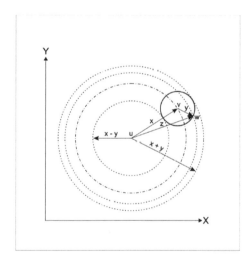

FIGURE 2.5: Construction of counter-example for Theorem 2.1.

Corollary 2.1. *If σ is monotonically increasing, that is, if $x > y$ implies $\sigma(x) \geq \sigma(y)$, $\forall x,y \in \mathbb{R}^+ \cup \{0\}$, then Condition (3) of MPT simplifies to*

$$\sigma(x) + \sigma(y) \geq \sigma(x+y).$$

□

Example 2.6. *Table 2.3 lists a number of MPTs. These follow from Theorem 2.1 and Corollary 2.1.*

TABLE 2.3: MPT examples.

MPT	Non − MPT
$\lceil x \rceil$	$\lfloor x \rfloor, round(x) = \lfloor x + 0.5 \rfloor$
ax^b, $a > 0$ and $0 < b \leq 1$	ax^b, $a > 0$ and $b > 0$
	x^2
$1 - exp(-mx)$, $m > 0$	
mx, $m > 0$	
$m_1 x/(m_2 + m_3 x)$, $m_1, m_2, m_3 > 0$	
$\ln(1 + x)$	

However, for a particular A and d, $\sigma(d)$ may still be a metric. For example, let $\sigma = \lfloor x \rfloor$, $A = \mathbb{R}^2$, and $d(x, y) = \lceil E_2(x, y) \rceil$. Clearly, $D(x, y) = \sigma(d(x, y)) = \lfloor \lceil E_2(x, y) \rceil \rfloor = \lceil E_2(x, y) \rceil$ is a metric even though $\lfloor . \rfloor$ is not an MPT.   □

MPTs are combined to form new MPTs as the following lemma states.

Lemma 2.2. *If g and h are MPT's, then $g \circ h$, $g + h$ and $\max(g, h)$ are also MPTs.*   □

Example 2.7. *Using $g(x) = \lceil x \rceil$ and $h(x) = 1 - exp(-x)$, $\sigma = g \circ h = \lceil 1 - exp(-x) \rceil$ is a normalized 0/1 approximation for any metric d. Since $\forall x, y \in A$, $0 \leq \sigma(d(x, y)) \leq 1$, we have*

$$\begin{aligned} \sigma(d(x, y)) &= 0 &\Longleftrightarrow& \quad x = y \\ &= 1 &\Longleftrightarrow& \quad x \neq y. \end{aligned}$$

□

Next we generalize by introducing MPTs with n arguments.

2.2.4.1   Generalized Metricity Preserving Transforms (GMPT)

Definition 2.15. *Let $\sigma_n : (\mathbb{R}^+ \cup \{0\})^n \to \mathbb{R}^+ \cup \{0\}$ be a transformation which combines n metrics $d_1, d_2, d_3, \cdots, d_n$ on A. Define $D(x, y) = \sigma_n(d_1(x, y), d_2(x, y), \ldots, d_n(x, y))$, $\forall x, y \in A$. σ_n is a **Generalized Metricity Preserving Transforms or GMPT** if D is a metric for every possible choice of metric d_i, $1 \leq i \leq n$ on any set A.*   □

Theorem 2.2 [61] presents the GMPT condition by extending Theorem 2.1.

Theorem 2.2. *A transformation* $\sigma_n : (\mathbb{R}^+ \cup \{ \nvdash \}^n) \to \mathbb{R}^+ \cup \{0\}$ *is metricity preserving iff* σ_n *satisfies the following conditions:*

1. σ_n *is total.*

2. $\sigma_n(x_1, x_2, \ldots, x_n) = 0$ *iff* $x_1 = x_2 = \cdots = x_n = 0$.

3. $\forall x_i, y_i \in \mathbb{R}^+ \cup \{0\}, 1 \leq i \leq n, \sigma_n(x_1, x_2, \ldots, x_n) + \sigma_n(y_1, y_2, \ldots, y_n)$
   $\geq \max_{z_i = |x_i - y_i|_{1 \leq i \leq n}}^{x_i + y_i} \{\sigma(z_1, z_2, \ldots, z_n)\}.$

Proof. Sufficiency follows from Theorem 2.1. To prove necessity, assume $A = \mathbb{R}^2$ and $d_1 = d_2 = d_3 = \cdots = d_n = E_2$. This leads to a similar contradiction as in Theorem 2.1. $\qquad \square$

Example 2.8. $\sigma_n(d_1, d_2, \ldots, d_n) = d_1 + d_2 + \cdots + d_n$ *and* $\sigma_n(d_1, d_2, \ldots, d_n) = \max(d_1, d_2, \ldots, d_n)$ *are GMPTs from Theorem 2.2. These are observed by [182]. Yet another GMPT*

$$\sigma_n(d_1, d_2, \ldots, d_n) = \left(\sum^n m_i \cdot d_i^2 / \sum^n m_i \right)^{1/2},$$

$m_i > 0, 1 \leq i \leq n$ *has been a popular choice for weighted Mahalanobis distance [123].* $\qquad \sqcup$

Example 2.9. *Consider the* i^{th} *maximum component function* $f_i(u)$ *of vector* $|\mathbf{u}|, \forall \mathbf{u} \in \mathbb{R}^n$ *(Definition 2.4). Clearly*

$$d_i(\mathbf{u}, \mathbf{v}) = \sum_{j=1}^{i} f_j(\mathbf{u} - \mathbf{v})$$

is a metric over $\mathbb{R}^n$. *This induces a class of generalized metrics*

$$d(\mathbf{u}, \mathbf{v}; M) = \max_{i=1}^{n} \{d_i(\mathbf{u}, \mathbf{v})/m_i\}$$

over $\mathbb{R}^n$, *where* $M = \{m_i : 1 \leq i \leq n, m_i \in \mathbb{R}^+\}$ *and* $1 \leq m_i \leq n$. *The direct proof for the metricity of* $d(M)$ *is quite involved. But*

$$\sigma_1(x) = x/m \quad and$$
$$\sigma_n(x_1, x_2, \ldots, x_n) = \max_{i=1}^{n} x_i$$

being GMPT's, the metricity of $d(M)$ *immediately follows, once the metricity of* $d_i's$ *are established.* $\qquad \square$

In a number of similar situations, identification of proper GMPT may save a lot of effort for the metricity proofs as we shall observe in this chapter.

2.3 Neighborhood Distances

We now present different distance functions based on neighborhood sets.

2.3.1 m-Neighbor Distance

From Euclidean geometry, we know that the length of the shortest path (that is, a straight line between two points) is given by the Euclidean distance. A similar result holds for $O(m)$-neighbor (Table 2.1) paths in digital n-D geometry. $O(m)$-paths are denoted by $\Pi(\mathbf{u}, \mathbf{v}; m : n)$. Examples in 2-D and 3-D are presented in Figs. 2.3 and 2.4.

We start with the closed functional form for this distance [60].

Definition 2.16. $\forall m, n \in \mathbb{N}$ and $\forall \mathbf{u}, \mathbf{v} \in \mathbb{Z}^n$, we define m-**neighbor distance** $d_m^n(\mathbf{u}, \mathbf{v})$ between $\mathbf{u}$ and $\mathbf{v}$ as

$$d_m^n(\mathbf{u}, \mathbf{v}) = \max(\max_{k=1}^n |u_k - v_k|, \left\lceil \frac{\sum_{k=1}^n |u_k - v_k|}{m} \right\rceil).$$

$\square$

Using the norm for this distance, we write $d_m^n(\mathbf{u}, \mathbf{0})$ as $d_m^n(\mathbf{u}) = \max(\max_{k=1}^n |u_k|, \left\lceil \frac{\sum_{k=1}^n |u_k|}{m} \right\rceil)$. Also, d_m^n is used in place of $d_m^n(\mathbf{u}, \mathbf{v})$ in cases where no confusion arises.

Theorem 2.3. $\forall m, n \in \mathbb{N}$, d_m^n is a metric over $\mathbb{Z}^n$.

Proof. The proof follows from MPT (Theorem 2.1) given that $\max_{k=1}^n |u_k|$ & $\sum_{k=1}^n |u_k|$ are metrics and scaling $(1/m)$, ceiling $(\lceil \cdot \rceil)$ and maximum (max) are MPTs (Table 2.3). An elaborate proof is given in [60]. $\square$

The above theorem suggests an infinite class of m-neighbor distances, each characterized by some $m \in \mathbb{N}$. This is an apparent contradiction as we know that for $n = 2$ and $n = 3$ there are only 2 (d_4 and d_8) and 3 (d_6, d_{18} and d_{26}) possible distances respectively. The following lemma removes this contradiction by showing that infinitely many of these distances are necessarily same and there are only n distances in n-D.

Lemma 2.3. $\forall m, n \in \mathbb{N}$, $m > n$ and $\forall \mathbf{x} \in \mathbb{Z}^n$, $d_m^n(\mathbf{x}) = d_n^n(\mathbf{x})$.

Proof. Show that $\forall m, m \geq n, d_m^n(\mathbf{x}) = \max_i^n |x_i|$. $\square$

Corollary 2.2. There exists exactly n number of m-neighbor distance functions in n-D space $\mathbb{Z}^n$, given by $d_m^n(\mathbf{u}, \mathbf{v}) = \max(d_n^n(\mathbf{u}, \mathbf{v}), \left\lceil \frac{d_1^n(\mathbf{u}, \mathbf{v})}{m} \right\rceil)$ for $1 \leq m \leq n$. $\square$

TABLE 2.4: m-Neighbor Norms in 2-D and 3-D.

n	m	Distance	$d_m^n(\mathbf{u})$
	1	City Block	$d_1^2 = |u_1| + |u_2|$
2	2	Chessboard	$d_2^2 = \max(|u_1|, |u_2|)$
	1	Grid	$d_1^3 = |u_1| + |u_2| + |u_3|$
3	2	d_{18}	$d_2^3 = \max(|u_1|, |u_2|, |u_3|, \left\lceil \frac{|u_1| + |u_2| + |u_3|}{2} \right\rceil)$
	3	Lattice	$d_3^3 = \max(|u_1|, |u_2|, |u_3|)$

Let us consider an example at this point.

Example 2.10. *Table 2.4 lists the m-Neighbor norms in 2-D and 3-D.* □

We know that in 2-D, *City Block distance* between two pixels is always greater than the *Chessboard distance*. A similar result holds in n-D. $\forall \mathbf{u} \in \mathbb{Z}^n$, the values of the m-Neighbor norms are ordered in a non-increasing sequence (for increasing m). That is:

$$d_1^n(\mathbf{u}) \geq d_2^n(\mathbf{u}) \geq \cdots \geq d_i^n(\mathbf{u}) \geq d_{i+1}^n(\mathbf{u}) \geq \ldots \geq d_n^n(\mathbf{u}).$$

The following is proved in [60].

Lemma 2.4. $\forall \mathbf{u} \in \mathbb{Z}^n$, $d_r^n(\mathbf{u}) \geq d_s^n(\mathbf{u})$, $\iff r \leq s$.

Lemma 2.5. $\forall \mathbf{x}, \mathbf{y} \in \mathbb{Z}^n$, $\mathbf{x}$ *and* $\mathbf{y}$ *are r-neighbors iff* $d_r^n(\mathbf{x}, \mathbf{y}) = 1$ *and* $d_s^n(\mathbf{x}, \mathbf{y}) > 1$, $\forall s, s < r$. □

Corollary 2.3. $\forall \mathbf{x}, \mathbf{y} \in \mathbb{Z}^n$ *are $O(r)$-adjacent neighbors iff* $d_r^n(\mathbf{x}, \mathbf{y}) = 1$. □

2.3.1.1 Length of Shortest Path

We adopt the generic shortest path Algorithm 2 and adapt it for $O(m)$-paths. Note the choice of the next point in this case with reference to the general discussions on such a choice in Section 2.2.2.1.

At termination, *length* gives the number of iterations of the while loop, which is the same as the number of points traced. In the following theorem, we show that this is also given by the $O(m)$-neighbor distance function. (For details of the proof, see [60]).

Theorem 2.4. d_m^n *gives the length of shortest $O(m)$-path between two points. That is,* $\forall \mathbf{u}, \mathbf{v} \in \mathbb{Z}^n$,

$$d_m^n(\mathbf{u}, \mathbf{v}) = |\Pi^*(\mathbf{u}, \mathbf{v}; m : n)|.$$

Proof. We use the path tracing Algorithm 3 for the proof. There are two parts:

Algorithm 3: TRACE(x, n, m): Trace an $O(m)$-path from **0** to **x**

Require: x $\in \Sigma_n$

   y $\leftarrow$ **x**

   length $\leftarrow 0$

   print x

   while y $\neq$ **0 do**

      for $1 \leq i \leq m$ **do**

         if $y_i \neq 0$ **then**

            $y_i \leftarrow y_i - 1$ // Choice of next point

         end if

      end for

      length $\leftarrow$ *length* $+ 1$

      print y

   end while

   print *length*

- Prove that Algorithm 3 traces a shortest $O(m)$-path.

- Prove that the length of the path is given by the expression of d_m^n.

Proof proceeds by induction on the path length and relies on the fact that if **w** is a point on a shortest path $\Pi^*(\mathbf{u}, \mathbf{v}; m : n)$, then $\Pi^*(\mathbf{u}, \mathbf{w}; m : n)$ and $\Pi^*(\mathbf{w}, \mathbf{v}; m : n)$ are both shortest paths between respective points and use the same sequence of points as in $\Pi^*(\mathbf{u}, \mathbf{v})$. $\square$

2.3.1.2 m-Neighbor Distance in Real Space

We relax the integral constraints from d_m^n to generalize it further over the n-D real space $\mathbb{R}^n$. We call it the **real m-neighbor distance** δ_m^n [66]. In δ_m^n, m is any positive real value less than n, and δ_m^n itself is allowed to be non-integral.

Definition 2.17. $\forall n$, and $\forall \mathbf{x}, \mathbf{y} \in \mathbb{R}^n, \delta_m^n : \mathbb{R}^n \times \mathbb{R}^n \to R^+ \cup \{0\}$ *is defined as:*

$$\delta_m^n(\mathbf{x}, \mathbf{y}) = \max(\max_{i=1}^{n} |x_i - y_i|, \frac{\sum_{i=1}^{n} |x_i - y_i|}{m}).$$

Clearly $\delta_m^n(\mathbf{x}, \mathbf{y}) = \delta_m^n(\mathbf{x} - \mathbf{y}, \mathbf{0}) = \delta_m^n(\mathbf{x} - \mathbf{y})$. $\square$

The following results are proved in [66]:

Theorem 2.5. δ_m^n *is a metric over* $\mathbb{R}^n$. $\square$

Theorem 2.6. $\forall \mathbf{x}, \mathbf{y} \in \mathbb{R}^n$

1. $\forall, m \geq n, \delta_m^n(\mathbf{x}, \mathbf{y}) = \delta_n^n(\mathbf{x}, \mathbf{y}) = \max_{i=1}^{n} |x_i - y_i|,$

2. $\forall, m \leq 1, \delta_m^n(\mathbf{x}, \mathbf{y}) = \sum_{i=1}^n |x_i - y_i|/m$, *and*

3. *if $m > m'$ then $\delta_m^n(\mathbf{x}, \mathbf{y}) \leq \delta_{m'}^n(\mathbf{x}, \mathbf{y})$. That is δ_m^n is a monotonically non-increasing function of m.*

□

2.3.2 t-Cost Distance

In this section we present a class of distance functions (to be precise $2^n - 1$ distinct functions) in n-D grid point space $\mathbb{Z}^n$ with non-unity (but integral) neighbor costs. We show in the Theorem 2.7 that only n of the possible $(2^n - 1)$ functions satisfy metric properties and derive the necessary and sufficient condition for the same. Subsequently, these n metrics are called *t-**cost distances*** in n-D. They represent the shortest path length for t-cost neighborhoods (Table 2.1). Under this neighborhood notion, two points in $\mathbb{Z}^n$ are neighbors when their corresponding hypervoxels share a hyperplane of any dimension. However, the cost associated with the neighborhood is at most t, $1 \leq t \leq n$, such that if two consecutive points on a shortest path share a hyperplane of dimension r, the distance between them is taken as $\min(t, n - r)$. For $\mathbf{u}, \mathbf{v} \in \mathbb{Z}^n$, t-cost paths are represented as $\Pi(\mathbf{u}, \mathbf{v}; t : n)$. A shortest path is illustrated in Fig. 2.6.

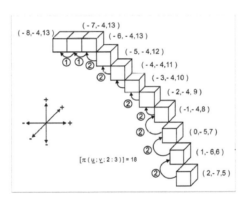

FIGURE 2.6: A minimal 2-cost path $\Pi^*(2 : 3)$ from (2,-7,5) to (-8,-4,13) in 3-D. The costs between adjacent points on the path are encircled in the figure. We have $|\Pi^*| = 8 \times 2 + 2 \times 1 = 18$. Also, $D_2^3((2, -7, 5), (-8, -4, 13)) = D_2^3((10, 3, 8)) = \max(10, 3, 8) + \max(\min(10, 3), \min(3, 8), \min(8, 10)) = 10 + 8 = 18$.

The t-cost distance provides a generalization for City Block and Chessboard distances in 2-D and grid and lattice distances in 3-D into n-D space. There are n distinct t-cost norms.

Definition 2.18. $\forall n, n \geq 1$ *we define a component-based distance function* $d_p, 1 \leq p \leq 2^n - 1$ *over* $\mathbb{Z}^n$ *as*

$$d_p(\mathbf{u}, \mathbf{v}) = \Sigma_{i=1}^n [b(i).f_i(\mathbf{u} - \mathbf{v})] \qquad \forall \mathbf{u}, \mathbf{v} \in \mathbb{Z}^n,$$

where

$$p = \Sigma_{i=1}^n [b(i).2^{n-i}] = [b(1), b(2), \ldots, b(n)]_2$$

and $0 \leq b(i) \leq 1, 1 \leq i \leq n$. $\square$

2.3.2.1 Necessary and Sufficient Condition for Metricity

In Theorem 2.7 [58], we present the necessary and sufficient condition on p, for d_p to be a metric. Before proceeding with the theorem, we state the condition for positive definiteness in Lemma 2.6.

Lemma 2.6. $\forall n, n \geq 1, d_p$ *is positive definite iff* $2^{n-1} \leq p \leq 2^{n-1}$. $\square$

Theorem 2.7. $\forall n, n \geq 1$ *and* $1 \leq p \leq 2^n - 1, d_p$ *is a metric iff* $p = [1^r 0^{n-r}]_2$ *for some* $r, 1 \leq r \leq n$. $\square$

Corollary 2.4. $\forall n, n \geq 1, d_p$ *is a non-metric iff* $1 \leq p \leq 2^{n-1} - 1$. $\square$

Corollary 2.5. *Define* **t-cost distance** *[58]* $D_t^n : \mathbb{Z}^n \times \mathbb{Z}^n \to P$ *for* $n \in N, t \in \{1, 2, 3, \ldots, n\}$ *as follows.* $\forall \mathbf{u}, \mathbf{v} \in \mathbb{Z}^n$,

$$\begin{aligned} D_t^n(\mathbf{u}, \mathbf{v}) &= d_p(\mathbf{u}, \mathbf{v}), \qquad \text{for } p = [1^t 0^{n-t}]_2 \\ &= \Sigma_{i=1}^t f_i(\mathbf{u} - \mathbf{v}). \end{aligned}$$

$\square$

From the preceding theorem D_t^n is a metric, $1 \leq t \leq n$, for all n. Actually, the n t-cost distances are the only metrics in the general definition of $(2^n - 1)$ distance functions.

Corollary 2.6. $\forall n, n \geq 1$, *the number of* d_p's *that are Metric, Non-Metric and semi-metric are given by* n *(Theorem 2.7),* $2^{n-1} - 1$*(Lemma 2.6) and* $(2^n - 1) - (2^{n-1} - 1) - n = 2^{n-1} - n$, *respectively.* $\square$

Note that if we allow p to be zero, then we get a trivial distance function d_0, defined as $d_0(\mathbf{u}) = 0, \forall \mathbf{u} \in \mathbb{Z}^n$. It violates definiteness and does not have any interesting property.

Now we illustrate the generalized distance, d_p's in 2- and 3-D.

Example 2.11. *Let* $n = 2$:

$$\begin{aligned} d_1(\mathbf{u}) &= \min(|u_1|, |u_2|) & & - \text{ } \textit{Non-metric.} \\ d_2(\mathbf{u}) &= D_1^2(\mathbf{u}) = \max(|u_1|, |u_2|) & & - \text{ } \textit{Chessboard distance.} \\ d_3(\mathbf{u}) &= D_2^2(\mathbf{u}) = d_1(\mathbf{u}) + d_2(\mathbf{u}) = |u_1| + |u_2| & & - \text{ } \textit{City Block distance.} \end{aligned}$$

TABLE 2.5: t-Cost norms in 2-D and 3-D.

n	t	$D_t^n(\mathbf{u})$
2	1	$\max(|u_1|, |u_2|)$
	2	$|u_1| + |u_2|$
3	1	$\max(|u_1|, |u_2|, |u_3|)$
	2	$\max(|u_1| + |u_2|, |u_2| + |u_3|, |u_3| + |u_1|)$
	3	$|u_1| + |u_2| + |u_3|$

Let $n = 3$:

$$d_1(\mathbf{u}) = \min(|u_1|, |u_2|, |u_3|) \qquad\qquad - \textit{Non-metric.}$$
$$d_2(\mathbf{u}) = \max(\min(|u_1|, |u_2|)), \min(|u_2|, |u_3|), \min(|u_1|, |u_2|)) \quad - \textit{Non-metric.}$$
$$d_3(\mathbf{u}) = d_1(\mathbf{u}) + d_2(\mathbf{u}) \qquad\qquad - \textit{Non-metric.}$$
$$d_4(\mathbf{u}) = D_1^3(\mathbf{u}) = \max(|u_1|, |u_2|, |u_3|) \qquad - \textit{Lattice distance.}$$
$$d_5(\mathbf{u}) = d_1(\mathbf{u}) + d_4(\mathbf{u}) \qquad\qquad - \textit{Semi-metric.}$$
$$d_6(\mathbf{u}) = D_2^3(\mathbf{u}) = d_2(\mathbf{u}) + d_4(\mathbf{u}) \qquad - \textit{A new metric.}$$
$$d_7(\mathbf{u}) = D_3^3(\mathbf{u}) = d_1(\mathbf{u}) + d_2(\mathbf{u}) + d_4(\mathbf{u}) - |u_1|, |u_2|, |u_3| \quad - \textit{Grid distance.}$$

$\square$

In Table 2.5, t-cost norms in 2-D and 3-D are listed.

2.3.2.2   Length of Shortest Path

Theorem 2.8. *D_t^n gives the length of the shortest t-cost path between two points. That is, $\forall \mathbf{u}, \mathbf{v} \in \mathbb{Z}^n$,*

$$D_t^n(\mathbf{u}, \mathbf{v}) = |\Pi^*(\mathbf{u}, \mathbf{v}; t : n)|.$$

Proof. Devise a shortest path algorithm for t-cost paths by adapting Algorithm 2 and then proceed as in the case of d_m^n (Section 2.3.1.1). For details refer to [58]. $\square$

2.3.2.3   Real-Valued Cost

An interesting and natural generalization to D_t^n demands the cost parameter t to be a real value where mapping of D_t^n is modified to $D_t^n : \mathbb{Z}^n \times \mathbb{Z}^n \to R^+ \cup \{0\}$. Following the technique of analysis adopted so far, we prove the property of D_t^n for real t as well. It may be noted that, though the t-cost neighborhood and the corresponding t-paths are defined primarily for integer costs, they have immediate extensions to the real case. However, the functional form of D_t^n undergoes change, as stated in the following lemma.

Lemma 2.7. $\forall n \in N$, $\forall t \in R^+$, $\forall \mathbf{x} \in \mathbb{Z}^n$,

$$D_t^n(\mathbf{x}) = \sum_{i=1}^{\lfloor t \rfloor} f_i(\mathbf{x}) + (t - \lfloor t \rfloor).f_{\lceil t \rceil}(\mathbf{x}).$$

$\square$

Note that if t is integral, that is $t = \lceil t \rceil$, the above reduces to Corollary 2.5. Moreover, D_t^n is a linear combination of $D_{\lfloor t \rfloor}^n$ and $D_{\lceil t \rceil}^n$. Hence we get,

Lemma 2.8. $\forall n \in N$, $\forall t \in R^+$, $\forall \mathbf{x} \in \mathbb{Z}^n$,

$$D_t^n(\mathbf{x}) = (t - \lfloor t \rfloor).D_{\lceil t \rceil}^n(\mathbf{x}) + (1 + \lfloor t \rfloor - t).D_{\lceil t \rceil}^n(\mathbf{x}).$$

$\square$

Using this result, we prove the result similar to Theorem 2.7.

Theorem 2.9. D_t^n *is a metric over* Z^n *and it gives the length of the shortest* t-*path between two points, where* t *is any positive real cost.* $\square$

2.3.2.4 t-Cost Distance in Real Space

Interestingly, the t-cost distance generalizes to the real n-D space.

Theorem 2.10. $\forall n \in N$, $\forall t \in R^+$, $D_t^n : \mathbb{R}^n \times \mathbb{R}^n \to \mathbb{R}^+ \cup \{0\}$ *is a metric.* $\square$

We present further generalizations of this result as weighted t-cost distance in Section 2.4.3.

2.3.3   Weighted Cost Distance

In general, computing the distance from each object grid point to each background grid point in an image leads to very high computational cost. So often the spatial consistency of a distance map is used to allow for propagation of local information. This is known as **chamfering** and is discussed in Chapter 5. When using the chamfer algorithm, only a small neighborhood of each grid point is considered. A weight, a local distance, is assigned to each grid point in the neighborhood. By propagating the local distances in the two-scan algorithm, the correct distance map is obtained. For example, City Block and Chessboard distances are obtained by using unit weights for the neighbors.

Borgefors et al. [22, 23, 24, 25, 202] worked extensively on computational aspects of digital distances using chamfering. Chamfering is used to compute the distance maps of the distances discussed so far. In addition, it opens the opportunity to define new measures through a linear combination of difference of coordinates of points (leading to different types of motion such as vertical and horizontal motions in 2-D) and local weights. These are known as **weighted cost distances** [37].

For digital distances, these weights are integers. The weighted distances are defined as follows in low dimensions:

In 2-D:

$$d^2_{<a,b>}(\mathbf{u}, \mathbf{v}) = a|u_1 - v_1| + (b - a)|u_2 - v_2|$$

where a and b are positive integers and $a \leq b \leq 2a$.

In 3-D:

$$d^3_{<a,b,c>}(\mathbf{u}, \mathbf{v}) = a|u_1 - v_1| + (b - a)|u_2 - v_2| + (c - b)|u_3 - v_3|$$

where a, b, and c are positive integers and $b \leq 2a$, $b \leq c$ and $a + c \leq 2b$.

The advantage of such distances includes ease of computation by chamfering, simple functional form, and the ability to approximate the Euclidean norm fairly accurately. In [37], Borgefors et al. present their chamfering algorithm as the **Weighted Distance Transform (WDT)**. They also prove various interesting results for face-centered-cubic (FCC) and body-centered-cubic (BCC) grids in 3-D.

2.3.4 Knight's Distance

In the game of chess, the move of a knight has been particularly interesting because of its tricky and non-proximal nature. If a knight is placed in the cell $\mathbf{x} \in \mathbb{Z}^2$, then in the next step it moves to any of the eight possible cells $\mathbf{y}$, where $(\mathbf{y} - \mathbf{x}) \in N(Knight) = \{(\pm 1, \pm 2), (\pm 2, \pm 1)\}$ (Fig. 2.1), where the cost δ is taken to be unity. The notions of Knight's path and path length naturally follow.

The **Knight's distance** $d_{Knight}(\mathbf{u}, \mathbf{v})$ [67] is defined as: $\forall \mathbf{u}, \mathbf{v} \in \mathbb{Z}^2$,

$$
\begin{aligned}
d_{Knight}(\mathbf{u}, \mathbf{v}) &= \max(\lceil x_1/2 \rceil, \lceil (x_1 + x_2)/3 \rceil) + (x_1 + x_2)) && if\ x \neq (1, 0), (2, 2), \\
&\quad - \max(\lceil x_1/2 \rceil, \lceil (x_1 + x_2)/3 \rceil)\ \mathrm{mod}\ 2 \\
&= 3 && if\ x = (1, 0), \\
&= 4 && if\ x = (2, 2),
\end{aligned}
$$

where

$$x_1 = \max(|u_1 - v_1|, |u_2 - v_2|),\quad x_2 = \min(|u_1 - v_1|, |u_2 - v_2|).$$

The distance between any two points is the number of steps taken to trace a shortest path from one point to another. Hence, at every step on a shortest path the larger coordinate (x_1) of $\mathbf{x}$ should be decreased by 2, whereas x_2 should be decreased by 1. This indicates the need for the first term $\max(\lceil x_1/2 \rceil, \lceil (x_1 + x_2)/3 \rceil)$ in d_{Knight}, where the ceiling function is introduced to get the integral number of steps. The second term in d_{Knight} is due to a parity correction necessary for the motion at the final step to be defined. For two points, namely (1, 0) and (2, 2), such a function does not hold due to the peculiar nature of the motion. Thus they are explicitly defined. The proofs of the functional form and the metric properties of d_{Knight} are given in [67].

Theorem 2.11. d_{Knight} *is a metric over* $\mathbb{Z}^2$.

Theorem 2.12. $\forall \mathbf{u}, \mathbf{v} \in \mathbb{Z}^2$, $d_{Knight}(\mathbf{u}, \mathbf{v})$ *is the length of a shortest Knight's path from* $\mathbf{u}$ *to* $\mathbf{v}$ *as defined by* $N(Knight)$.

Proof. Adopt Algorithm 2 for $N(Knight)$ and proceed as in d_m^n.   □

The Knight's move is non-proximal and somewhat atypical in digital geometry. Yet, it has been extended in [72] as a **Super-Knight's Distance** with a class of neighborhood sets $N(Super - Knight) = \{(\pm p, \pm q), (\pm p, \pm q)\}$ where $p, q \in \mathbb{P}$ and $p \geq q \geq 1$. $d_{Super-Knight}$ is a metric under certain conditions (see Exercise 3) on $N(Super - Knight)$ [72].

2.4   Path-Dependent Neighborhoods and Distances

In the last section, we relaxed various neighborhood conditions from $O(m)$-neighborhood to get new distances besides d_m^n. Specifically, we got t-cost distances by relaxing the unity cost criteria and the Knight's distance (albeit in 2-D) by diluting the proximal condition. In this section, we introduce a new form of relaxation in terms of path dependence of neighborhoods while reverting back to neighbors being separated by unity costs and being proximal. We start with an example in 2-D.

Example 2.12. *In 2-D, City Block distance uses $O(1)$- or 4-neighbors while Chessboard distance uses $O(2)$- or 8-neighbors. The first one produces a diamond as a disk that is too small for a circle while the second produces a square that is too large. Rosenfeld and Pfaltz [182] suggested a compromise between them as* **octagonal distance**, *where a path starts with an $O(1)$ neighbor but then $O(1)$- and $O(2)$-neighbors are used alternately along the path.*

Such a path from (0,0) to (9,5) is shown in Fig. 2.7.

Octagonal distance results in octagonal disks that better approximate circles, Hence the name.   □

In [59], the notion of path dependence has been extended further by allowing for arbitrary sequences of neighborhoods for arbitrary dimensions. We call them **hyperoctagonal distances** and present their properties.

2.4.1   Hyperoctagonal Distance

We first need to formalize the mathematics behind sequences and provide for necessary characterizations.

Definition 2.19. *A finite sequence that consists of the first n natural numbers is called a* **Neighborhood Sequence** *(or N-Sequence) in n-D. It is*

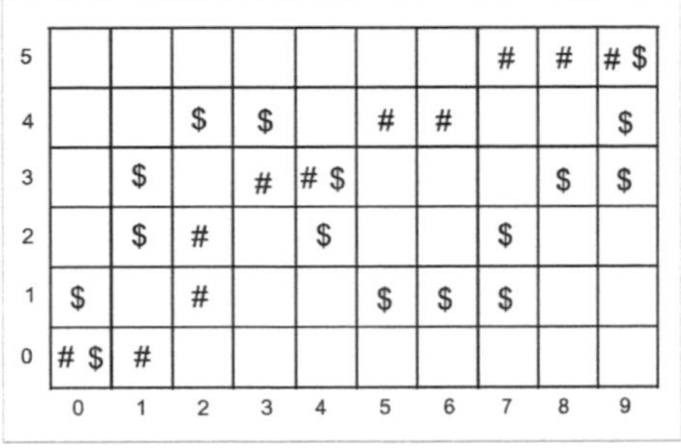

	0	1	2	3	4	5	6	7	8	9
5								#	#	# $
4			$	$		#	#			$
3		$		#	# $				$	$
2		$	#		$		$			
1	$		#			$	$	$		
0	# $	#								

FIGURE 2.7: Two paths from $(0,0)$ to $(9,5)$ using octagonal distance. The path Π marked with $ has a length $|\Pi|=15$ and the path Π^* marked with # has a length $|\Pi^*|=10$. Along either path, the adjacency relation alternates between $O(1)$-neighbor and $O(2)$-neighbor. Clearly $|\Pi^*|$ has the minimal length.

represented as

$$B = \{b(i) : i = 1, 2, \cdots, p; b(i) \in \{1, 2, \cdots, n\}\}$$

*where $p = |B|$ is called the **period** or **length** of the sequence.* □

A few N-Sequences in low dimensions are shown below:

n	B	p	n	B	p	n	B	p
2	{1}	1	3	{1}	1	4	{1}	1
2	{2}	1	3	{2}	1	4	{2}	1
2	{1,2}	2	3	{3}	1	4	{3}	1
2	{2,1}	2	3	{1,2}	2	4	{4}	1
2	{1,2,2}	3	3	{2,3}	2	4	{1,2}	2
2	{1,2,2,2}	4	3	{1,3}	2	4	{2,3}	2
2	{1,2,1,2,2}	5	3	{1,2,3}	3	4	{1,2,4}	3
2	{2,2,1,1,1}	5	3	{2,3,1,2,2,3}	6	4	{1,2,2,3,4}	5

Definition 2.20. *Given an N-Sequence B, the notion of an $N(B)$-path $\Pi(\mathbf{u}, \mathbf{v}; B)$ between $\mathbf{u}, \mathbf{v} \in \mathbb{Z}^n$ extends naturally from the definitions of digital paths (Section 2.2.2) using neighborhood set $N(B)$. On such a path, neighborhood relations are cyclically used from the N-Sequence B.*
The minimal path is denoted by $\Pi^*(\mathbf{u}, \mathbf{v}; B)$. □

Definition 2.21. *The **Sum Sequence** $F = \{f(1), f(2), \cdots, f(p)\}$ of an N-Sequence $B = \{b(1), b(2), \cdots, b(p)\}$ is defined as*

$$f(i) = \sum_{j=1}^{i} b(j) \quad \forall i, 1 \leq i \leq p.$$

☐

By definition, $f(0) = 0$.

Definition 2.22. *A* **sorted N-Sequence** $S(B)$ *of an N-Sequence B is a reordering of the elements of B in non-decreasing order. That is,*

$$S(B) = \{s(1), s(2), \cdots, s(p)\}$$

where

$$s(i) \le s(i+1), \forall i, 1 \le i \le p - 1.$$

☐

If an N-Sequence is intrinsically ordered, that is, $B = S(B)$, then it is alternately represented as:

$$B = [\alpha_1, \alpha_2, \cdots, \alpha_n] = \{1^{\alpha_1}, 2^{\alpha_2}, \cdots, n^{\alpha_n}\}$$

where α_i denotes the number of consecutive i's in the sequence. Clearly, $\forall i, 1 \le i \le n$, $0 \le \alpha_i \le p$, and $\sum_{i=1}^{n} \alpha_i = p$.

Definition 2.23. *A* **subsequence** $B(i, j)$ *of an N-Sequence B is formed by taking j elements from B starting from the i-th element. That is,*

$$\forall i \forall j, 1 \le i, j \le p, \quad B(i, j) = \{b(i), b(i+1), \cdots, b(p), b(1), b(2), \cdots, b(k)\}$$

where

$$k = \begin{cases} (i+j) - (p+1), & i+j > p, \\ i+j-1, & i+j \le p. \end{cases}$$

☐

Definition 2.24. *The* **wave front set** $WF(B)$ *of an N-Sequence B is defined recursively as:*

$$\begin{aligned} WF(B(1, i)) &= W(i), 1 \le i \le p \\ W(0) &= \{\mathbf{0}\} \\ W(i) &= \{\mathbf{w} : \mathbf{w} = \mathbf{u} + \mathbf{v} \wedge \mathbf{w} \in \Sigma_n \wedge \\ & \quad \textstyle\sum_{j=1}^{n} u(j) = b(i) \wedge \forall j, 0 \le u(j) \le 1 \wedge \mathbf{v} \in W(i-1)\}. \end{aligned}$$

Finally, $WF(B) = WF(p)$. ☐

Definition 2.25. *For two sequences* $\mathbf{u}, \mathbf{v} \in \mathbb{Z}^n$, $\mathbf{u}$ **component-wise dominates** $\mathbf{v}$ *iff:*

$$\mathbf{u} \ge_c \mathbf{v} \quad or \quad u(i) \ge v(i), \quad \forall i, 1 \le i \le n.$$

☐

Definition 2.26. *X and Y are two N-Sequences with* $|X| = |Y|$. *X* **dominates** *Y (written as* $X \succ Y$*) iff* $\forall \mathbf{u} \in WF(Y), \exists \mathbf{v} \in WF(X)$ *such that* $\mathbf{v} \ge_c \mathbf{u}$. ☐

Definition 2.27. *An* N-Sequence *B is* **well-behaved (wb)** *iff*

$$B(i, j) \succ B(1, j) \quad \forall i \forall j, 1 \leq i, j \leq p.$$

$\square$

From [59], we present the functional form for $d(\mathbf{u}, \mathbf{v}; B)$, the length of the shortest $N(B)$-path, as a function of $\mathbf{u}$, $\mathbf{v}$, and B.

Theorem 2.13. *The length of the minimal path from $\mathbf{u}$ to $\mathbf{v}$ determined by B is denoted by $d(\mathbf{u}, \mathbf{v}; B) = |\Pi^*(\mathbf{u}, \mathbf{v}; B)|$ and is given by:*

$$d(\mathbf{u}, \mathbf{v}; B) = \max_{i=1}^{n} d_i(\mathbf{u}, \mathbf{v})$$

where

$$
\begin{aligned}
d_i(\mathbf{u}, \mathbf{v}) &= p \left\lfloor \frac{a_i}{f_i(p)} \right\rfloor + h(z_i; B_i) \\
&= \sum_{j=1}^{p} \left\lfloor \frac{a_i + g_i(j)}{f_i(p)} \right\rfloor \\
a_i &= \sum_{j=1}^{n-i+1} x(j) \\
\mathbf{x} &= (x_1, x_2, \cdots, x(n))
\end{aligned}
$$

such that $\mathbf{x}$ is formed by sorting $|u_i - v_i|$, $1 \leq i \leq n$, in non-increasing order, that is, $x_i \geq x_j$ for $i < j$:
$B_i = \{b_i(1), b_i(2), \cdots, b_i(p)\}$ *such that $\forall i, 1 \leq i \leq n$,*

$$
b_i(j) = \begin{cases} b(j), & b(j) < n - i + 2, \\ n - i + 1, & \text{otherwise} \end{cases}
$$

$$
f_i(j) = \begin{cases} \sum_{k=1}^{j} b_i(k), & 1 \leq j \leq p, \\ 0, & j = 0 \end{cases}
$$

$$
\begin{aligned}
z_i &= a_i \bmod f_i(p) \\
h(z_i; B_i) &= \min\{k : f_i(k) \geq z_i\}; \text{ that is,} \\
f_i(h(z_i; B_i) - 1) &< z_i \leq f_i(h(z_i; B_i)); \text{ and} \\
g_i(j) &= f_i(p) - f_i(j-1) - 1, 1 \leq j \leq p.
\end{aligned}
$$

Proof. The proof is rather complicated [59]. It proceeds as in the case of d_m^n (Section 2.3.1.1). $\square$

2.4.1.1 Necessary and Sufficient Condition for Metricity

Unfortunately, for all B's the corresponding $d(B)$'s do not satisfy the metric (specifically, triangular) properties. For example, in 2-D, $d(\{1, 2\})$ (octagonal distance) is a metric [182]. But $d(\{2, 1\})$ is not a metric as it violates the triangle inequality for the points $(0,0)$, $(1,1)$ and $(2,2)$ with $d((0,0), (1,1); \{2,1\})=1$ and $d((1,1), (2,2); \{2,1\})=1$ but $d((0,0), (2,2); \{2,1\})=3$ $(>1+1)$.

In the following theorem (from [59]), we present a necessary and sufficient condition on a B for the $d(B)$ to be a metric.

TABLE 2.6: Functional forms of $d(\mathbf{x}; B)$'s in 2-D [71], p = 1, 2, 3, 4. For compactness we use $\alpha_1 = \max(|x_1|, |x_2|), \alpha_2 = |x_1| + |x_2|$.

p	Well-Behaved B	$d(\mathbf{x}; B)$
1	$\{1\}$	α_2 : Simple
	$\{2\}$	α_1 : Simple
2	$\{1,2\}$	$\max(\lceil 2\alpha_2/3 \rceil, \alpha_1)$: Simple
3	$\{1,1,2\}$	$\max(\lceil 3\alpha_2/4 \rceil, \alpha_1)$: Simple
	$\{1,2,2\}$	$\max(\lceil 3\alpha_2/5 \rceil, \alpha_1)$: Simple
4	$\{1,1,1,2\}$	$\max(\lceil 4\alpha_2/5 \rceil, \alpha_1)$: Simple
	$\{1,1,2,2\}$	$\max(\lfloor (\alpha_2+1)/6 \rfloor + \lfloor (\alpha_2+3)/6 \rfloor + \lfloor (\alpha_2+4)/6 \rfloor + \lfloor (\alpha_2+5)/6 \rfloor, \alpha_1)$
	$\{1,2,1,2\}$	$\max(\lceil 2\alpha_2/3 \rceil, \alpha_1)$: Simple; same as d(1, 2)
	$\{1,2,2,2\}$	$\max(\lceil 4\alpha_2/7 \rceil, \alpha_1)$: Simple

Theorem 2.14. $d(B)$ *is a metric iff B is well-behaved.*   □

The general functional form of a $d(B)$ is complex. However, it usually takes a simple form when the corresponding $d(B)$ is a metric, as the following example illustrates in low dimensions.

Example 2.13. *In Tables 2.6 and 2.7, we present the functional forms of $d(B)$'s in 2-D and 3-D for short sequences.*

Theorems 2.13 and 2.14 present strong characterizations for N-Sequence distances in n-D that generalize the results for $O(m)$-neighbor distances in d_m^n and also generate a set of new metrics. Their further characterizations in terms of hyperspheres are presented in Section 2.5.5 where we observe that the hyperspheres of these distances asymptotically approach shapes of hyperoctagons, justifying their nomenclature as **hyperoctagonal distances**.

2.4.2   Octagonal Distances in 2-D

Hyperoctagonal distances presented in the last section take interesting forms when we restrict them to 2-D. These generalize the *octagonal distance* originally proposed by Rosenfeld and Pfaltz [182].

First, we specialize Theorem 2.13 (from [56]) to get a simpler form for a $d(B)$.

Theorem 2.15. *In 2-D, $d(\mathbf{x}; B) = |\Pi^*(\mathbf{x}; B)|$ and is given by:*

$$d(\mathbf{x}; B) = \max\{d_1(\mathbf{x}), d_2(\mathbf{x})\}$$

where

$$
\begin{aligned}
d_1(\mathbf{x}) &= p + \sum_{j=1}^{p} \left\lfloor \frac{x_1 + x_2 - f(j-1) - 1}{f(p)} \right\rfloor \\
d_2(\mathbf{x}) &= \max(x_1, x_2).
\end{aligned}
$$

     □

TABLE 2.7: Functional forms of $d(\mathbf{x}; B)$'s in 3-D [71], p = 1, 2, 3. For compactness, we use $\alpha_1 = \max(|x_1|, |x_2|, |x_3|), \alpha_2 = \max(|x_1| + |x_2|, |x_2| + |x_3|, |x_3| + |x_1|), \alpha_3 = |x_1| + |x_2| + |x_3|$.

p	Well-Behaved B	$d(\mathbf{x}; B)$
1	{1}	α_3
	{2}	$\max(\lfloor (\alpha_3 + 1)/2 \rfloor, \alpha_1)$
	{3}	α_1
2	{1,2}	$\max(\lfloor (2\alpha_3 + 2)/3 \rfloor, \alpha_1)$
	{1,3}	$\max(\lfloor (\alpha_3 + 3)/4 \rfloor + \lfloor (\alpha_3 + 2)/4 \rfloor, \lfloor (2\alpha_2 + 2)/3 \rfloor, \alpha_1)$
	{2,3}	$\max(\lfloor (2\alpha_3 + 4)/5 \rfloor, \alpha_1)$
3	{1,1,2}	$\max(\lfloor (3\alpha_3 + 3)/4 \rfloor, \alpha_1)$
	{1,1,3}	$\max(\lfloor (\alpha_3 + 4)/5 \rfloor + \lfloor (\alpha_3 + 3)/5 \rfloor + \lfloor (\alpha_3 + 2)/5 \rfloor, \lfloor (3\alpha_2 + 3)/4 \rfloor, \alpha_1)$
	{1,2,2}	$\max(\lfloor (3\alpha_3 + 4)/5 \rfloor, \alpha_1)$
	{1,2,3}	$\max(\lfloor (\alpha_3 + 2)/3 \rfloor + \lfloor (\alpha_3 + 4)/6 \rfloor, \lfloor (3\alpha_2 + 4)/5 \rfloor, \alpha_1)$
	{1,3,3}	$\max(\lfloor (\alpha_3 + 6)/7 \rfloor + \lfloor (2\alpha_3 + 4)/7 \rfloor, \lfloor (3\alpha_2 + 4)/5 \rfloor, \alpha_1)$
	{2,2,3}	$\max(\lfloor (\alpha_3 + 6)/7 \rfloor + \lfloor (\alpha_3 + 4)/7 \rfloor + \lfloor (\alpha_3 + 2)/7 \rfloor, \alpha_1)$
	{2,3,3}	$\max(\lfloor (\alpha_3 + 7)/8 \rfloor + \lfloor (\alpha_3 + 5)/8 \rfloor + \lfloor (\alpha_3 + 2)/8 \rfloor, \alpha_1)$

Distances obtained from the above expression are presented in Table 2.6.

Next we present a lemma from [56] that simplifies the well-behaved condition for metricity.

Lemma 2.9. *In 2-D an N-Sequence B is well-behaved iff*

$$f(i) + f(j) \quad \leq \quad \begin{cases} f(i+j), & i+j \leq p, \\ f(p) + f(i+j-p), & i+j > p. \end{cases}$$

□

Consider $B = \{2, 1\}$. Here $f(1) = 2$, but $f(1) + f(1) = 4 > f(1 + 1) = f(2) = 1$. So $\{2, 1\}$ is not well-behaved and $d(\{2, 1\})$ is not a metric as was concluded earlier.

All of these $d(B)$ distances are called **octagonal distances** as their disks, $H(B; r) = \{\mathbf{x} : d(\mathbf{x}; B) \leq r\}$ take the shape of octagons with vertices at $\{(\pm r, \pm m(r)), (\pm m(r), \pm r)\}$ where $\forall x, 0 \leq x \leq m(r), d((x, r); B) = d((r, x); B) = r$ and $\forall x, x > m(r), d((x, r); B) = d((r, x); B) > r$. The diagram in Fig. 2.8 shows the geometric structure of $m(r)$'s.

From [56] we have

Theorem 2.16. *For a well-behaved B (metric $d(B)$)*

$$m(r) = \lfloor r/p \rfloor \cdot (f(p) - p) + f(r \bmod p) - (r \bmod p)$$

□

Note that $m(r)$ is a function of B, and in specific cases may degenerate to 0 or to r giving rise to diamonds or squares.

We present a generalization of this result in n-D in Section 2.5.5 on hyperspheres of hyperoctagonal distances.

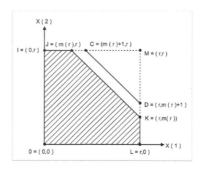

FIGURE 2.8: Vertices of the octagon of a metric $d(B)$ (using wb B).

2.4.2.1   Best Simple Octagonal Distance

We find from Table 2.6 that the summation part of $d(B)$ is often a fairly complex integer function. For example, if $B = \{1, 1, 2, 2\}$, then the sum is $\lfloor(\alpha + 1)/6\rfloor + \lfloor(\alpha + 3)/6\rfloor + \lfloor(\alpha + 4)/6\rfloor + \lfloor(\alpha + 5)/6\rfloor$, for $\alpha = x_1 + x_2$, whereas for $B = \{1, 2\}$, it is $\lfloor(\alpha + 1)/3\rfloor + \lfloor(\alpha + 2)/3\rfloor = \lceil 2\alpha/3 \rceil$. This is not a coincidence. For some well-behaved B's the sum turns out to be a single ceiling function. Such distances are easy to handle and efficient to perform computations with. These are called **simple octagonal distances** [55]. The following theorem from [55] shows that for every p and $p \le f(p) \le 2p$, there exists a unique B that defines a simple $d(B)$.

Theorem 2.17. $\forall \mathbf{x} \in \Sigma_2$, $d(\mathbf{x}; B)$ *is simple and of the form* $\max(x_1, x_2, \lceil (x_1 + x_2)/m \rceil)$ *iff* $b(i) = \lfloor if(p)/p\rfloor - \lfloor (i-1)f(p)/p\rfloor$, $1 \le i \le p$, *where* $1 < m < 2$, $m = f(p)/p$, *and* $f(p)$ *and* p *are relatively prime. In addition, for* $m = 1$, $B = \{1\}$ *and* $d(\mathbf{x}; B) = x_1 + x_2$, *and for* $m = 2$, $B = \{2\}$ *and* $d(\mathbf{x}; B) = \max(x_1, x_2)$ *are also simple.*   □

Example 2.14. *For* $B = \{1, 1, 2, 1, 2\}$, $d(\mathbf{x}; B) = \max(x_1, x_2, \lceil 5(x_1 + x_2)/7 \rceil)$ *is simple and has special significance for its excellent accuracy in approximation.*

Examples of a few other simple octagonal distances are presented in Table 2.6. These are marked as "Simple" in the table.   □

Minimization of the average absolute (normalized) and average relative errors of these simple distances with regard to the Euclidean norm is used to identify the best approximate digital distances in 2-D digital geometry. It is shown in [55] that the neighborhood sequences $\{2\}$, $\{1, 2\}$, $\{1, 1, 2\}$, and $\{1, 1, 2, 1, 2\}$ have special significance in distance measurement in digital geometry as they are simple, and no octagonal distance of reasonable neighborhood structure is expected to offer a better accuracy of approximation.

2.4.3 Weighted t-Cost Distance

In Section 2.3.2 it was shown that the t-cost distance $D_t^n(\mathbf{u})$ is a metric in $\mathbb{Z}^n$. Further, it was presented in Section 2.3.2.4 that it is also a metric in $\mathbb{R}^n$. In this section it is generalized by composing various t-cost norms with a set of positive weights. This is known as the **weighted t-cost norm** [150] $WD^n(\mathbf{x}, \mathbf{y}; \mathbf{w}) : \mathbb{R}^n \times \mathbb{R}^n \to \mathbb{R}^+ \cup \{0\}$ and is defined as follows:

Definition 2.28. $\forall n, n \geq 1$ *and* $\forall \mathbf{x}, \mathbf{y} \in \mathbb{R}^n$

$$WD^n(\mathbf{u}; \mathbf{w}) = \max_{1 \leq t \leq n} \{w_t . D_t^n(\mathbf{u})\}$$

where $\mathbf{u} = |\mathbf{x} - \mathbf{y}|$ *and* $\mathbf{w} = \{w_t : w_t \in \mathbb{R}^+, 1 \leq t \leq n\}$. *That is, w_t's are non-negative real constants.* □

It may be noted that distances computed by $WD^n(\mathbf{u})$ are real-valued. A modification to its form for integral distance values could be made in various ways like:

$$\widetilde{WD^n}(\mathbf{u}; \mathbf{w}) = \max_{1 \leq t \leq n} \{\lceil w_t . D_t^n(\mathbf{u}) \rceil\}$$

The metric property of $WD^n(\mathbf{u}; \mathbf{w})$ is given in the following theorem.

Theorem 2.18. $WD^n(\mathbf{u}; \mathbf{w})$ *is a metric in n-D real space $\mathbb{R}^n$.*

Proof. The proof follows from the facts that every $D_t^n(\mathbf{u})$ is a metric, weights w_t's are non-negative and the composing function max is an MPT. □

Corollary 2.7. $\widetilde{WD^n}(\mathbf{u}; \mathbf{w})$ *is a metric in n-D real space $\mathbb{R}^n$.* □

As an m-neighbor norm, $d_m^n(\mathbf{u}) = \max\left(D_1^n(\mathbf{u}), \left\lceil \frac{D_n^n(\mathbf{u})}{m} \right\rceil\right)$, WD^n specializes to d_m^n for $w_1 = 1$, $w_n = \frac{1}{m}$ and all other weights being zero. Also it becomes the usual t-cost norm D_t^n, when $w_t = 1$ and all other weights are zero. Hence, WD^n provides a generalization with infinite possibilities including both d_m^n and D_t^n norms.

In [167], d_m^n and D_t^n are generalized by the class of **t-cost-m-Neighbor distance**. WD^n presents another perspective for generalizing them.

2.5 Hyperspheres of Digital Distances

In the last section we studied a number of distance functions in n-D for various neighborhood set notions. Conditions for their metricity were established and the relationships to shortest paths elucidated. We have also observed that many of them preserve a certain order of their computed distance values.

The length of the shortest path, however, provides only one type of measure defined by such distance functions. Considering the hyperspheres formed by $d(N(\cdot))$, we get two more measurable quantities, namely the volume of the sphere and the area of its surface. In discrete point space, the meaning of such hyperspheres and their corresponding volumes and surface areas are analyzed in this section.

A hypersphere H of a given radius r around the origin $\mathbf{0}$ is the set of points whose distance measure from $\mathbf{0}$ is less than or equal to r. The set of points whose distance measure from $\mathbf{0}$ is exactly equal to r is defined as the surface S of the sphere. The number of points in H gives us the volume of the hypersphere, and that of S, the surface area.

For example, in 2-D, the disk (of *radius r*) of City Block distance is an isothetic diamond with vertices at $\{(\pm r, 0), (0, \pm r)\}$. It has an area of $2r^2 + 2r + 1$ and a perimeter of $4r$. Similarly, the disk of Chessboard distance is isothetic square with vertices at $\{(\pm r, \pm r)\}$, area $4r^2 + 4r + 1$, and perimeter $8r$. Also, the diamond is totally contained within the square.

Likewise in 3-D, lattice, d_{18} and grid distances define three convex digital solids with the following properties (Table 2.8):

TABLE 2.8: Properties of hyperspheres of common distances in 2-D and 3-D.

Distance	Vertices	Perimeter / Surface Area	Area / Volume
City Block	$\{(\pm r, 0), (0, \pm r)\}$	$4r$	$2r^2 + 2r + 1$
Chessboard	$\{(\pm r, \pm r)\}$	$8r$	$4r^2 + 4r + 1$
Lattice	$\{(\pm r, 0, 0), (0, \pm r, 0), (0, 0, \pm r)\}$	$24r^2 + 2$	$18r^3 + 12r^2 + 6r + 1$
d_{18}	$\{(\pm r, \pm r, 0), (\pm r, 0, \pm r), (0, \pm r, \pm r)\}$	$20r^2 - 4r + 2$	$\frac{20}{3}r^3 + 8r^2 + \frac{10}{3}r + 1$
Grid	$\{(\pm r, \pm r, \pm r)\}$	$4r^2 + 2$	$\frac{4}{3}r^3 + 2r^2 + \frac{8}{3}r + 1$

In this section we present closed-form expressions for volumes and surfaces of hyperspheres. Using absolute and relative volumetric and surface error measures, we also study how well these digital hyperspheres approximate the Euclidean hypersphere.

2.5.1 Notions of Hyperspheres

Definition 2.29. $S(N(\cdot); r)$ *is the* **hypersurface** *of radius r in n-D for neighborhood set $N(\cdot)$. It is the set of n-D grid points that lie exactly at a distance r, $r \geq 0$, from the origin when $d(N(\cdot))$ is used as the distance.*

$$S(N(\cdot); r) = \{\mathbf{x} : \mathbf{x} \in \mathbb{Z}^n, d(\mathbf{x}; N(\cdot)) = r\}$$

The **surface area** $surf(N(\cdot); r) = \|S(N(\cdot); r)\|$ *of a hypersurface $S(N(\cdot); r)$ is defined as the number of points in $S(N(\cdot); r)$.* $\square$

In the digital space $surf(N(\cdot); r)$ is often a polynomial in r of degree $n-1$ with rational coefficients.

Definition 2.30. $H(N(\cdot); r)$ *is the* **hypersphere** *of radius* r *in* n-*D for neighborhood Set* $N(\cdot)$. *It is the set of* n-*D grid points that lie within at a distance* r, $r \geq 0$, *from the origin when* $d(N(\cdot))$ *is used as the distance.*

$$H(N(\cdot); r) = \{\mathbf{x} : \mathbf{x} \in \mathbb{Z}^n, 0 \leq d(\mathbf{x}; N(\cdot)) \leq r\}$$

The **volume** $vol(N(\cdot); r) = ||H(N(\cdot); r)||$ *of a hypersphere* $H(N(\cdot); r)$ *is defined as the number of points in* $H(N(\cdot); r)$. □

In the digital space $vol(N(\cdot); r)$ is often a polynomial in r of degree n with rational coefficients.

Note that

$$H(N(\cdot); r) = \bigcup_{s=0}^{r} S(N(\cdot); s) \quad and \quad vol(N(\cdot); r) = \sum_{s=0}^{r} surf(N(\cdot); r).$$

Definition 2.31. *The* **shape feature** $\psi_n(N(\cdot))$ *of a hypersphere in* n-*D is defined as:*

$$\psi_n(N(\cdot)) = \lim_{r \to \infty} \frac{(surf(N(\cdot); r)^n}{(vol(N(\cdot); r))^{n-1}}$$

□

$vol(N(\cdot); r)$ and $surf(N(\cdot); r)$ being polynomials in r of degree n and $n-1$, respectively, $\psi_n(N(\cdot))$ is a dimension-less quantity characterized by the parameters of the neighborhood set $N(\cdot)$. Further, if $vol(N(\cdot); r)$ and $surf(N(\cdot); r)$ contain the term for the highest power of r alone, the limit in the above definition becomes redundant.

Though the above definition of the shape feature is provided in general for n-D, it finds major use up to three dimensions, only where the surfaces and volumes (and therefore *shapes*) can be physically visualized.

Definition 2.32. *An* $\mathbf{x} \in \mathbb{Z}^n$ *is said to be a* **vertex** *or* **corner** *of* $H(N(\cdot); r)$ *if* $\mathbf{x} \in H(N(\cdot); r)$ *and*

$$x_i = \max\{y : d((x_1, x_2, \cdots, x_{i-1}, y, 0, 0, \cdots, 0); r) = r, 1 \leq i \leq n.$$

In other words, $\forall 1 \leq i \leq n$,

$$d((x_1, x_2, \cdots, x_{i-1}, x_i, 0, 0, \cdots, 0); r) \quad = \quad r \; and$$
$$d((x_1, x_2, \cdots, x_{i-1}, x_i + 1, 0, 0, \cdots, 0); r) \quad > \quad r.$$

□

Note that, by symmetry, if $\mathbf{x}$ is a vertex of $H(N(\cdot); r)$, then all points belonging to $\phi(\mathbf{x})$ are vertices of $H(N(\cdot); r)$.

2.5.2 Euclidean Hyperspheres

To compare the digital hyperspheres of various neighborhood sets with the hyperspheres $H_E^n(r)$, $r \geq 0$, of Euclidean norm in n-D, we note the following:

Definition 2.33. *The* **Euclidean surface area, volume,** *and* **shape feature** *measures are given as follows:*

Volume:	$V_n^E(r)$	$=$	$\{\mathbf{x} : \mathbf{x} \in \mathbb{Z}^n, E_n(\mathbf{x}) \leq r\}$	$= L_n r^n$
Surface Area:	$S_n^E(r)$	$=$	$\{\mathbf{x} : \mathbf{x} \in \mathbb{Z}^n, E_n(\mathbf{x}) = r\}$	$= \frac{d}{dr} V_n^E(r) = n L_n r^{n-1}$
Shape Feature:	ψ_n^E	$=$	$\frac{(S_n^E(r))^n}{(V_n^E(r))^{n-1}}$	$= n^n L_n$

where

$$L_n = \frac{\pi^{\lfloor n/2 \rfloor}}{\prod_{i=0}^{\lceil n/2 \rceil - 1} \left(\frac{n}{2} - i\right)}$$

□

The comparison is quantified in terms of the following error measures:

Definition 2.34. *The* **error measures** *between Euclidean and digital hyperspheres are defined as follows:*

Surface Error:	$E_S^n(N(\cdot))$	$=$	$\lim_{r \to \infty} \left\| \frac{S_n^E(r) - surf(N(\cdot);r)}{r^{n-1}} \right\|$
Volume Error:	$E_V^n(N(\cdot))$	$=$	$\lim_{r \to \infty} \left\| \frac{V_n^E(r) - vol(N(\cdot);r)}{r^n} \right\|$
Shape Feature Error:	$E_\psi^n(N(\cdot))$	$=$	$\left\| \psi_n^E - \psi_n(N(\cdot)) \right\|$
Absolute Volumetric Error:	$A_V(N(\cdot);r)$	$=$	$\left\| V_n^E(r) - vol(N(\cdot);r) \right\|$
Relative Volumetric Error:	$R_V(N(\cdot);r)$	$=$	$\frac{A_V(N(\cdot);r)}{V_n^E(r)}$
Absolute Surface Error:	$A_S(N(\cdot);r)$	$=$	$\left\| S_n^E(r) - surf(N(\cdot);r) \right\|$
Relative Surface Error:	$R_S(N(\cdot);r)$	$=$	$\frac{A_S(N(\cdot);r)}{S_n^E(r)}$

□

Now we present expressions for $surf(N(\cdot);r)$ and $vol(N(\cdot);r)$ for various neighborhood sets and compare these quantities against the Euclidean hyperspheres in the sense of the above error measures.

2.5.3 Hyperspheres of m-Neighbor Distance

First we explore the properties of hypersurfaces and hyperspheres of $O(m)$-neighbor distance d_n^m from [70]. We use the following notation:

$$S(m, n; r) \implies S(N(\cdot); r), \quad surf(m, n; r) \implies surf(N(\cdot); r) \ and$$

$$H(m, n; r) \implies H(N(\cdot); r), \quad vol(m, n; r) \implies vol(N(\cdot); r).$$

We present a few results that are used in the measurement Theorem 2.19. These follow from binomial expansion and enumerations for solutions of simple integer equations.

Lemma 2.10. $\forall r, t \in \mathbb{N}$ *the following holds:* $h_n(r,s) =$

$$(-1)^n \sum_{i=\lceil s/r \rceil}^{n} \binom{n}{i} (-2)^i \times \left[\sum_{j=0}^{\lfloor s/(r+1) \rfloor} \binom{i}{j} \binom{s+i-1-(r+1)j}{i-1} (-1)^j \right]$$

where $h_n(r,s) = $ *the number of distinct ways to select* $\mathbf{x}$ *from* $\mathbb{Z}^n$ *to satisfy the equation*

$$\sum_{i=1}^{n} |x_i| = s, 0 \le x_i \le r, 1 \le i \le n.$$

$h_n(r,s)$ *is a polynomial of degree* $n-1$ *in* s *with rational coefficients.* $\qquad \square$

Corollary 2.8. *Special cases of* $h_n(r,s)$:

$$h_n(r,s) = \begin{cases} 0 & s < 0, \\ 1 & s = 0, \\ (-1)^n \sum_{t=1}^{n} \binom{n}{t} (-2)^t \binom{t-1+s}{s} & r \ge s. \end{cases}$$

$\qquad \square$

Now we present the surface area theorem from [70].

Theorem 2.19. $\forall m, n \in \mathbb{N}, m \le n,$ *and* $r \in \mathbb{N}, surf(m,n;r) =$

$$\sum_{k=1}^{\lfloor m(r-1)/r \rfloor} \left[\sum_{s=0}^{\lceil (m-k)r-m \rceil} h_{n-k}(r-1,s) \right] \binom{n}{k} . 2^k + \sum_{s=m(r-1)+1}^{mr} h_n(r,s)$$

where $h_n(r,s)$ *is as defined in Lemma 2.10.*

Proof. $surf(m,n;r)$ is the number of distinct solutions in $\mathbf{x} \in \mathbb{Z}$ for $d_m^n(\mathbf{x}) = r$. The counting is done with the help of the above lemmas. $\qquad \square$

Corollary 2.9. $surf(m,n;r)$ *and* $vol(m,n;r)$ *are polynomials in* r *with rational coefficients of degree* $n-1$ *and* n, *respectively.* $\qquad \square$

Example 2.15. *Let us take an example to illustrate the theorem. Consider* $n = 3, m = 2, r = 2$ *to list the points on the surfaces:*

$$
\begin{aligned}
S(2,3;2) &= \{(\pm 2, 0, 0), (0, \pm 2, 0), (0, 0, \pm 2), \\
&\quad (\pm 2, \pm 1, 0), (\pm 2, 0, \pm 1), (\pm 1, \pm 2, 0), \\
&\quad (\pm 1, 0, \pm 2), (0, \pm 1, \pm 2), (0, \pm 2, \pm 1), \\
&\quad (\pm 2, \pm 2, 0), (\pm 2, 0, \pm 2), (0, \pm 2, \pm 2), \\
&\quad (\pm 2, \pm 1, \pm 1), (\pm 1, \pm 2, \pm 1), (\pm 1, \pm 1, \pm 2), \\
&\quad (\pm 1, \pm 1, \pm 1)\} \\
S(2,3;1) &= \{(\pm 1, 0, 0), (0, \pm 1, 0), (0, 0, \pm 1) \\
&\quad (\pm 1, \pm 1, 0), (\pm 1, 0, \pm 1), (0, \pm 1, \pm 1)\} \\
S(2,3;0) &= \{(0, 0, 0)\}
\end{aligned}
$$

Next we verify the polynomials derived above.

$$||S(2,3;2)|| = 74 = 2 - 4*2 + 20*2^2 = surf(2,3;2)$$
$$||S(2,3;1)|| = 18 = 2 - 4 + 20 = surf(2,3;1)$$
$$||H(2,3;2)|| = 1 + ||S(2,3;1)|| + ||S(2,3;2)|| = 1 + 18 + 74$$
$$= 93 = 1 + \frac{10}{3}*2 + 8*2^2 + \frac{20}{3}*2^3$$
$$= vol(2,3;2),$$
$$||H(2,3;1)|| = 1 + ||S(2,3;1)|| = 1 + 18$$
$$= 19 = 1 + \frac{10}{3} + 8 + \frac{20}{3}$$
$$= vol(2,3;1).$$

$\square$

TABLE 2.9: Volume and surface polynomials [70] for hyperspheres ($n = 1, 2, ..., 5$).

Params		Coefficients in $surf(m,n;r)$					Coefficients in $vol(m,n;r)$						Euclidean	
n	m	r^0	r^1	r^2	r^3	r^4	r^0	r^1	r^2	r^3	r^4	r^5	$S_n^E(r)$	$V_n^E(r)$
1	1	2	-	-	-	-	1	2	-	-	-	-	0	$2r$
2	1	0	4	-	-	-	1	2	2	-	-	-	$2\pi r$	πr^2
2	2	0	8	-	-	-	1	4	4	-	-	-		
3	1	2	0	4	-	-	1	$\frac{8}{3}$	2	$\frac{4}{3}$	-	-	$4\pi r^2$	$\frac{4}{3}\pi r^3$
3	2	2	-4	20	-	-	1	$\frac{10}{3}$	8	$\frac{20}{3}$	-	-		
3	3	2	0	24	-	-	1	6	12	8	-	-		
4	1	0	$\frac{16}{3}$	0	$\frac{8}{3}$	-	1	$\frac{8}{3}$	$\frac{10}{3}$	$\frac{4}{3}$	$\frac{2}{3}$	-	$2\pi^2 r^3$	$\frac{1}{2}\pi^2 r^4$
4	2	0	16	-16	32	-	1	$\frac{16}{3}$	8	$\frac{32}{3}$	8	-		
4	3	0	$\frac{32}{3}$	-8	$\frac{184}{3}$	-	1	4	$\frac{50}{3}$	28	$\frac{46}{3}$	-		
4	4	0	16	0	64	-	1	8	24	32	16	-		
5	1	2	0	$\frac{20}{3}$	0	$\frac{4}{3}$	1	$\frac{46}{15}$	$\frac{10}{3}$	$\frac{8}{3}$	$\frac{2}{3}$	$\frac{4}{5}$	$\frac{8}{3}\pi^2 r^4$	$\frac{8}{15}\pi^2 r^5$
5	2	2	$\frac{-32}{3}$	52	$\frac{-88}{3}$	36	1	$\frac{62}{15}$	$\frac{40}{3}$	44	32	$\frac{36}{5}$		
5	3	2	$\frac{7}{3}$	60	$\frac{-176}{3}$	124	1	$\frac{46}{5}$	$\frac{50}{3}$	32	$\frac{142}{3}$	$\frac{124}{3}$		
5	4	2	-8	$\frac{196}{3}$	-8	$\frac{476}{3}$	1	$\frac{18}{5}$	$\frac{80}{3}$	212	$\frac{232}{3}$	$\frac{476}{15}$		
5	5	2	0	80	0	160	1	10	40	80	80	32		

Reprinted from *Information Sciences*, 50(1990), P. P. Das and B. N. Chatterji, Hyperspheres in Digital Geometry, 73–91, Copyright (1990), with permission from Elsevier.

From Corollary 2.9 and the Table 2.9 we observe the following:

1. Similar to the Euclidean norm, the surface and volume expressions in n-D digital space are polynomials in r of degree $n-1$ and n, respectively.

2. Unlike the Euclidean measures that involve π, the coefficients in these

polynomials are rational. This is due to the fact that in discrete cases, all operations are performed with integers and hence in no way can an irrational coefficient arise.

3. These polynomials are order preserving over m and r as the underlying distance measure d_m^n maintains an order.

Corollary 2.10. *The following hold for surfaces and volumes:*

1. $H(m, n; r) \subset H(m, n; r + 1)$ *and* $vol(m, n; r) < vol(m, n; r + 1)$.

2. $S(m, n; r) \cap S(m, n; r + 1) = \phi$, *but* $surf(m, n; r) < surf(m, n; r + 1)$.

3. $H(m, n; r) \subset H(m + 1, n; r)$ *and* $vol(m, n; r) < vol(m + 1, n; r)$.

4. $surf(m, n; r) < surf(m + 1, n; r)$, *but* $S(m, n; r)$ *and* $S(m + 1, n; r)$ *are unrelated.*

$\square$

$vol(m, n; r)$, being a polynomial of degree n in r (Corollary 2.9), its coefficients are computed from the values of volume for $n + 1$ distinct values of $r = 0, 1, 2, ..., n$ and solving for $n+1$ simultaneous equations in $n+1$ coefficients using matrix methods. [70] illustrates this approach in depth.

2.5.3.1   Vertices of Hyperspheres

In Table 2.8, we present the vertices of the hyperspheres of d_m^n in 2-D and 3-D. The following theorem generalizes the result.

Theorem 2.20. *The vertices of* $H(m, n; r)$ *are given by* $\phi(\mathbf{x})$ *where*

$$\mathbf{x} = (\underbrace{r, r, \cdots, r}_{m}, \underbrace{0, 0, \cdots, 0}_{n-m})$$

and $\phi(\cdot)$ *is the* 2^n *symmetry function (Definition 2.2).* $\square$

2.5.3.2   Errors in Surface and Volume Estimations

We present the error between digital and Euclidean hyperspheres from [70].

Theorem 2.21. $\forall n, n \geq 1$ *and neighborhood* $m \leq n$, *the volumetric errors between Euclidean and digital hyperspheres satisfy the following:*

- *Absolute Error is unbounded and*

- *Relative Error is bounded*

Proof. From Corollary 2.9, we get $A_V(m, n; r) =$

$$\left| V_n^E(r) - vol(m, n; r) \right| = \left| (L_n - v_n) r^n - \sum_{i=0}^{n-1} v_i r^i \right| \approx \left| (L_n - v_n) r^n \right|, r \to \infty$$

Hence, the absolute error is unbounded.

$$R_V(m, n; r) = \frac{A_V(m, n; r)}{V_n^E(r)}, r \geq 1$$

$$= \left| \sum_{i=0}^{n} \frac{v_i}{L_n} r^{i-n} - 1 \right| = \left| \sum_{i=0}^{n} \frac{v_i}{L_n} r^{-i} - 1 \right| < \sum_{i=0}^{n} \left| \frac{v_i}{L_n} \right| r^{-i} + 1$$

$$< \sum_{i=0}^{n} \frac{|v_i|}{L_n} + 1$$

Hence, the relative error is bounded. Note that v_i's are dependent on m and this upper bound is used to select an optimal m for a given n to get the *best* m-Neighbor distance in the volumetric sense. $\square$

Similar results follow for errors $A_S(m, n; r)$ and $R_S(m, n; r)$ in surface area.

2.5.3.3 Hyperspheres of Real m-Neighbor Distance

Next we extend the notion of $H(m, n; r)$ for real m-Neighbor distance δ_m^n. Besides the expression of volume like before, the explorations in the real (continuous) space offer nice results in terms of the vertices of the hyperspheres and their inscribed and circumscribed Euclidean hyperspheres.

We start with the definitions of hypersphere and its volume in the continuous case.

Definition 2.35. *An* **m-hypersphere** $H(m, n; r)$ *in n-D, of radius r and center* $\mathbf{0}$, *is defined as a subset of* $\mathbb{R}^n$ *as follows:*

$$H(m, n; r) = \{ \mathbf{x} : \mathbf{x} \in \mathbb{R}^n, \ \delta_m^n(\mathbf{x}) \leq r \}.$$

H being 2^n-*symmetric, it is sufficient to restrict our attention to the all-positive hyperoctant* $\mathbb{R}_+^n$ *where* $x_i \geq 0, \forall i, 1 \leq i \leq n$. $\square$

Definition 2.36. *The* **volume** $\|H(m, n; r)\|$ *of an m-hypersphere is defined as:* $\|H(m, n; r)\| =$

$$\int_{\delta_m^n(\mathbf{x}) \leq r} d\mathbf{x}, \quad \text{where } d\mathbf{x} = dx_1 dx_2 \ldots dx_n \text{ and } \mathbf{x} = (x_1, x_2, \cdots, x_n).$$

$\square$

The expression for $||H(m, n; r)||$ as presented in Theorem 2.22 is derived in [66] with the use of the inclusion–exclusion principle from combinatorics.

Theorem 2.22. $\forall n \in \mathbb{N}, 0 < m \leq n, r \geq 0$, *the volume of the m-hypersphere* $H(m, n; r)$ *is given by* $||H(m, n; r)|| = v_n(m) \cdot r^n$, *where*

$$v_n(m) = (2^n/n!) \cdot \sum_{j=0}^{\lceil m \rceil - 1} \binom{n}{j} \cdot (m - j)^n \cdot (-1)^j.$$

[66] also establishes the vertices of the hypersphere as:

Theorem 2.23. *The vertices of the polytope* $H(m, n; r)$ *are* $\phi(\mathbf{x})$, *where*

$$\mathbf{x} = (\underbrace{r, r, \cdots, r}_{\lfloor m \rfloor}, (m - \lfloor m \rfloor)r, \underbrace{0, 0, \cdots, 0}_{n - \lceil m \rceil})$$

and $\phi(\cdot)$ *is the* 2^n *symmetry function of n-D point. Given* $\mathbf{x} \in (\mathbb{R}^+)^n$, $\phi(\mathbf{x})$ *gives the set of points in* $\mathbb{R}^n$ *obtained by the reflection and permutation of* $\mathbf{x}$.

The Euclidean hyperspheres that can be **inscribed within** or **circumscribed around** these m-hyperspheres demonstrate interesting properties. While the former is called an **insphere**, the latter is termed as a **circumsphere**. We define the *inradius* r_I and the *circumradius* r_C, respectively, as:

$$\begin{aligned} r_I &= \max\{r' : H_E^n(r') \subseteq H(m, n; r)\} \text{ and} \\ r_C &= \min\{r' : H_E^n(r') \supseteq H(m, n; r)\} \end{aligned}$$

where $H_E^n(r')$ is a Euclidean hypersphere of radius r' with the center at the origin. Note that r_I and r_C both are functions of m, n, and r.

Clearly, $H_E^n(r_I)$ and $H_E^n(r_C)$ are the insphere and circumsphere of $H(m, n; r)$, respectively. They touch the $H(m, n; r)$ at the furthest inner points $\mathbf{t}_I$ and nearest outer points $\mathbf{t}_C$, where $\delta_m^n(\mathbf{t}_I) = \delta_m^n(\mathbf{t}_C) = r$ and $E_n(\mathbf{t}_I) = r_I$ and $E_n(\mathbf{t}_C) = r_C$. In the next theorem we present the expressions for these quantities.

Theorem 2.24. $\forall n, n \geq 1, 0 < m \leq n$ *we have*

1. $r_I = \min(1, m/\sqrt{n}) \cdot r$ *and* $r_C = \sqrt{(\lfloor m \rfloor + (m - \lfloor m \rfloor)^2)} \cdot r$

$$\begin{aligned} \mathbf{t}_I &\in \phi(r, 0, 0, \cdots, 0) \quad for \ m \leq \sqrt{n} \\ &\in \phi(mr/n, mr/n, \cdots, mr/n) \quad for \ m \geq \sqrt{n} \ and \end{aligned}$$

2.

$$\mathbf{t}_C \in \phi(\underbrace{r, r, \cdots, r}_{\lfloor m \rfloor}, (m - \lfloor m \rfloor)r, \underbrace{0, 0, \cdots, 0}_{n - \lceil m \rceil})$$

where $\phi(\cdot)$ *is the* 2^n *symmetry function of an n-D point.* $\square$

Example 2.16. *Let $n = 2$. There are two ranges for m.*
Case 1: $\lceil m \rceil = 1$ or $0 < m \le 1$:

$\delta_m^2((x_1, x_2)) = (x_1 + x_2)/m$ and $H(m, 2; r) = \{(x_1, x_2) : x_1, x_2 \in \mathbb{R}$ and $(x_1 + x_2) \le r\}$. $H(m, 2; r)$, therefore, is a diamond with vertices $\{(\pm mr, 0), (0, \pm mr)\}$. By 2^2 symmetry, this is a triangle in the first quadrant with vertices $\{(0, 0), (mr, 0), (0, mr)\}$. So, $\|H(m, 2; r)\| = 4 * \frac{1}{2} * mr * mr = (2m^2)r^2$ and $v_2(1, m) = 2m^2$.

Case 2: $\lceil m \rceil = 2$ or $1 < m \le 2$:

$\delta_m^2((x_1, x_2)) = \max(\max(x_1, x_2), (x_1 + x_2)/m)$ and $H(m, 2; r) = \{(x_1, x_2) : x_1, x_2 \in \mathbb{R}$ and $\max(\max(x_1, x_2), (x_1 + x_2)/m) \le r\}$. $H(m, 2; r)$, therefore, is an octagon with vertices $\{(\pm r, \pm (mr - r)), (\pm (mr - r), \pm r)\}$. By 2^2 symmetry, this is a pentagon in the first quadrant with vertices $\{(0, 0), (r, 0), (r, mr - r), (mr - r, r), (0, r)\}$. Using inclusion–exclusion, $\|H(m, 2; r)\| = 4 * (\frac{1}{2} * (mr)^2 - 2 * \frac{1}{2} * (mr - r)^2) = 2 * (m^2 - 2m^2 + 4m - 2) * r^2 = (-2m^2 + 8m - 4)r^2$ and $v_2(2, m) = -2m^2 + 8m - 4$.

□

2.5.4 Hyperspheres of t-Cost Distance

We start with the hyperspheres for the generalized cost-composite distance functions d_p (Section 2.3.2).

Definition 2.37. *Hypersphere of a d_p is defined as:*

$$HS_p^n(r) = \{\mathbf{x} : \mathbf{x} \in \mathbb{Z}^n, 0 \le d_p(\mathbf{x}) \le r\}, \qquad r \ge 0.$$

□

Clearly the hyperspheres of non-metric d_p's are infinite. Hence,

Lemma 2.11. $\forall n, n \ge 1, 1 \le p \le 2^{n-1} - 1, HS_p(r)$ *is an infinite set for any $r \ge 0$.* □

However, whenever d_p is a metric (or semi-metric), we observe interesting shapes of these digital hyperspheres. Clearly in 2-D, $HS_2^3(r)$ (for Chessboard distance) is a *square* and $HS_3^2(r)$ (for City Block distance) is *diamond*. $HS_1^2(r)$, on the other hand, is in the shape of a cross extending to infinity on either arm (see Fig. 2.9).

In 3-D, $HS_4^3(r)$ is a *cube* and $HS_7^3(r)$ is a *octahedron*. The shape of $HS_5^3(r)$ is a combination of digital pyramids and cubes and $HS_5^3(r)$ is a peculiar solid for a semi-metric d_5. Check the shapes of $HS_6^3(6)$ and $HS_5^3(4)$ in Fig. 2.10. $HS_p^3(r)$ for $p = 1, 2$, and 3 are quite intricate infinite 3-D solids.

Volumes and surface areas of $HS_p^n(r)$ would be interesting to compute

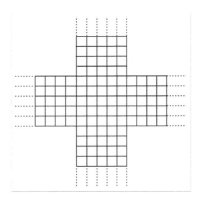

FIGURE 2.9: $HS_1^2(2)$ or *circle* of d_1, in 2-D for radius = 2. HS, extends to infinity on all four directions.

for $2^{n-1} \le p \le 2^n - 1$ (as d_p is a non-metric for $1 \le p \le 2^{n-1} - 1$; from Lemma 2.11 these quantities would be unbounded). Unfortunately, there are no general results for these in the literature. So only the results of 2- and 3-D are presented here.

Lemma 2.12. *In 2-D*

$$
\begin{aligned}
S_2(r) &= 8r \\
S_3(r) &= 4r
\end{aligned}
\qquad
\begin{aligned}
V_2(r) &= 4r^2 + 4r + 1 \\
V_3(r) &= 2r^2 + 2r + 1
\end{aligned}
$$

and in 3-D,

$$
\begin{aligned}
S_4(r) &= 24r^2 + 2. \\
S_5(r) &= 12r^2 + 6, \quad r\ odd \\
&= 12r^2 + 2, \quad r\ even \\
S_6(r) &= 6r^2 \qquad\quad r\ odd \\
&= 6r^2 + 2, \quad r\ even \\
S_7(r) &= 4r^2 + 2. \\
\\
S_p(r) &= 1. \qquad\qquad r = 0.
\end{aligned}
\qquad
\begin{aligned}
V_4(r) &= 8r^3 + 12r^2 + 6r + 1 \\
V_5(r) &= 4r^3 + 6r^2 + 6r + 3 \qquad r\ odd \\
&= 4r^3 + 6r^2 + 6r + 1 \qquad r\ even \\
V_6(r) &= 2r^3 + 3r^2 + 2r \qquad\quad r\ odd \\
&= 2r^3 + 3r^2 + 2r + 1 \qquad r\ even \\
V_7(r) &= 4/3r^3 + 2r^2 + 8/3r + 1
\end{aligned}
$$

$\square$

2.5.5 Hyperspheres of Hyperoctagonal Distances

In this section we explore the properties of the hyperspheres of hyperoctagonal distances $d(B)$ (Section 2.4.1). Hence, the neighborhood set is taken

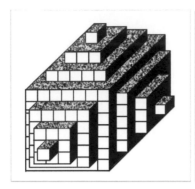 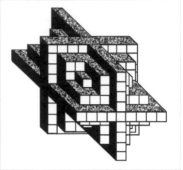

Reprinted from *Information Sciences* 59(1992), P. P. Das, J. Mukherjee and B. N. Chatterji, The t-Cost Distance

in Digital Geometry, 1–20, Copyright (1992), with permission from Elsevier.

FIGURE 2.10: $HS_6^3(6)$ or *sphere* of d_6, in 3-D for radius $= 6$ and $HS_5^3(4)$ or *sphere* of d_5, in 3-D for radius $= 4$.

as the neighborhood sequence B. We use the following notation:

$$H(B;r) \implies H(N(\cdot);r).$$

2.5.5.1   Vertices of Hyperspheres and Approximations

The vertices of a $H(B)$ are given in the following theorem from [64].

Theorem 2.25. *For a well-behaved B, an $H(B;r)$ has its corners at $\phi(\mathbf{x})$ (where $\phi(\cdot)$ is the 2^n symmetry function in Definition 2.2), with $\mathbf{x}$ computed as follows:*

$$x_i = \lfloor r/p \rfloor \cdot (f_i(p) - f_{i-1}(p)) + f_i(r \bmod p) - f_{i-1}(r \bmod p), 1 \le i \le n$$

where

$$
\begin{aligned}
Length: \quad & p \;=\; |B| \\
N - Sequence: \quad & B \;=\; \{b(1), b(2), \cdots, b(p)\} \\
Trimmed\ B: \quad & B_i \;=\; \{b_i(1), b_i(2), \cdots, b_i(p)\}, \\
& \qquad b_i(j) = \min(b(i), i),\ \forall i, 1 \le i \le p \\
Sum\ Sequence: \quad & F_i \;=\; \{f_i(1), f_i(2), \cdots, f_i(p)\}, \\
& \qquad f_i(j) = \textstyle\sum_{k=1}^{j} b_i(k), \forall i, 1 \le i \le p\ and \\
& \quad F_0 = \{0, 0, \cdots, 0\}\ and\ f_i(0) = 0
\end{aligned}
$$

$\square$

It may be noted that for $r < p$, some of the vertices merge to form degenerate circles and spheres in both 2-D and 3-D. The following bounds hold for all N-Sequences.

Corollary 2.11. $\forall n,$ $B,$ *and* $r,$ $x_1 = r,$ *and* $0 \le x_i \le r,$ $\forall i, 1 \le i \le n.$ ☐

We now illustrate examples in 2-D and 3-D.

Example 2.17. *Let* $n = 2,$ $B = \{1, 1, 2, 1, 2\},$ $p = |B| = 5$ *and* $r = 4.$ *Hence, the vertices of* $H(\{1, 1, 2, 1, 2\}; 4)$ *are given as:*

$$
\begin{aligned}
B_2 &= \{1, 1, 2, 1, 2\}, B_1 = \{1, 1, 1, 1, 1\} \text{ and} \\
F_2 &= \{1, 2, 4, 5, 7\}, F_1 = \{1, 2, 3, 4, 5\}
\end{aligned}
$$

$$
\begin{aligned}
x_1 &= \lfloor r/5 \rfloor \cdot (f_1(5) - f_0(5)) + f_1(r \bmod 5) - f_0(r \bmod 5) \\
&= \lfloor r/5 \rfloor \cdot (5 - 0) + (r \bmod 5 - 0) & = r \\
x_2 &= \lfloor r/5 \rfloor \cdot (f_2(5) - f_1(5)) + f_2(r \bmod 5) - f_1(r \bmod 5) \\
&= \lfloor r/5 \rfloor \cdot (7 - 5) + (f_2(r \bmod 5) - r \bmod 5) & = \lfloor 2r/5 \rfloor
\end{aligned}
$$

With $r = 4,$ *the vertices are* $\{(\pm 4, \pm 1), (\pm 1, \pm 4)\}.$ ☐

Example 2.18. *Let* $n = 3,$ $B = \{1, 1, 3\},$ $p = |B| = 3$ *and* $r = 4.$ *Hence, the vertices of* $H(\{1, 1, 3\}; 4)$ *are given as:*

$$
\begin{aligned}
B_3 &= \{1, 1, 3\}, B_2 = \{1, 1, 2\}, B_1 = \{1, 1, 1\} \text{ and} \\
F_3 &= \{1, 2, 5\}, F_2 = \{1, 2, 4\}, F_1 = \{1, 2, 3\}
\end{aligned}
$$

$$
\begin{aligned}
x_1 &= \lfloor r/3 \rfloor \cdot (f_1(3) - f_0(3)) + f_1(r \bmod 3) - f_0(r \bmod 3) \\
&= \lfloor r/3 \rfloor \cdot (3 - 0) + (r \bmod 3 - 0) & = r \\
x_2 &= \lfloor r/3 \rfloor \cdot (f_2(3) - f_1(3)) + f_2(r \bmod 3) - f_1(r \bmod 3) \\
&= \lfloor r/3 \rfloor \cdot (4 - 3) + (r \bmod 3 - r \bmod 3) & = \lfloor r/3 \rfloor \\
x_3 &= \lfloor r/3 \rfloor \cdot (f_3(3) - f_2(3)) + f_3(r \bmod 3) - f_2(r \bmod 3) \\
&= \lfloor r/3 \rfloor \cdot (5 - 4) + (r \bmod 3 - r \bmod 3) & = \lfloor r/3 \rfloor
\end{aligned}
$$

With $r = 4,$ *the vertices are* $\{(\pm 4, \pm 1, \pm 1), (\pm 1, \pm 4, \pm 1), (\pm 1, \pm 1, \pm 4)\}.$ ☐

Now we consider two special cases:

- Hyperoctagonal distances specialize to m-neighbor distance when $p = 1.$ Hence, $\forall n, n \ge 1, \forall m, 1 \le m \le n$ and $B = \{m\}$

$$
\mathbf{x} = (\underbrace{r, r, \cdots, r}_{m}, \underbrace{0, 0, \cdots, 0}_{n-m}).
$$

  This was presented earlier in Theorem 2.20 for d_m^n (over $\mathbb{Z}^n$) and in Theorem 2.23 for δ_m^n (over $\mathbb{R}^n$). Hence, the above theorem is a true generalization of the same.

- Hyperoctagonal distances specialize to octagonal distance when $n = 2.$ Hence, for $\mathbf{x} \in Z^2$: $x_1 = r$ and $x_2 = \lfloor r/p \rfloor \cdot (f(p) - p) + f(r \bmod p) - (r \bmod p).$

  This was presented earlier in Theorem 2.16.

2.5.5.2   Special Cases of Hyperspheres in 2-D and 3-D

Next we take a deeper look into the disks (or spheres) in 2-D and 3-D for a subclass of hyperoctagonal distances where the N-Sequence B is intrinsically sorted (see Definition 2.22), that is, $B = S(B)$. Examples of these disks and spheres are shown in Figs. 2.11 and 2.12, respectively.

Under this condition, we show that the vertices are nicely approximated and compact forms for the perimeter, area, volume, and shape feature exist.

Approximations of Vertices in 2-D and 3-D

We start by approximating the vertices in 2-D (digital disks) and 3-D (digital spheres). The approximations are used to compute various properties of the digital disks (spheres), and further to identify the best digital approximate to the Euclidean metric. We present these approximate expressions for the coordinates of the vertices in Theorems 2.26 [151] and Theorem 2.27 [154].

B being sorted, in 2-D, we write B as a doublet, $B=[\alpha_1, \alpha_2] = \{1^{\alpha_1}, 2^{\alpha_2}\}$ and $\alpha_1 + \alpha_2 = p$. If we approximate the vertex $\mathbf{x}(r) \in \Sigma_n$, of $H(B;r)$, by $\tilde{\mathbf{x}}(r)$ where $\tilde{\mathbf{x}} = \left(r, \frac{r\alpha_2}{p}\right)$, the difference between $\mathbf{x}(r)$ and $\tilde{\mathbf{x}}(r)$ is bounded by the following theorem from [151].

Theorem 2.26. *For any $r > 0$ and for any sorted B, we have*
1. $r - x_1(r) = 0$;

2. $0 \le \left\{ \frac{r\alpha_2}{p} - x_2(r) \le \alpha_1 - \frac{\alpha_1^2}{p} \right\} \le \begin{cases} \frac{p}{4} & p \text{ even} \\ \frac{(p^2-1)}{4p} & p \text{ odd} \end{cases}$ □

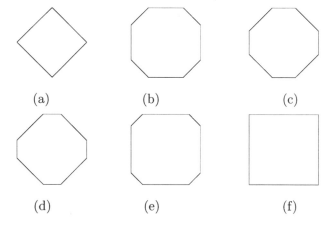

(a)   (b)   (c)

(d)   (e)   (f)

FIGURE 2.11: Digital circles of 2-D octagonal distances for sorted N-Sequences: (a) $\{1\}$, (b) $\{1, 2\}$, (c) $\{1, 1, 2\}$, (d) $\{1, 1, 1, 2\}$, (e) $\{1, 2, 2\}$, and (f) $\{2\}$.

Similarly, in 3-D, we write a sorted B as a triplet, $B=[\alpha_1, \alpha_2, \alpha_3] = \{1^{\alpha_1}, 2^{\alpha_2}, 3^{\alpha_3}\}$ and $\alpha_1 + \alpha_2 + \alpha_3 = p$. If we approximate the vertex $\mathbf{x}(r) \in \Sigma_n$, of $H(B;r)$, by $\tilde{\mathbf{x}}(r)$ where $\tilde{\mathbf{x}} = \left(r, \frac{r(\alpha_2+\alpha_3)}{p}, \frac{r\alpha_3}{p} \right)$, the difference between $\mathbf{x}(r)$ and $\tilde{\mathbf{x}}(r)$ is bounded by the following theorem from [154].

Theorem 2.27. *For any $r > 0$ and for any sorted B, we have*

1. $r - x_1(r) = 0$;

2. $0 \leq \left\{ \begin{array}{l} \frac{r(\alpha_2+\alpha_3)}{p} - x_2(r) \leq \alpha_1 - \frac{\alpha_1^2}{p} \\[2mm] \frac{r\alpha_3}{p} - x_3(r) \leq \alpha_3 - \frac{\alpha_3^2}{p} \end{array} \right\} \leq \left\{ \begin{array}{ll} \frac{p}{4} & p \text{ even} \\[2mm] \frac{(p^2-1)}{4p} & p \text{ odd} \end{array} \right.$ □

Perimeter, Area, Volume and Shape Features

Next we present a few theorems (from [128]) related to the computation of the geometric features of the digital discs (and spheres) when the N-Sequence is intrinsically sorted.

Theorem 2.28. *The perimeter P and area A of a digital disk of radius r for an octagonal distance in 2-D are given by*

$$
\begin{array}{rcccl}
S(\{1^{\alpha_1}, 2^{\alpha_2}\}; r) & = & P(\beta; r) & = & 4rP(\beta) \text{ and} \\
H(\{1^{\alpha_1}, 2^{\alpha_2}\}; r) & = & A(\beta; r) & = & r^2 F(\beta)
\end{array}
$$

where $P(\beta) = (2 - \sqrt{2})\beta + \sqrt{2}$, $F(\beta) = 2 + 4\beta - 2\beta^2$ and $\beta = \frac{\alpha_2}{\alpha_1+\alpha_2} = \frac{\alpha_2}{p}$ □

Theorem 2.29. *The volume V and the surface area A of the polyhedron of radius r for a 3-D octagonal metric are given by:*

$$
\begin{array}{rcccl}
S(\{1^{\alpha_1}, 2^{\alpha_2}, 3^{\alpha_3}\}; r) & = & A(\beta; r) & = & 4r^2 G(\beta,\gamma) \text{ and} \\
H(\{1^{\alpha_1}, 2^{\alpha_2}, 3^{\alpha_3}\}; r) & = & V(\beta; r) & = & \frac{4}{3}r^3 T(\beta,\gamma)
\end{array}
$$

where

$T(\beta,\gamma) = \{1 + 3\beta + 3\gamma + 6\beta\gamma + 3\beta^2 - 6\gamma^2 + 3\beta\gamma^2 - 6\beta^2\gamma - 2\beta^3 + \gamma^3\},$

$G(\beta,\gamma) = \{\beta^2(3-2\sqrt{3}) + \gamma^2(\sqrt{3}-3) + \beta\gamma(2\sqrt{3}-6\sqrt{2}+6) + \beta(2\sqrt{3}) + \gamma(6\sqrt{2}-4\sqrt{3}) + \sqrt{3}\}$

$\beta = \frac{\alpha_2+\alpha_3}{\alpha_1+\alpha_2+\alpha_3} = \frac{\alpha_2+\alpha_3}{p}$ *and* $\gamma = \frac{\alpha_3}{\alpha_1+\alpha_2+\alpha_3} = \frac{\alpha_3}{p}$. □

It may be noted that for $m = 0$, $n = 0$, an octagonal face in the digital sphere degenerates to a point, a rectangular one degenerates to a straight line segment, and a hexagon degenerates to a triangle. Similarly, for $m \neq 0$, $n = 0$, octagon degenerates to a square, a rectangle degenerates to a straight line, and a hexagon degenerates to a triangle, and for $m = 1$, $n = 1$, the octagon degenerates to a square, a rectangle degenerates to a straight line, and the hexagon degenerates to a point. But in all such cases, the expression given in Theorem 2.29 holds.

Next we compute the shape feature based on the above measurements. In 2-D, the shape feature is defined as:

$$
\psi_2(\beta) = \frac{(\text{perimeter})^2}{(\text{area})} = \frac{(4rP(\beta))^2}{r^2 F(\beta)} = \frac{16(\beta(2-\sqrt{2})+\sqrt{2})^2}{2+4\beta-2\beta^2}
$$

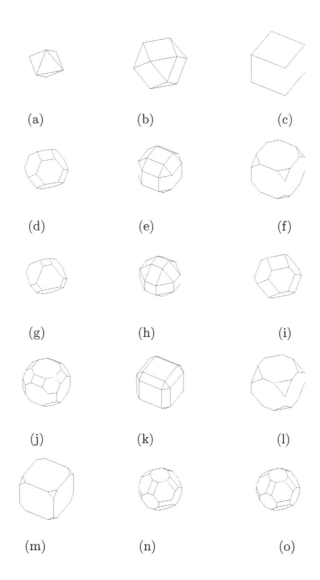

FIGURE 2.12: Digital spheres of 3-D octagonal distances for sorted N-Sequences: (a) $\{1\}$, (b) $\{2\}$, (c) $\{3\}$, (d) $\{1,2\}$, (e) $\{1,3\}$, (f) $\{2,3\}$, (g) $\{1,1,2\}$, (h) $\{1,1,3\}$, (i) $\{1,2,2\}$, (j) $\{1,2,3\}$, (k) $\{1,3,3\}$, (l) $\{2,2,3\}$, (m) $\{2,3,3\}$, (n) $\{1,1,1,2,2,3\}$, and (o) $\{1,2,2,2,3,3\}$.

where $\beta = \frac{\alpha_2}{\alpha_1 + \alpha_2} = \frac{\alpha_2}{p}$ and $P(\cdot)$ and $F(\cdot)$ are defined in Theorem 2.28. In 3-D it is defined as:

$$\psi_3(\beta, \gamma) \quad = \quad \frac{\text{area}^3}{\text{volume}^2} \quad = \quad \frac{(4r^2 G(\beta,\gamma))^3}{(\frac{4}{3}r^3 T(\beta,\gamma)^2} \quad = \quad \frac{6(G(\beta,\gamma))^3}{(T(\beta,\gamma))^2}$$

where $\beta = \frac{\alpha_2 + \alpha_3}{\alpha_1 + \alpha_2 + \alpha_3} = \frac{\alpha_2 + \alpha_3}{p}$, $\gamma = \frac{\alpha_3}{\alpha_1 + \alpha_2 + \alpha_3} = \frac{\alpha_3}{p}$, and $G(\cdot, \cdot)$ and $T(\cdot, \cdot)$ are defined in Theorem 2.29.

2.5.6 Hyperspheres of Weighted t-Cost Distance

The generalized computations of the vertices and volume of a hypersphere $H(\mathbf{w}; r)$ of *weighted t-cost norms* $WD^n(\mathbf{w})$ has so far been elusive for arbitrary cost vectors $\mathbf{w}$. However, there are interesting results for well-behaved cost vectors. A cost vector $\mathbf{w}$ is said to be *well-behaved* if the weights are ordered in non-increasing manner. That is, $w_1 \geq w_2 \geq \cdots \geq w_n \geq 0$. The corresponding distance is known as a **well-behaved weighted t-cost distance**.

For a well-behaved $\mathbf{w}$, we first compute the vertices of a hypersphere $H(\mathbf{w}; r)$ following Lemma 3 in [150]. The lemma has some limitations, as it does not apply the constraint on non-increasing order of the coordinates of the vertices, which is assumed in the proposition. Hence we present its modified version in the following theorem. The proof follows the arguments presented in [150].

Theorem 2.30. *For a well-behaved* $\mathbf{w}$, *vertices of* $H(\mathbf{w}; r)$ *are given by* $\phi(\mathbf{u})$ *where* $\mathbf{u} \in \Sigma_n$ *and*

$$u_i = \begin{cases} \frac{r}{w_i} & \text{for } i = 1. \\ \min\left(\frac{r}{w_i} - \sum_{j=1}^{i-1} u_j, u_{i-1} \right) & 2 \leq i \leq n \end{cases}$$

$\square$

The result reported previously (Lemma 3 of [150]) is a special case of the above lemma. This is stated in the following corollary.

Corollary 2.12 (Lemma 3 of [150]). *If* $\frac{1}{w_1} \geq \left(\frac{1}{w_2} - \frac{1}{w_1} \right) \geq \cdots \geq$ $\left(\frac{1}{w_{n-1}} - \frac{1}{w_n} \right) \geq 0$, *the vertices of* $H(\mathbf{w}; r)$ *are given by* $\phi(\mathbf{u})$ *where* $\mathbf{u} \in \Sigma_n$ *and*

$$u_i = \begin{cases} \frac{r}{w_i} & \text{for } i = 1. \\ \left(\frac{1}{w_i} - \frac{1}{w_{i-1}} \right) r & 2 \leq i \leq n \end{cases}$$

$\square$

The above corollary holds for inverse square root weights, that is, $w_t = \frac{1}{\sqrt{t}}$. Hence, the vertices of the hypersphere $H_{isr}^n(r)$ of the **inverse square root weighted t-cost norm** $(WD_{isr}^n(\cdot))$ are obtained as:

Corollary 2.13. *The vertices of $H_{isr}^n(r)$ are $\phi(\mathbf{u})$, where $\mathbf{u} \in \Sigma_n$ and*

$$u_i = \begin{cases} r & \text{for } i = 1. \\ (\sqrt{i} - \sqrt{i-1})r & 2 \leq i \leq n \end{cases}$$

$\square$

Shapes of a digital circle and sphere in 2-D and 3-D for inverse square root weighted t-cost norms are shown in Figs. 2.13 (a) and (b), respectively.

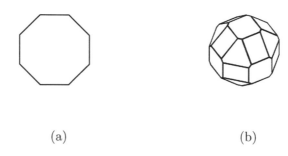

(a)   (b)

FIGURE 2.13: (a) A digital circle and (b) a digital sphere of inverse square root weighted t-cost norm.

Using the results from [154], different measures related to the shape of the disks of the *well-behaved weighted t-cost norm* in 2-D and 3-D are stated now. We restrict to the (normalized) ordered set of weights with $w_1 = 1$.

Theorem 2.31. *The perimeter and area of the circle of radius r for a weighted t-cost norm in 2-D with the ordered set of weights as $\{1, w_2\}$, such that $0 < w_2 \leq 1$, are $4rP(\beta)$ and $r^2 F(\beta)$, where $\beta = \frac{1}{w_2} - 1$, and*

$$P(\beta) = (2 - \sqrt{2})\beta + \sqrt{2},$$

and,

$$F(\beta) = 2 + 4\beta - 2\beta^2.$$

$\square$

Theorem 2.32. *The surface area and volume of the sphere of radius r for a weighted t-cost norm in 3-D with the ordered set of weights as $\{1, w_2, w_3\}$, such that $0 < w_3 \leq w_2 \leq 1$, are $4r^2 G(\beta, \gamma)$ and $\frac{4}{3}r^3 T(\beta, \gamma)$, where $\beta = \frac{1}{w_2} - 1$, $\gamma = \frac{1}{w_3} - \frac{1}{w_2}$, and,*

$$G(\beta, \gamma) = (3 - 2\sqrt{3})\beta^2 + (\sqrt{3} - 3)\gamma^2 + (2\sqrt{3} - 6\sqrt{2} + 6)\beta\gamma + 2\sqrt{3}\beta + (6\sqrt{2} - 4\sqrt{3})\gamma + \sqrt{3},$$

and,

$$T(\beta, \gamma) = 1 + 3\beta + 3\gamma + 6\beta\gamma + 3\beta^2 - 6\gamma^2 + 3\beta\gamma^2 - 6\beta^2\gamma - 2\beta^3 + \gamma^3.$$

$\square$

From the above theorems, we state the measures of disks of inverse square root weighted t-cost norm in 2-D and 3-D in the following corollaries.

Corollary 2.14. *In 2-D, the perimeter and area of $H_{isr}^2(r)$ are given by $6.6274r$ and $3.313708r^2$, respectively. $H_{isr}^2(r)$ is a regular octagon.* ☐

Corollary 2.15. *In 3-D, the surface area and volume of $H_{isr}^3(r)$ are given by $14.3319r^2$ and $4.7773r^3$, respectively.* ☐

For inverse square root weighted t-cost norm the n-D hypersphere $H_{isr}^n(r)$ encloses the n-D Euclidean hypersphere $H_E^n(r)$. Also $H_E^n(r)$ touches all the center points of the faces (hyperplane of dimension $n - 1$) of $H_{isr}^n(r)$. For example, in 2-D the regular octagon of radius r encloses the Euclidean circle of the same radius (refer to Fig. 2.14), which touches each side of the regular octagon at its midpoints, that is, the signed permutation sets $\phi((r, 0))$ and $\phi((\frac{r}{\sqrt{2}}, \frac{r}{\sqrt{2}}))$.

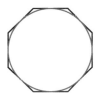

FIGURE 2.14: Circle of inverse square root weighted t-cost norm in 2-D (that is, a regular octagon) enclosing the Euclidean circle of the same radius.

This containment of the Euclidean hypersphere within that of the inverse square root weighted t-cost norm of the same radius is stated in the following theorem [150].

Theorem 2.33. *$H_E^n(r)$ is enclosed within $H_{isr}^n(r)$. Also $H_E^n(r)$ touches $H_{isr}^n(r)$ exactly at $3^n - 1$ points, which are centers (that is, $\phi(\mathbf{v}_{r,t})$, $1 \le t \le n$) of the hyperfaces (of dimension $n - 1$) of the hypersphere.* ☐

For example, in 2-D and 3-D, the Euclidean sphere touches the circle and sphere of the inverse square root weighted t-cost norms at 8 and 26 points, respectively. The points of contact in 2-D are already described before and demonstrated in Fig. 2.14. In 3-D, for a radius of r, these points belong to the sets $\phi((r, 0, 0))$, $\phi((\frac{r}{\sqrt{2}}, \frac{r}{\sqrt{2}}, 0))$, and $\phi((\frac{r}{\sqrt{3}}, \frac{r}{\sqrt{3}}, \frac{r}{\sqrt{3}}))$.

2.5.6.1 Proximity to Euclidean Hyperspheres

In [154], a geometric approach is adopted to measure the proximity of the octagonal distance functions to Euclidean norms in 2-D and 3-D. In this method, measures related to the perimeter, area, and shape of octagonal disks in 2-D are compared with those of disks in the Euclidean space. In 3-D, surface area, volume, and shape of the spheres are compared. For comparing the degree of approximation to the Euclidean norm, the error measures from Definition 2.34 are used [154].

Error measures in 2-D: For $w_1 = 1$, and $\beta = \frac{1}{w_2} - 1$,

$$Perimeter\ Error = E_S^2(\beta) \quad = \quad \left| \frac{2\pi r - 4rP(\beta)}{r} \right| = 2\left| \pi - 2P(\beta) \right|,$$

$$Area\ Error = E_V^2(\beta) \quad = \quad \left| \frac{\pi r^2 - r^2 F(\beta)}{r^2} \right| = \left| \pi - F(\beta) \right|,\ and$$

$$Shape\ Feature\ Error = E_\psi^2(\beta) \quad = \quad \left| \pi - \frac{(\beta(2 - \sqrt{2}) + \sqrt{2})^2}{2 + 4\beta - 2\beta^2} \right| = \left| \pi - S(\beta) \right|.$$

where $S(\beta) = \frac{(\beta(2-\sqrt{2})+\sqrt{2})^2}{2+4\beta-2\beta^2}$.

Error measures in 3-D: For $w_1 = 1$, $\beta = \frac{1}{w_2} - 1$, and $\gamma = \frac{1}{w_3} - \frac{1}{w_2}$,

$$Surface\ Area\ Error \ = E_S^3(\beta, \gamma) = \left| \pi - (G(\beta, \gamma)) \right|,$$

$$Volumetric\ Error \ = E_V^3(\beta, \gamma) = \left| \pi - T(\beta, \gamma) \right|,\ and$$

$$Shape\ Feature\ Error \ = E_\psi^3(\beta, \gamma) = \left| \pi - \frac{G(\beta, \gamma)^3}{T(\beta, \gamma)^2} \right|.$$

The functional forms of $P(\beta)$ and $F(\beta)$ are described in Theorem 2.31, whereas $T(\beta, \gamma)$ and $G(\beta, \gamma)$ are defined in Theorem 2.32. In Tables 2.10 and 2.11 we present values of corresponding error measures for some of the representative octagonal distances approximated by weighted t-cost distance functions. This set includes distance functions, which are reported [154] to be of good approximation of Euclidean norm. For examples, in 2-D the distance functions defined by N-Sequences of $\{1, 1, 2\}$ and $\{1, 1, 1, 2\}$ provide very low values of error measures. In Table 2.10, we also report the corresponding error measures for *inverse square root weighted t-cost norm* $(WD_{isr}^2(\cdot))$ in 2-D. It has been observed that all the error measures have low values with this norm. Similar observations are also drawn from the results presented in Table 2.11. In particular, in both tables, *the error values related to the shape of the circle and sphere are the smallest among the distance functions chosen for comparison.*

TABLE 2.10: Geometric error measures in 2-D. Smallest error measures are highlighted in bold font.

B	$E_p^{(2)}$	$E_a^{(2)}$	$E_s^{(2)}$
$\{1,2\}$	1.717	0.858	0.858
$\{1,1,2\}$	0.155	**0.030**	0.189
$\{1,1,1,2\}$	**0.041**	0.267	0.247
$\{1,2,2\}$	0.936	0.636	0.307
$\{1,2,2,2\}$	1.131	0.733	0.405
WD_{isr}^2	0.344	0.172	**0.172**

Reprinted from *Pattern Recognition Letters*, 32(2011), J. Mukherjee, On approximating Euclidean metrics by weighted t-cost distances in arbitrary dimension, 824–831, Copyright (2011), with permission from Elsevier.

TABLE 2.11: Geometric error measures in 3-D. Smallest error measures are highlighted in bold font.

B	$E_a^{(3)}$	$E_v^{(3)}$	$E_s^{(3)}$
$\{1,2\}$	0.206	0.142	1.028
$\{2,3\}$	2.541	2.733	2.176
$\{1,3\}$	0.912	1.108	0.548
$\{1,1,3\}$	**0.180**	**0.044**	0.472
$\{2,2,3\}$	2.295	2.562	1.797
$\{1,1,1,2,3\}$	0.241	0.114	0.509
$\{1,1,1,1,2,2,3\}$	0.246	0.083	0.598
WD_{isr}^2	0.441	0.441	**0.441**

Reprinted from *Pattern Recognition Letters*, 32(2011), J. Mukherjee, On approximating Euclidean metrics by weighted t-cost distances in arbitrary dimension, 824–831, Copyright (2011), with permission from Elsevier.

2.6   Error Estimation and Approximation of Euclidean Distance

Image processing and other related applications over the digital space $\mathbb{Z}^n$ require a quantification of the continuous space $\mathbb{R}^n$. Thus, for all types of Euclidean measures we need some discrete approximation applicable in grid point space. Among such measures in Euclidean space, the distance measure is most important. Common approximations of this Euclidean distance, E_n include $trunc(E_n)$ or $\lfloor E_n \rfloor$, $round(E_n)$ or $\lfloor E_n + 0.5 \rfloor$, $(E_n)^2$, and $\lceil E_n \rceil$. Unfortunately, the first three of these violate the requirements of a metric. And the last one, though a metric in the topological sense, suffers from high computational cost and the lack of a suitable neighborhood definition in $\mathbb{Z}^n$ for well-formed shortest path structures and algorithms. Hence, attempts are made to use digital metrics to approximate E_n in such cases. Rosenfeld and Pfaltz [182] show that in 2-D, the octagonal distance provides a good approximation to E_2. Using variable neighborhood sets, Yamashita and Ibaraki present an algorithm in [221] to formulate an approximate digital function.

In this section, we explore the potential of the distance functions discussed in this chapter, especially m-Neighbor distance d_m^n and t-cost distance D_t^n as an approximation for E_n in n-D grid point space. Depending on the neighborhood set (characterized by its parameters, like m for m-Neighbor distance d_m^n or t for t-cost distance), we have several choices for digital distances in n-D. Any one of them could be used to approximate E_n. The best choice would clearly be the one with minimum error. Thus, a measurement of errors between $d(N\cdot)$ and E_n would indicate the goodness of approximation of such distances. Our main focus, in this section, is to investigate the properties of errors between $d(N(\cdot))$ and E_n.

Naturally, there are two obvious choices for error measures—the *absolute error* and the *relative* or *proportional error*. Unfortunately, the absolute error is often not useful (unless it is constrained within a finite subset $\mathbb{Z}^n$), as it grows in an unbounded manner as the image dimensions increase. On the other hand, the proportional error (the ratio between $d(N(\cdot))$ and E_n) is usually bounded. Since any relative error is a linear function of the proportional error, we conclude that all types of relative errors are also bounded. Thus, we concentrate on various kinds of relative errors. In some cases, however, we also use the absolute error with a given image dimension to understand how the approximations work within finite limits.

Using such measures, we choose an appropriate distance from among $d(N(\cdot))$ distances in n-D (that is, make an optimal choice of the parameters of the neighborhood set), which produces the least error to approximate E_n. This choice necessitates the computation of the maximum of relative error between $d(N(\cdot))$ and E_n. In this section, we present various bounds for these errors to make optimal choices of parameters for these distances.

2.6.1 Notions of Error

We first formally define the notions of errors.

Definition 2.38. *Errors are computed between pairs of points in n-D as follows:* $\forall \mathbf{u}, \mathbf{v} \in \mathbb{Z}^n$

$a(\mathbf{u}, \mathbf{v}) =$ **absolute error** *between $d(\mathbf{u}, \mathbf{v}; N(\cdot))$ and Euclidean distance $E_n(\mathbf{u}, \mathbf{v})$: $a(\mathbf{u}, \mathbf{v}) = |E_n(\mathbf{u}, \mathbf{v}) - d(\mathbf{u}, \mathbf{v}; N(\cdot))|$*

$r(\mathbf{u}, \mathbf{v}) =$ **relative error** *between $d(\mathbf{u}, \mathbf{v}; N(\cdot))$ and Euclidean distance $E_n(\mathbf{u}, \mathbf{v})$: $r(\mathbf{u}, \mathbf{v}) = \frac{a(\mathbf{u}, \mathbf{v})}{E_n(\mathbf{u}, \mathbf{v})}$*

Using translation invariance, we often deal with $a(\mathbf{x})$ and $r(\mathbf{x})$. □

Definition 2.39. $\mathbf{REL}_i(N(\cdot), n)$ *represents the* **maximum of relative error of type** *i, $0 \leq i \leq 4$ for a given $N(\cdot)$ in n-D:*

$$\mathbf{REL}_0(N(\cdot), n) = \max_{\mathbf{u} \in \mathbb{Z}^n} r(\mathbf{u})$$

$$\mathbf{REL}_1(N(\cdot), n) = \max_{\mathbf{u} \in \mathbb{Z}^n} \frac{a(\mathbf{u})}{d(N(\cdot); \mathbf{u})}$$

$$\mathbf{REL}_2(N(\cdot), n) = \max_{\mathbf{u} \in \mathbb{Z}^n} \frac{a(\mathbf{u})}{\max\{d(N(\cdot); \mathbf{u}), E_n(\mathbf{u})\}}$$

$$\mathbf{REL}_3(N(\cdot), n) = \max_{\mathbf{u} \in \mathbb{Z}^n} \frac{a(\mathbf{u})}{|d(N(\cdot); \mathbf{u}) + E_n(\mathbf{u})|}$$

$$\mathbf{REL}_4(N(\cdot), n) = \max_{\mathbf{u} \in \mathbb{Z}^n} \frac{a(\mathbf{u})}{\sqrt{d(N(\cdot); \mathbf{u})^2 + E_n^2(\mathbf{u})}}$$

Note that unlike $\mathbf{REL}_{0,1}(N(\cdot), n)$, the last three relative errors $\mathbf{REL}_{2-4}(N(\cdot), n)$ are normalized:

$$\mathbf{REL}_i \leq 1, \qquad 2 \leq i \leq 4.$$

If the neighborhood set $N(\cdot)$ is characterized by a parameter λ, the optimal choice of this parameter minimizing some maximum error (as above) is denoted by λ_i^{opt}. That is,

$$\mathbf{REL}(\lambda_i^{opt}, n) = \min_\lambda \mathbf{REL}_i(\lambda, n).$$

□

For example, for m-Neighbor distance $\lambda = m$, or for t-cost distance $\lambda = t$. The range of λ is decided by the type of the parameter and lies between 1 and n for m and t. Wider ranges are also possible for parameters of other $N(\cdot)$'s.

We often write $\mathbf{REL}_0$ simply as $\mathbf{REL}$ if the context is clear.

2.6.2 Error of m-Neighbor Distance

It has been shown in [69] that the maximum relative error $\mathbf{REL}(m, n)$ is bounded by a real constant, whereas, unfortunately, the absolute error has no upper bound. Hence, only the relative error is used as a measure of goodness of digital approximation to Euclidean distance.

We quote the theorems below. The proofs are from [69].

Theorem 2.34. *The absolute error $a(\mathbf{u})$ is unbounded, that is, for all $M \in R^+, \exists \mathbf{u} \in \Sigma_n$ such that $a(\mathbf{u}) = |E_n(\mathbf{u}) - d_m^n(\mathbf{u})| > M$.*   □

Theorem 2.35. $\forall m, n,\ 1 \leq m \leq n$,

$$\mathbf{REL}_0(m, n) = \max_{u \in \Sigma_n} r(\mathbf{u}) < \frac{n}{m}.$$

□

Corollary 2.16. $\forall n, \forall m, 1 \leq m \leq n, \mathbf{REL}_{1-4}(\mathbf{m}, \mathbf{n})$ *are all bounded. The bounds are as follows:*

$$\mathbf{REL}_1(m, n) \quad < \quad \max\left\{\sqrt{m} - 1, 1 - \frac{n}{n+m}\right\}$$

$$\mathbf{REL}_2(m, n) \quad < \quad \max\left\{1 - \frac{1}{\sqrt{m}}, 1 - \frac{n}{n+m}\right\} \qquad < \quad 1$$

$$\mathbf{REL}_3(m, n) \quad < \quad \max\left\{\frac{\sqrt{m}-1}{\sqrt{m}+1}, \frac{n}{n+2m}\right\} \qquad < \quad 1$$

$$\mathbf{REL}_4(m, n) \quad < \quad \max\left\{\frac{\sqrt{m}-1}{\sqrt{m}+1}, \frac{n}{\sqrt{n^2+2nm+2m^2}}\right\} \quad < \quad 1$$

□

The upper bound on $\mathbf{REL}_0(m, n)$ as obtained in Theorem 2.35 is rather loose and tends to suggest that d_m^n improves as m increases and equals n. However, further analysis shows that this is not true, and $\mathbf{REL}_0(m, n)$ indeed minimizes for an m between 1 and n. The mathematics for it is complicated [69]. Based on the analysis the authors present Table 2.12 for optimal choice of m in low dimensions.

2.6.3   Error of Real m-Neighbor Distance

The above analysis of relative errors between d_m^n extends naturally to the case of δ_m^n using real m. In addition, the absolute normalized error is also bounded.

Definition 2.40. *The maxima of absolute (normalized) error between δ_m^n and E_n is defined as: $A(m, n) =$*

$$\max_{\mathbf{x} \in R^n, |x_i| \leq M} |E_n(\mathbf{x}) - \delta_m^n(\mathbf{x})|/M = \max_{0 \leq x_i \leq M} |E_n(\mathbf{x}) - \delta_m^n(\mathbf{x})|/M$$

A is independent of the point spread parameter M, where the spread is $[0, M]^n$. Hence, it is not a parameter in $A()$.

TABLE 2.12: Optimal neighborhood value for least relative error.

n	m_0^{opt}	m_1^{opt}	m_2^{opt}	m_3^{opt}	m_4^{opt}
1	1	1	1	1	1
2	2	1	1	1	1
3	2	2	2	2	2
4	2	2	2	2	2
5	2	2	2	2	2
6	2	2	2	2	2
7	3	2	2	2	2
8	3	2	2	2	2
9	3	2	2	2	2
10	3	2	2	2	2

Reprinted from *Information Sciences* 59(1992), P. P. Das, J. Mukherjee and B. N. Chatterji, The t-Cost Distance in Digital Geometry, 1–20, Copyright (1992), with permission from Elsevier.

The maxima of relative error between δ_m^n and E_n is defined as: $R(m,n) =$

$$\max_{\mathbf{x} \in R^n} |E_n(\mathbf{x}) - \delta_m^n(\mathbf{x})|/E_n(\mathbf{x}).$$

$\square$

The following theorem is proved in [66].

Theorem 2.36.

$$
\begin{aligned}
A(m,n) &= \max(n/m - \sqrt{n}, \sqrt{\lfloor m \rfloor + (m - \lfloor m \rfloor)^2} - 1) \\
&= \max(\sqrt{n}(r/r_I - 1), r_C/r - 1) \\
R(m,n) &= \max(\sqrt{n}/m - 1, 1 - 1/\sqrt{(\lfloor m \rfloor + (m - \lfloor m \rfloor)^2)}) \\
&= \max(r/r_I - 1, 1 - r/r_C)
\end{aligned}
$$

where r_I and r_C are the radii of the inscribed and circumscribed hyperspheres, respectively (Section 2.5.3.3). $\square$

2.6.4 Error of t-Cost Distance

In this section we explore the suitability of a t-cost distance in n-D for the approximation of E_n. A measurement of error between D_t^n and E_n has been presented from [58].

We quote the theorems below. The proofs are from [58].

Theorem 2.37. *Absolute error $a(\mathbf{u})$ is unbounded, that is, $\forall M$, $M \in R^+$, $\exists \mathbf{u} \in \mathbb{Z}^n$ such that $a(\mathbf{u}) = |E_n(\mathbf{u}) - D_t^n(\mathbf{u})| > M$.* $\square$

Lemma 2.13. *Proportional error is bounded over n-D space, that is, $\forall n \geq 1$, $1 \leq t \leq n$ $\exists C_1, C_2 \in R^+$ such that $\forall \mathbf{u} \in \mathbb{Z}^n - \{0\}$*

$$C_1 . D_t^n(\mathbf{u}) \leq E_n(\mathbf{u}) \leq C_2 . D_t^n(\mathbf{u}).$$

$\square$

Theorem 2.38. $\forall n$, $n \geq 1$, $1 \leq t \leq n$,

$$\mathbf{REL}(t, n) = \max_{\mathbf{u} \in \mathbb{Z}^n}\{r(u)\} = \max(\sqrt{t} - 1, 1 - t/\sqrt{n}).$$

$\square$

Optimal Choice of Cost

For every n, there are n different choices of the associated cost t, $1 \leq t \leq n$. Every cost gives a corresponding maximum error $\mathbf{REL}(t, n)$. We want to choose cost t^{opt} in such a way that it minimizes the maxima of relative error. That is,

$$\mathbf{REL}(t^{opt}, n) = \min_{t=1}^{n} \mathbf{REL}(t, n).$$

This choice of cost clearly gives the best D_t^n. Interestingly, for all n the value of t^{opt} lies between 1 and 3. Hence the theorem from [58].

Theorem 2.39. $\forall n$, $n \geq 1$, we have $1 \leq t^{opt} \leq 3$. $\quad\square$

Proof. Clearly, $\mathbf{REL}(t, n)$ minimises for the solution of $\sqrt{t} - 1 = 1 - t/\sqrt{n}$. Here the right-hand side is < 1. Hence $t < 4$. $\quad\square$

Example 2.19. *For $n = 2$, $\mathbf{REL}(1, 2) = 1 - 1/\sqrt{2} = 0.2929 < 0.4142 = \sqrt{2} - 1 = \mathbf{REL}(2, 2)$. Thus D_1^2 or Chessboard is better than D_2^2 or City Block in 2-D.*

For $n=3$, $\mathbf{REL}(1, 3) = 1 - 1/\sqrt{3} = 0.4226$, $\mathbf{REL}(2, 3) = \sqrt{2} - 1 = 0.4142$ and $\mathbf{REL}(3, 3) = \sqrt{3} - 1 = 0.7321$. Thus, the newly introduced metric D_2^3 is better than both D_1^3 or lattice distance and D_3^3 or grid distance in 3-D. $\quad\square$

2.6.4.1 Error of t-Cost Distance for Real Costs

Finally, we present a result of the error estimate for real costs.

Theorem 2.40. $\forall n \in N$, $\forall t \in R^+$,

$$\mathbf{REL}(t, n) \leq (t - \lceil t \rceil).\mathbf{REL}(\lfloor t \rfloor, n) + (1 + \lceil t \rceil - t).\mathbf{REL}(\lceil t \rceil, n).$$

$\square$

2.7   Summary

In this chapter, we introduced the generalized notions of neighborhoods, paths, and distances in n-D digital geometry [71, 54, 221] and provided a generic framework for tracing shortest paths and proving the metricity conditions [61]. We studied a number of classes of digital distances under this framework; notably, m-Neighbor [60], t-cost [58], hyperoctagonal [59], weighted cost [37] and weighted t-cost [150] distances. Properties of the hyperspheres [70, 58] of these distances in terms of the vertices, volumes, surface areas, and shape features were explored in depth and the hyperspheres were geometrically and algebraically compared with their Euclidean counterparts. Estimations of errors in distance measures [69, 55] between digital distances and the Euclidean norm were presented to identify *good* digital measures. A number of interesting and useful special cases of distances and approximations in 2-D and 3-D were presented in the form of Knight's [67, 72], octagonal [56, 55, 63] and weighted t-cost [150] distance. Results for more classes of distances like t-cost-m-Neighbor distances [167] and anisotropic neighborhood [62, 141] distances in 2-D are given as exercises (Section 2.7) below.

Though the shortest path algorithms for most of these distances are outlined (primarily for path construction and metricity proofs), we do not discuss the (chamfer) algorithms for computing the distance transform for them. These are taken up in Chapter 6.

For lack of space, most of the results in this chapter are presented without proof. The reader is encouraged to refer to the details in the references in the text.

Exercises

1. Generalize Algorithm 2 to work for any pair of points $\mathbf{u}, \mathbf{v} \in \mathbb{Z}^n$. *Note that the algorithm as presented works only when the destination point is* $\mathbf{0}$, *and the source point* $\mathbf{u}$ *has all positive coordinates and the coordinate values are ordered, that is* $\mathbf{u} \in \Sigma_n$.

2. Formulate the shortest path algorithms for t-cost, Knight's, octagonal and hyperoctagonal distances by adopting Algorithm 2.

3. Super-Knight's distance (Section 2.3.4) is defined by the neighborhood sets $N(Super - Knight) = \{(\pm p, \pm q), (\pm p, \pm q)\}$ where $p, q \in \mathbb{P}$ and $p \geq q \geq 1$. Show that [72] $d_{Super-Knight}$ is a metric iff $N(Super - Knight)$ is well-behaved, that is, $(p + q) \bmod 2 = 1$ and $gcd(p, q) = 1$.

4. As an exception to symmetric and isotropic (Section 2.2.1.1) neighborhood sets, many anisotropic yet symmetric neighborhoods in 2-D also define metrics [62, 141].

N_1	=	$\{(\pm 1, 0), (0, \pm 1)\}$	$d_{N_1}(\mathbf{x})$	=	$|x_1| + |x_2|$
N_{2A}	=	$\{(\pm 1, 0), \pm(1, 1)\}$	$d_{N_{2A}}(\mathbf{x})$	=	$\max(|2x_1 - x_2|, |x_2|)$
N_{2B}	=	$\{(\pm 1, 0), \pm(1, -1)\}$	$d_{N_{2B}}(\mathbf{x})$	=	$\max(|2x_1 + x_2|, |x_2|)$
N_{2C}	=	$\{(0, \pm 1), \pm(1, 1)\}$	$d_{N_{2C}}(\mathbf{x})$	=	$\max(|2x_2 + x_1|, |x_1|)$
N_{2D}	=	$\{(0, \pm 1), \pm(1, -1)\}$	$d_{N_{2D}}(\mathbf{x})$	=	$\max(|2x_2 - x_1|, |x_1|)$
N_{3A}	=	$\{(\pm 1, 0), (0, \pm 1), \pm(1, 1)\}$	$d_{N_{3A}}(\mathbf{x})$	=	$\max(|x_1|, |x_2|, |x_1 - x_2|)$
N_{3B}	=	$\{(\pm 1, 0), (0, \pm 1), \pm(1, -1)\}$	$d_{N_{3B}}(\mathbf{x})$	=	$\max(|x_1|, |x_2|, |x_1 + x_2|)$
N_{4A}	=	$\{(\pm 1, 0), (\pm 1, \pm 1)\}$	$d_{N_{4A}}(\mathbf{x})$	=	$\max(2\lceil(|x_1| - |x_2|)/2\rceil, 0) + |x_2|$
N_{4B}	=	$\{(0, \pm 1), (\pm 1, \pm 1)\}$	$d_{N_{4B}}(\mathbf{x})$	=	$\max(2\lceil(|x_2| - |x_1|)/2\rceil, 0) + |x_1|$
N_5	=	$\{(\pm 1, 0), (0, \pm 1), (\pm 1, \pm 1)\}$	$d_{N_5}(\mathbf{x})$	=	$\max(|x_1|, |x_2|)$

(a) Prove that the above $d's$ are metrics and give the lengths of the shortest paths for the corresponding neighborhood sets.

(b) Study the disks of these distances.

(c) Two metrics d_1 and d_2 over $\mathbb{Z}^2$ are *isomorphic* if there exists a bijection $\lambda : \mathbb{Z}^2 \to \mathbb{Z}^2$ such that $\forall \mathbf{x}, \mathbf{y} \in \mathbb{Z}^2$, $d_1(\lambda(\mathbf{x}), \lambda(\mathbf{y})) = d_2(\mathbf{x}, \mathbf{y})$ and $d_2(\lambda^{-1}(\mathbf{x}), \lambda^{-1}(\mathbf{y})) = d_1(\mathbf{x}, \mathbf{y})$. Prove that the following sets of metrics are isomorphic to each other: $\{d_{N_{2A}}, d_{N_{2B}}, d_{N_{2C}}, d_{N_{2D}}\}$, $\{d_{N_{3A}}, d_{N_{3B}}\}$ and $\{d_{N_{4A}}, d_{N_{4B}}\}$.

5. The $t-cost-m-Neighbor$ (or tCmN) [167] neighborhood set $N(t, m, n)$, is defined $\forall n, m, t \in \mathbb{P}$, $1 \le m \le n$, $1 \le t \le n$ as $\{w : w \in \{0, \pm 1\}^n, h_n(w) \le m\}$ with an associated cost function $\delta(w) = min(t, h_n(w))$ where $h_n(\cdot)$ is the component sum function (Definition 2.5). It induces tCmN paths $\pi(\mathbf{u}, \mathbf{v}; t, m : n)$ and the tCmN norm defined as $\forall \mathbf{x} \in \mathbb{Z}^n$,

$$d(\mathbf{x}; t, m : n) = \max_{i=0}^{t} S_i(\mathbf{x}),$$
$$S_i(\mathbf{x}) = \sum_{j=1}^{i} f_j(\mathbf{x}) + [minimum(1, (t - i)/(m - i)) \sum_{j=i+1}^{n} f_j(\mathbf{x})],$$
$$0 \le i \le n,$$

where

$$minimum(1, (t - i)/(m - i)) = \begin{cases} (t - i)/(m - i), & 0 \le (t - i)/(m - i) < 1, \\ 1, & (t - i) \ge (m - i). \end{cases}$$

and $f_i(\cdot)$ is the component function (Definition 2.4).

Note that, for $i = m$, $minimum(1, (t - i)/0) = 1$ and $0 \le minimum(1, (t - i)/(m - i)) \le 1$.

Prove the following for the tCmN neighborhood [167] $\forall n, 1 \le m \le n$, $\forall t, 1 \le t \le n$:

(a) $d(t, m : n)$ is a metric over $\mathbb{Z}^n$.

(b) $\forall \mathbf{x} \in \mathbb{Z}^n$, $d(\mathbf{x}; t, m : n) = |\pi^*(\mathbf{x}; t, m : n)|$.

(c) $d(t, m : n)$ is a generalization of m-Neighbor [60] and t-cost [58] distances. That is, $d_m^n = d(1, m : n)$ and $D_t^n = d(t, n : n)$.

6. Let us consider two additional well-behaved conditions for an N-Sequence—one stronger and one weaker.

   - An N-Sequence B is *strongly well-behaved* iff $S(B(i,j)) \geq_c S(B(1,j))$, $\forall i \forall j, 1 \leq i, j \leq p$.
   - An N-Sequence B is *weakly well-behaved* iff

   $$f(i) + f(j) \leq \begin{cases} f(i+j), & i+j \leq p, \\ f(p) + f(i+j-p), & i+j > p. \end{cases}$$

   Prove the following for hyperoctagonal distances from [59, 56, 65]:

   (a) If a B is strongly well-behaved, it is well-behaved.
   (b) If a B is strongly well-behaved, $d(B)$ is a metric.
   (c) If a B is well-behaved, it is weakly well-behaved.
   (d) If a B is *not* weakly well-behaved, $d(B)$ cannot be a metric.
   (e) In 2-D, a B is strongly well-behaved iff it is weakly-well-behaved.

7. Let $S(p) = \{B : |B| = p\}$ be the set of all N-Sequences of length p in 2-D. Define a relation R over $S(p)$ as: $\forall B_1, B_2 \in S(p)$, $B_1 \ R \ B_2$ iff $f_1(i) \leq f_2(i)$, $\forall 1 \leq i \leq p$ where $f(i)$ is the sum sequence. Prove (Theorem 3 in [63]) that $\forall p \geq 1$, $< S(p), R >$ is a distributive lattice with the least element $S_0 = \{\{1\}\}$ and the greatest element $S_1 = \{\{2\}\}$.

8. From Corollary 2.9 we know that $surf(m, n; r) = \sum_{i=0}^{n-1} s_i r^i$, the surface area of the hypersphere $H(m, n; r)$ of d_m^n is a polynomial of degree $n-1$ in r with rational coefficients (s_i). Devise a direct method to compute these coefficients for any given m and n without using the expression presented in Theorem 2.19.

   Hint: Numerically evaluate $surf(m, n; r)$ for $r = 0, 1, \cdots, n-1$ and equate with the polynomial above to get n simultaneous equations in n unknown $s_i's$. Solve these equations by matrix method using rational algebra. For details see [70].

   Repeat the above for $vol(m, n; r)$ and verify Table 2.9.

9. Using the approach from Exercise 8, compute a table like Table 2.9 for surface and volume expressions for hyperspheres of t-cost distance D_t^n for $n = 1, 2, \cdots, 5$. Assume that surface and volume expressions are polynomials with rational coefficients.

10. Compute the corners of hyperspheres for the following N-Sequences:

    (a) Show that for $n = 2$ and $B = \{1\}$, vertices of $H(B; r)$ are $\phi((r, 0))$.
    (b) Show that for $n = 3$ and $B = \{1, 3, 3\}$, vertices of $H(B; r)$ are $\phi((r, \lfloor 2r/3 \rfloor, \lfloor 2r/3 \rfloor))$.

11. Compute the approximate vertices of the disk of radius 20 centered at the point $(-30, 40)$ with the distance function $\{1, 1, 2\}$ in 2-D integral coordinate space. Find the maximum possible deviation of these coordinate points from their true positions. Compute its area and perimeter errors in percentage with respect to the Euclidean disk of the same radius.

12. Compute the volume of a digital sphere of radius 10 in the metric space of the weighted distance function, where weights are given by $\{1, \frac{1}{2}, \frac{1}{3}\}$. Show that the distance is the same as the 26-neighbor distance function in 3-D.

13. Represent an m-neighbor distance in n-D as a weighted t-cost distance function and compute the vertices of its hypersphere of radius r using the Theorem 2.30.

14. Prove that the hypersphere of the inverse square root weighted t-cost distance function encloses a Euclidean hypersphere of the same dimension and radius.

15. $\forall n, n \geq 1, 1 \leq p < q \leq n$, and the following holds [69]: $\forall (\mathbf{u}) \in \Sigma_n,$

$$d_q^m(\mathbf{u}) \leq d_p^m(\mathbf{u}) \leq \lceil q/p \rceil d_q^m(\mathbf{u}).$$

16. Prove the following bounds for m-Neighbor distances in 3-D [68]:

$$
\begin{array}{llll}
d_1^3(\mathbf{x}) & \geq & E_3(\mathbf{x}) & \\
d_2^3(\mathbf{x}) & > & E_3(\mathbf{x}), & iff \quad \mathbf{x} = \quad (\pm 1, \pm 1, \pm 1) \\
 & = & E_3(\mathbf{x}), & iff \quad \mathbf{x} \in \quad \{(\pm k, 0, 0), (0, \pm k, 0), (0, 0, \pm k), \\
 & & & \quad (\pm 2, \pm 2, \pm 1), (\pm 2, \pm 1, \pm 2), (\pm 1, \pm 2, \pm 2)\}, k \geq 0 \\
 & < & E_3(\mathbf{x}), & Otherwise \\
d_3^3(\mathbf{x}) & \leq & E_3(\mathbf{x}) &
\end{array}
$$

Chapter 3

Digitization of Straight Lines and Planes

Line drawings represent a large class of pictorial data used in such diverse areas as cartography, computer graphics, alphanumeric text, engineering drawing, and so on. Straight lines are particularly important because of their simplicity and their ability to approximate any curve to any desired accuracy. More than two thousand years ago, Euclid introduced straight lines through an axiomatic approach and made them the basic elements in his geometry. It is no wonder that studies in discrete geometry concentrated on defining and characterizing discrete counterparts of this vital concept.

With the rapid growth of three-dimensional scene analysis, the study of the geometry of 3-D digital images has raised considerable interest. In 3-D, straight lines and planes are the most elementary geometric objects. Hence, the properties of their discrete forms are fundamental to this geometry and are discussed in this chapter. These include characterization of straight lines and planar segments in 2-D and 3-D digital grids. We also discuss how we can obtain their algebraic description in the Euclidean space, and estimate their lengths and areas in respective dimensions.

3.1   2-D Discrete Straight Line Segments

A discrete straight line segment (*DSLS*) is defined as a set of discrete points obtained from the digitization of some **continuous straight line segment (CSLS)** from (x_1, y_1) to (x_2, y_2) in $\mathbb{R}^2$.

There are various digitization schemes available in the literature. We discuss three such schemes for 2-D *CSLS*.

(a) Nearest integral coordinates (NIC): In this digitization scheme, we consider the digital image of a point p belonging to a straight line in a Euclidean space, which has integral coordinates with the nearest integers to the real coordinates of p.

(b) Inner digitization: Let l be a straight line. Whenever l crosses a coordinate line, the nearest digital point to the right of the point of crossing a digital grid line with respect to the direction of traversing l becomes a point of $D(l)$, the digital representation of l. We assume that the slope of the line l lies between 0 and 1.

(c) Object Boundary Quantization (OBQ): Intuitively, whenever a closed figure is placed on a square array of points, the grid points that are inside the figure and nearest to the contour of the figure form the OBQ image. Since a *CSLS* is not a closed figure, its OBQ is obtained after closing it by drawing a vertical or horizontal line from the DSLS to either the X or Y axis. It can be immediately understood that there is a main grid direction such that there exists only one digitization point in every column in the direction, if the equation of the line is $y = mx + c$ where $0 \leq m \leq 1$ and $0 \leq c < 1$. That direction turns out to be the direction of the X - axis. In this case, the OBQ image of the *CSLS* may be given by $(i, \lfloor mi + c \rfloor)$, where i is an integer that ranges from x_1 to x_2. For an illustration, consider the *CSLS* in Fig. 3.1. The *CSLS* is AB and the shaded part is considered to be its interior.

We may note that the inner digitization is the same as the Object Boundary Quantization (OBQ) in one direction of traversal of the straight line l.

A *DSLS* is represented by a chain code (refer to Section 1.3.3.2), which is denoted by the arrowhead line in Fig. 3.1. It is always assumed that the *DSLS* starts at the point $(0, \lfloor c \rfloor)$ when the equation of the straight line is $y = mx + c$. The chain code in this case is 1011101110. A block of continuous 0s or 1s in chain code is called its **run** and the number of symbols in a *run* is called its **run length**.

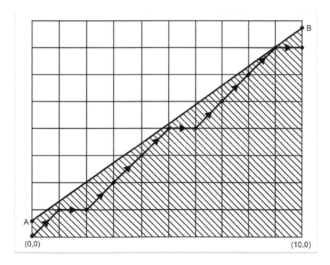

FIGURE 3.1: The OBQ of a continuous straight line and its digitization.

A set of digital points D is called *straight* if it is the digitization of a *CSLS*. Since the concept of digital straightness was introduced in the mid-1970s, many characterizations of digital straight lines have been formulated, and many algorithms for determining whether a digital arc is straight have been defined [88, 215, 216, 179, 109, 102, 174]. We mention one of the first results spelled out by Rosenfeld [176], which is known as the celebrated *chord property*. This property is a bijective relation between discrete arcs and discrete straight line segments. A *discrete arc* in 2-D (also called a *simple arc* in Section 1.2.1 of Chapter 1) is a sequence of discrete points p_i, $0 \leq i \leq n$ such that each p_i, except p_0 and p_n, has exactly two 8-neighbours p_{i-1} and p_{i+1}. p_0 and p_n have single 8-neighbour p_1 and p_{n-1}, respectively.

Theorem 3.1. *Let G be a discrete arc and $S \subseteq G$. S is a DSLS if and only if S is a discrete arc having the chord property. S has the chord property if for all discrete points $\mathbf{p}, \mathbf{q} \in \mathbf{S}$ and for every point (x, y) belonging to the CSLS joining $\mathbf{p}$ and $\mathbf{q}$, there exists some $(u, v) \in S$ such that*

$$max\{|x - u|, |y - v|\} < 1$$

$\square$

The chord property effectively means that all *CSLSs* connecting two arbitrary points of a *DSLS* lie *closer* to the discrete points of the *DSLS* than any other point in the digital plane. An example illustrating the chord property is given in Fig. 3.2.

In this figure, the points in the *DSLS* are represented by the round dots. The *CSLSs* joining any two points of the *DSLS* are also shown in the figure. If we consider any such *CSLS*, then we note that all points of that *CSLS* are within a unit distance from some point of the *DSLS* in the X or Y direction.

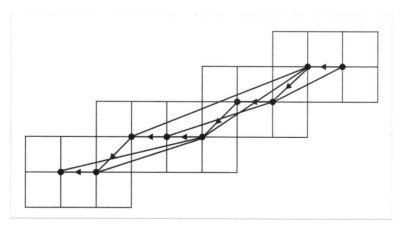

FIGURE 3.2: Chord property of a *DSLS*.

Considering the possible run lengths and runs of runs in the principal direction, Rosenfeld showed that there can be only two run lengths in the chain code of a *DSLS*. These run lengths can differ by at most one, and one of them occurs singly (also refer to Theorems 4.3 and 4.2 in Chapter 4). A number of algorithms for generating digital straight line segments have been developed since then [34, 5, 6].

Apart from characterizing whether a given set of 2-D digital points is a *DSLS*, there are two related problems.

1. To reconstruct a *DSLS* that gives us a *CSLS* whose digitization may have produced it.

2. To compute the domain of a *DSLS*. The domain of a *DSLS* denotes the set of all *CSLSs* whose digitization may have yielded the given *DSLS*.

In our subsequent discussion, we use the function $dist(p, q)$ as a generic notation for expressing distance between two geometric entities, where any one of them could be a point, straight line, *DSLS*, digital planar segment (DPS), and plane. We apply the common notion of our distances in Euclidean geometry in respective cases. For example, $dist(p, l)$ between a point p and a straight line l denotes the perpendicular distance from p to l. Similarly for $dist(D, l)$ between a *DSLS* D and a straight line l, denotes the maximum of $dist(p, l)$ such that $p \in D$. We define the following important constructs for a digitization scheme.

Definition 3.1. *Given a CSLS l, whose digitization produces a DSLS D, l is called the* **pre-image** *of D, and D is called the* **digital image** *of l.* □

Definition 3.2. *A support line (SL)(or support) of a DSLS is a continuous line going through at least one point of the DSLS so that all other points of the DSLS lie on one side of the SL.* □

Definition 3.3. *Given a DSLS D, and a CSLS l that lies on a support of D and $dist(D, l) < 1$, l is a pre-image of D. Such a support of D is called the nearest support of a DSLS.* □

3.1.1 Computation of Support Line of DSLS

One of the most popular schemes of 2-D *DSLS* was presented by Kim and Anderson [4]. In this section, we describe the concepts and results of their work. Let us assume that a 2-D *DSLS* is obtained by the inner digitization scheme and that the slope m of the corresponding *CSLS* lies between 0 and 1.

A *DSLS* D has two kinds of pre-images, one lying above and the other below D as the pre-image (i.e., *CSLS* l) can be traversed in two different directions in inner digitization. To find the nearest support of a *DSLS*, the following result is useful.

Lemma 3.1. *If D is a DSLS, then there exists a CSLS l such that D is the digital image of l and l passes through at least two points of D [4].* □

It is also easy to observe that such a pre-image of D is periodic in a sense. This we state in the following theorem.

Theorem 3.2. *Let $D = \{d_1, d_2, ..., d_n\}$ be a DSLS and let l be a CSLS whose digital image is D such that l contains at least two points of D. If d_i and d_j are successive such points, then d_i and d_j determine a period on D. That is, if $k = j - i$, then $dist(d_h, l) = dist(d_h + k, l)$ for $1 \leq h \leq n - k$ [4].* □

The following theorem helps us to find a pre-image of D that passes through at least two points of D.

Theorem 3.3. *If l is a CSLS that is a pre-image of DSLS D and l contains two points of D, then l is a segment of one of the nearest support of D.* □

Thus, we can make the following statement.

Corollary 3.1. *If D is a DSLS, then there is at most one line above (below) D that contains two points of D and also contains a CSLS that is a pre-image of D.* □

All these results help us to find the nearest support of *DSLS* D by finding a pre-image of D that passes through two points of D. Lemma 3.1 guarantees the existence of at least one such pre-image, and Corollary 3.1 tells us that there can be at most one above and one below D. We could, for example, examine each edge of $H(D)$, the convex hull of D, since any support that contains two points of D must contain an edge of $H(D)$.

But there may be many candidates if D is large. The following theorem allows us to limit the search to one possible candidate on each side of D.

Theorem 3.4. *Let D be a DSLS and let $P = \{v_1, v_2, .., v_j., ..., v_n\}$ be the vertices of $H(D)$ listed in counter-clockwise order, where v_1 is the leftmost point and v_j is the rightmost point of D. If line segment (v_i, v_{i+1}) lies on a pre-image of D, then the horizontal length of $(v_i, v_{i+1}), 1 \leq i \leq j$ is greater than the horizontal length of all of D to the left of v_i or to the right of v_{i+1}. In particular, (v_i, v_{i+1}) is the longest edge of $H(D)$ below D. Similarly, if there is an edge of $H(D)$ above D that lies on a pre-image of D, then it is the longest edge of $H(D)$ above D [4].* □

For finding the convex hull of a planar set of digital points, a variation of Graham's method [169] can be used. Once the convex hull is obtained, the nearest support of a digital line segment can be algorithmically computed. Such an algorithm has a time complexity of $O(n)$. It is possible to have two nearest support lines, one above the convex hull of the set of digital points and another nearest support line below the said set. We also note that the nearest support line above the digital point set, D, is also a pre-image of D if an OBQ digitization scheme is used.

Since there are many *CSLSs* leading to the same *DSLS*, another pertinent problem is to find the entire set of real line segments corresponding to a given chain code. Using the nearest upper support and nearest lower support of the convex hull of D, Kim and Anderson devised an algorithm to compute a polygon so that any *CSLS* inside that polygon is a pre-image of D.

3.1.2 Number Theoretic Characterization and Domain of DSLS

The structure of *straight strings* (which are chain codes of *DSLSs*) is also closely related to the number theoretical properties of the slope of that straight line. Dorst introduced a convenient diagrammatic representation, called a *spirograph*, for a *CSLS* to study the structure of a straight string [77]. We discuss here a closed form algebraic characterization of *DSLSs* [78] invented by Dorst and Smeulders. They considered the chain code of a *CSLS*, C, obtained by the OBQ digitization of a straight line $y = mx + c$ from $x = 0$ to $x = n$, where $0 \leq m < 1$ and $0 \leq c < 1$.

The main theorem of [78] is stated below.

Theorem 3.5. *The chain code of a straight line segment C in the standard situation can be mapped bijectively onto the quardruple (n, q, p, s) defined by*

$$n = \text{the number of elements in } C$$

$$q = \min_k \{k \in \{1, 2, .., n\} | k = n \text{ or } \forall i \in \{1, 2, .., n-k\} : C_{i+k} = C_i\}$$

$$p = \sum_{i=1}^{q} C_i$$

$$s \in \{0, 1, 2, ..., q-1\} \text{ and } \forall i \in \{1, 2, ..., q\} :$$
$$C_i - \lfloor (p/q)(i - s) \rfloor - \lfloor (p/q)(i - s - 1) \rfloor$$
where C_i is the i-th element of C.

□

Intuitively, q represents the periodicity of the given chain code of a *DSLS*. $\frac{p}{q}$ represents the best rational approximation of the slope and s denotes the phase shift and arises due to the intercept c.

The proof of this theorem is quite elaborate and employs involved algebraic and number theoretic manipulations. Thus, the proof is not discussed in this book and can be found in [78, 76].

As an example, consider the *DSLS* of the *CSLS* given by

$$y = \frac{1}{\sqrt{2}}x + \frac{1}{2} \text{ from x} = 0 \text{ to x} = 9.$$

The *DSLS* has the chain code 101110111. Its periodicity q is 4, the number of 1's in a period, p, is 3, and phase shift equivalent s is 1. See Fig. 3.3 for an illustration.

Thus, a straight string is represented by a unique 4-tuple (n, q, p, s) without any loss of information. This set of 4-tuple is called a faithful characterization of a $DSLS$. As $C_i = \lfloor \frac{p}{q}(i-s) \rfloor - \lfloor \frac{p}{q}(i-s-1) \rfloor$, we can easily see that among other lines, the line

$$L : y = \frac{p}{q}(x-s) + \lceil \frac{sp}{q} \rceil$$

is one pre-image of C. Note that L is analogous to the nearest upper support line of C as described by Kim and Anderson [4]. We also observe that if we shift L vertically upward by an amount $\frac{1}{q}$, then the new line L' hits some points outside C. All lines that can be drawn in this corridor between L and L' have the same digital image and hence the same chain code C. The line L passes through at least one point $(s, \lceil sp/q \rceil)$ belonging to the given digital set. Let P be the leftmost point on L, which is also a member of the digitization, and R be the rightmost such point. Similarly, L' passes through one or more grid points that are outside the given digitization. Let Q and S be such points on L' which are leftmost and rightmost, respectively. Dorst carried out the entire analysis in the (x, y)-plane using n inequalities $mi + c - 1 < y_i \leq mi + c$ arising out of the equations $y_i = \lfloor mi + c \rfloor$ for $i = 0$ to n. He showed that the feasible set of values for m ranges from the slope of the line QR to the slope of the line PS (refer to Fig. 3.3). For a given value of m, the span of c was also derived. The domain of D, which is the set of tuples (m, c) that satisfies the given n equations, was formulated in [78] . For an illustration, consider Fig. 3.3. The $DSLS$ is 101110111. The points P, R and Q, S are shown. Any straight line lying completely within the shaded region produces the given $DSLS$.

Inspired by McIlroy [139] to use the Farey series in characterizing the domain of a given $DSLS$ D, Dorst reviewed his proof for domain to achieve an elegant solution in [76]. In [76], Dorst carried out the analysis in the parametric (c, m)-plane. He also defined a *line parameter transformation* (LPT). Using this transformation, a line $y = mx + c$ is converted to a point (c, m). Also, a point (x, y) gets transformed to a line $c = -mx + y$ and vice versa. The transformation preserves the incidence relationship.

For $CSLS$s that are in the standard situation (i.e., $0 \leq m \leq 1, 0 \leq c < 1$) and are enclosed in $x = 0$ to $x = n$, it is sufficient to consider the grid points (i, j) such that $0 \leq j \leq i \leq n$ (see Fig. 3.4). Let us consider a $CSLS$ L. If we change the parameter (c, m) of L, the corresponding $DSLS$ $D(L)$ would not necessarily change. $D(L)$ changes only when L crosses a grid point which is *critical*. In the dual plane L is a point, and change in (c, m) of L means a movement of that point. When L sweeps over a critical point in the primal plane, the point representing L traverses the LPT image of the critical point. Hence, Dorst's idea is to partition the (c, m)-plane by drawing LPT images of the critical points (which in this case are all the grid points) into several facets, which are shown in Figs. 3.5 (a) for $n = 3$ and 3.5 (b) for n = 4. He quantitatively characterized the facets so constructed and proved that they are triangular or quadrilateral and each facet can be identified by four unique integers. Finally, he established that each facet stands for the domain of a

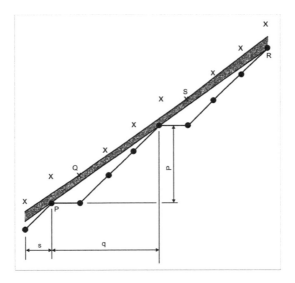

FIGURE 3.3: Faithful characterization of a *DSLS* [78]. *PR* is the support line of the *DSLS*. Any *CSLS* through the shaded region is a reconstruction. Note that the upper boundary of the shaded region is formed by three lines: *RQ* (extended), *QS*, and *PS* (extended). The lower boundary is similar.

Reprinted from *Pattern Recognition Letters*, 12(1991), S. Chattopadhyay et. al., A new method of analysis for discrete straight lines, 747-755, Copyright (1991), with permission from Elsevier.

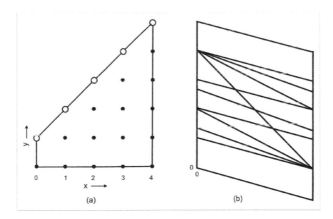

FIGURE 3.4: (a) The trapezium in the digital plane through which any *CSLS* (in the standard situation) of length 4 must pass. (b) LPT images of all digital points in (a).

particular straight string of length n. It is most interesting to note that the domain is in general a quadrilateral with vertices at $LPT(PR)$, $LPT(QR)$, $LPT(QS)$, and $LPT(PS)$ where P, R and Q, S are the points in the primal plane as discussed earlier. The domains for $n = 3$ and 4 are shown in Figs. 3.5(a) and 3.5(b).

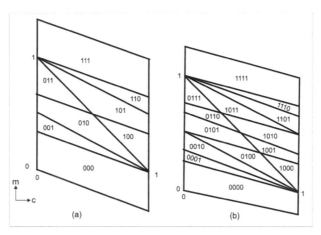

FIGURE 3.5: (a) Domains of all *DSLSs* of length 3. (b) Domains of all *DSLSs* of length 4.

3.2   Iterative Refinement: An Algebraic Characterization

In this section, we have considered that a *CSLS* $l : y = mx + c, 0 \le m < 1, 0 \le c < 1$ is digitized using OBQ from $x = 0$ to $x = n$ to obtain a *DSLS* $D(l) = D$. $D(L)$ is mathematically defined as follows.

$$D = D(L) = \{(i, y_i) | i \text{ is an integer and } y_i = \lfloor (mi + c) \rfloor, 0 \le i \le n\}$$

In an alternative notation, we list $(n + 1)$ y values in increasing values of x from 0 to n, i.e., alternatively we represent D as $\{y_i | 0 \le i \le n\}$.

3.2.1   Gradient and Intercept Estimation

The inequalities derived from the definition of the floor function and the $n + 1$ equations $y_i = \lfloor m_o i + c_o \rfloor$ for $i = 0$ *to* n obtained from D_o can be used to obtain an algorithm called *LR* that refines the upper and lower bounds of m and c iteratively. We present the algorithm in the following theorem [38].

In the following, we use m_o and c_o to denote possible original values of m and c, respectively, so that digitization of $l : y = m_o x + c_o$ yields D_o. That is, $D(l : y = m_o x + c) = D_o$.

Theorem 3.6. *(The LR Algorithm) If the upper and lower bounds of m and c are defined by the following algorithm: [38]*
For $k \geq 0$,

$$c_l^{k+1} = \max_i\{y_i - m_u^k.i\}, 0 \leq i \leq n,$$

$$c_u^{k+1} = \min_i\{y_i + 1 - m_l^k.i\}, 0 \leq i \leq n,$$

$$m_l^{k+1} = \max_i\{(y_i - c_u^k)/i\}, 1 \leq i \leq n, and$$

$$m_u^{k+1} = \min_i\{(y_i + 1 - c_l^k)/i\} 1 \leq i \leq n$$

and $c_l^0 = y_0, c_u^0 = y_0 + 1, m_l^0 = -(1/n), m_u^0 = (n+1)/n$, then, there exist c_l, c_u, m_l, m_u such that,

$$\lim_{k\to\infty} m_l^k = m_l, \lim_{k\to\infty} c_u^k = c_u, \lim_{k\to\infty} m_u^k = m_u, \lim_{k\to\infty} c_l^k = c_l \dots\dots(a)$$

$$c_l < c_o < c_u \text{ and } m_l < m_o < m_u \dots\dots(b).$$

Proof: From D_o we get, $y_i = \lfloor m_o i + c_o \rfloor \ \forall i, 0 \leq i \leq n$.
Using the definition floor function and rearranging terms, we can write

$$(y_i + 1 - c_o)/i > m_o \geq (y_i - c_o)/i.$$

Let c_l^k and c_u^k be such that $c_l^k < c_o < c_u^k$. So, from the above inequalities we get,

$$(y_i + 1 - c_l^k)/i > m_o \geq (y_i - c_u^k)/i.$$

Thus, $m_u^{k+1} = min_i((y_i + 1 - c_l^k)/i) > m_o > max_i((y_i - c_u^k)/i) = m_l^{k+1}$.

Similarly, we can show that $c_l^{k+1} < c_o < c_u^{k+1}$ holds if we have $m_l^k < m_o < m_u^k$.

The fact that $c_l^k < c_o < c_u^k$ and $m_l^k < m_o < m_u^k$ hold, $\forall k, k \geq 0$ can be easily proved by induction on k. The inductive basis holds for k = 0 as

$$c_l^0 = y_0 \leq c_o < \lfloor c_o + 1 \rfloor = y_0 + 1 = c_u^0 \text{ and}$$

$$m_l^0 = -1/n < 0 \leq m_o \leq 1 < (n+1)/n = m_u^0.$$

Now, let us assume that $c_l^k < c_o < c_u^k$ and $m_l^k < m_o < m_u^k$ hold for some k, $k > 0$.
From the above derivation we have

$$c_l^k < c_o < c_u^k \text{ implies } m_l^{k+1} < m_o < m_u^{k+1} \text{and}$$

$$m_l^k < m_o < m_u^k \text{ implies } c_l^{k+1} < c_o < c_u^{k+1}.$$

Hence, $c_l^k < c_o < c_u^k$ and $m_l^k < m_o < m_u^k$ for all $k \geq 0$.

The convergence of the iterative scheme is proved again by induction on k. We first show that for all $k \geq 0$, $c_l^{k+1} \geq c_l^k$ and $m_u^{k+1} \leq m_u^k$. The inductive basis holds for $k = 0$. As,

$$c_l^1 = \max_i \{y_i - m_u^0.i) \geq y_0 = c_l^0$$

$$m_u^1 = \min_i \{(y_i + 1 - c_l^0)/i) \leq ((y_n - y_0) + 1)/n \leq (n+1)/n = m_u^0.$$

It is easy to see that,

$$(i) c_l^{k+1} \geq c_l^k \text{ implies } m_u^{k+2} \leq m_u^{k+1} \text{ for } k \geq 0, and$$

$$(ii) m_u^{k+1} \leq m_u^k \text{ implies } c_l^{k+2} \geq c_l^{k+1} \text{ for } k \geq 0.$$

Hence,

$$c_l^0 \leq c_l^1 \leq c_l^2 \leq c_l^3 \leq ...c_l^k \leq c_l^{k+1}... < c_o \text{ and}$$
$$m_u^0 \geq m_u^1 \geq m_u^2 \geq m_u^3 \geq ...m_u^k \geq m_u^{k+1}... > m_o.$$

Now, if we get any k so that $\forall k', k' \geq k$, $c_l^{k'} = c_l^{k'+1}$, then there exists some j, so that $\forall j', j' \geq j, m_u^{j'} = m_u^{j'+1}$. In this case, the iteration ends at a fixed point. Otherwise, it forms a non-decreasing (or non-increasing) sequence upper bounded (lower bounded) by a constant value. Hence, the limit always exists and $\lim_{k \to \infty} c_l^k = c_l$ and $\lim_{k \to \infty} m_u^k = m_u$. Numerically, we can compute c_l and m_u by selecting a small quantity $\epsilon > 0$ and terminating when $c_l^k - c_l^{k-1} < \epsilon$ and $m_u^{k-1} - m_u^k < \epsilon$.

Similarly, we can prove that $\lim_{k \to \infty} m_l^k = m_l$ and $\lim_{k \to \infty} c_u^k = c_u$.

Hence, part (a). Part (b) actually follows from the first inductive proof and part (a). $\square$

Example 3.1. *Let $n = 4$ and D_o be (0,1,1,1,2) corresponding to the line $L_o : y = 0.4x + 0.6$.*

The convergence of c_l, c_u, m_l, and m_u can be verified. It may be observed that while c_u^k and m_l^k attain fixed points after only one iteration, c_l^k and m_u^k form monotone sequences. It can be checked that $\max_i \{y_i - m_u^k.i\}$ occurs at $i = 1$ and $y_i = 1$. Similarly, $\min_i \{(y_i + 1 - c_l^k)/i\}$ occurs at $i = 3, y_i = 1$. Thus, c_l and m_u follow the mutually recursive equations: $c_l^k = 1 - m_u^{k-1}$ and $m_u^k = (2 - c_u^{k-1})/3$.

3.2.2   Reconstruction of a DSLS

We first show that the rectangle (say R_{ul}) defined on the $c - m$ plane by the points (c_u, m_l) and (c_l, m_u) as diagonally opposite vertices, properly contains the domain of D_o. Subsequently, it is shown that R_{ul} is the smallest such rectangle, and if $R_{ul} \neq \phi$, then there exists at least one continuous line L such that $D\{L\} = D_o$. This we state in the following theorems.

Lemma 3.2. *Using the equations of Theorem 3.6, we can see that g and h may be defined so that $y_g = m_l g + c_u$ and $y_h + 1 = m_l h + c_u$. Then, $g > h$* [38]. $\square$

This lemma defines a relation between two support (extreme) grid points that define the bounds m_l and c_u. In the context of m_l and c_u these points are denoted as (g, y_g) and $(h, y_h + 1)$. A similar pair of support points for m_u and c_l are referred to as $(h', y_{h'} + 1)$ and $(g', y_{g'})$. For the D_o in Example 3.1, $g = 4, g' = 1, h = 0$, and $h' = 3$.

Theorem 3.7. *If for any $L : y = mx + c, D = D(L) = D_o$ then, $m_l < m < m_u$ and $c_l < c_o < c_u$. In other words, $Domain(D_o) \subseteq R_{ul}$* [38]. $\square$

The previous theorem states that R_{ul} indeed contains the domain of D_o. Before proving the tightness of the rectangle, we explore one more property of their bounds.

Definition 3.4. *Digitization D of a straight line $L : y = mx + c$ is said to be less than or equal to D_o (denoted as $D \leq D_o$) if and only if, $\forall i, 0 \leq i \leq n, \lfloor mi + c \rfloor \leq y_i = \lfloor m_o i + c_o \rfloor$. $D \geq D_o$ is defined similarly.* $\square$

Theorem 3.8. *If $m_l < m < m_u$ and $c_l < c < c_u$, then either $D \leq D_o$ or $D \geq D_o$* [38]. $\square$

The above theorem directly leads us to a binary search-like algorithm to estimate an m (or c) given a c (or m) in R_{ul} once the tightness of the rectangle R_{ul} is established. In doing so, we need the following definition.

Definition 3.5. *The (c, m)-bounds: m_l, c_u, m_u, c_l are defined to be consistent if and only if $m_l < m_u$ and $c_l < c_u$.* $\square$

Note that the following theorem uses the fact that the chain code under *standard situation* contains 0s and 1s only.

Theorem 3.9. *If (c, m)-bounds are consistent, then there exists at least one pre-digitized line L, such that $D(L) = D_o$.* [38].

Proof: First we highlight two simple properties of the lines $L_{lu}: y = m_l x + c_u$ and $L_{ul}: y = m_u x + c_l$. Let L_{lu} and L_{ul} intersect at $M = (x, y)$ (Fig. 3.6). Then:
Property 1: $0 < x \leq n$.
Property 2: There exists no grid point in the open triangles ABM and MDE.
Note that these properties directly follow from the consistency of the bounds.

From the above properties, we can say that $y_i = \lfloor m_l i + c_u \rfloor, x \leq i < n$ and $y_i = \lfloor m_u i + c_l \rfloor, 0 < i < x$. Thus, if we draw a straight line $FG(y = mx + c)$ through M (Fig. 3.6), then $y_i = \lfloor mi + c \rfloor$ if $i \neq x$. Now two cases may arise. If

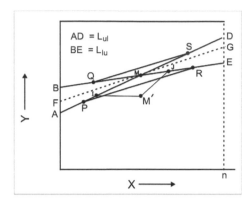

FIGURE 3.6: Construction of a straight line which after digitization produces D_o (Theorem 3.9). The nearest support line PR is also illustrated.

Reprinted from *Pattern Recognition Letters*, 12(1991), S. Chattopadhyay et al., A new method of analysis for discrete straight lines, 747-755, Copyright (1991), with permission from Elsevier.

x is not an integer, $D(FG) = D_o$. The same is true if (x, y) are integers and $y = y_x$. So, we deal with the case where x is an integer and $y = y_x + 1$.

Consider a point M' unit ordinate below M (Fig. 3.6).

Construct a line $M'I$ with slope zero and a line $M'J$ with slope $\pi/4$. The two intersections are obtained on two sides of M'.

No grid point exists in the triangle $MM'I$ or $MM'J$ as we are considering $0-1$ chain codes only. Now, it is possible to draw a *CSLS* L that lies in the region traced by $BMDEJM'IA$ so that $D(L) = D_o$. □

Thus, if we take any $m, m_l < m < m_u$, we may draw at least one *CSLS* L that lies in the region $BMDEJM'IA$ producing the same digitization as D_o. Numerically this line can be found by a simple *binary-search*-like algorithm that uses Theorem 3.8 to terminate. Similarly, we can compute one m for reconstruction given a c, such that $c_l < c < c_u$. So, we can say that R_{ul} is the smallest rectangle containing the domain of D_o.

3.2.3   Computation of Precise Domain of a DSLS

The reconstruction algorithm of the last section computes at least one straight line with the given digitization. We are more interested in all values of c, m parameters that allow for reconstruction. Next, we discuss computation of the domain of D_o.

We note that little modifications of the iteration equations of the last section directly lead to the computation of $Domain(D_o)$, of which R_{ul} is only a tight bound. We present it in the next theorem after necessary lemmas.

Lemma 3.3. *Suppose that D denoted by y_i from $0 \le i \le n$ is a DSLS. For* *any $0 \leq i, j \leq n$, the following hold: [38]*

$$(i-j)c_u \leq (iy_j - jy_i + i)\ldots\ldots\ldots(a)$$
$$(i-j)c_l \leq (iy_j - jy_i + i)\ldots\ldots\ldots(b)$$
$$(i-j)m_u \geq (y_i - y_j - i)\ldots\ldots\ldots(c)$$
$$(i-j)m_l \geq (y_i - y_j - i)\ldots\ldots\ldots(d).$$

□

Lemma 3.4. *The following hold:*

(A) $c_l < c < c_u$ if and only if $m_l^*(c) < m_u^*(c))$, where

$$m_l^*(c) = \max_i (y_i - c)/i \text{ and } m_u^*(c) = \min_i (y_i + 1 - c)/i$$

(B) $m_l < m < m_u$ if and only if $c_l^*(m) < c_u^*(m)$, where

$$c_l^*(m) = \max_i (y_i - mi) \text{ and } c_u^*(m) = \min_i (y_i + 1 - mi)$$

Proof: Part(A): Let $c_l < c < c_u$. Assume $m_l^*(c) \geq m_u^*(c)$. Clearly, there exists $p, q > 0$ such that $m_l^*(c) = (y_p - c)/p$ and $m_u^*(c) = (y_q + 1 - c)/q$. So, $m_l^*(c) \geq m_u^*(c)$ implies $(y_p - c)/p \geq (y_q + 1 - c)/q$. Rearranging terms we gets $(p-q)c \geq (py_q - qy_p + p)$.

Thus, from Lemma 3.3, $(p-q)c \geq (p-q)c_u$ and $(p-q)c \geq (p-q)c_l$. If $p > q$ then $c \geq c_u$ and if $p < q$, then $c \leq c_l$. $p = q$ is ruled out as it implies that $0 \geq p > 0$.

Hence $c_l < c < c_u$ implies $m_l^*(c) < m_u^*(c)$.

Conversely, let $m_l^*(c) < m_u^*(c)$. Select any m such that $m_l^*(c) \leq m < m_u^*(c)$. From the definition of $m_l^*(c)$ and $m_u^*(c)$ we get,

$$\forall i, m_i + c - 1 < y_i \leq m_i + c.$$

That is, $y_i = \lfloor mi + c \rfloor$. In other words, $D(L(c, m)) = D_o$. Hence, from Theorem 3.7, we get $c_l < c < c_u$.

Part(B) can be proved analogously. □

Theorem 3.10. *(The domain theorem). The domain Domain(D_o) of a DSLS D_o is given by the following [38].*

$$Domain(D_o) = \bigcup_{c_l < c < c_u} [m_l^*(c), m_u^*(c)); \text{ or}$$

$$Domain(D_o) = \bigcup_{m_l < m < m_u} [c_l^*(m), c_u^*(m))$$

For a given c, $[m_l^*(c), m_u^*(c))$ of a *DSLS* defines a line interval in the (c, m)-plane. Taking union of these intervals for $c_l < c < c_u$, we get a region in the (c, m)- plane. Thus, $Domain(D_o)$ is a region in the (c, m)-plane.

Proof: From Theorem 3.7, $Domain(D_o) \subseteq R_{ul}$. Again from Lemma 3.4, for given $c, c_l < c < c_u$, and for all $m, m \in [m_l^*(c), m_u^*(c)), (c, m) \in Dom(D_o)$, if $m < m_l^*(c)$, then $m < (y_p - c)/p$ for some p. In other words, $y_p > mp + c$ or $y_p > \lfloor mp + c \rfloor$. Thus, $m < m_l^*(c)$ implies that $(c, m) \notin Dom(D_o)$. Similarly, $m \geq m_u^*(c)$ implies that $(c, m) \notin Dom(D_o)$. Hence, for any $c, c_l < c < c_u$, $(c, m) \in Dom(D_o)$ if and only if $m \in [m_l^*(c), m_u^*(c))$.   □

Corollary 3.2. *R_{ul} is the smallest rectangle in the (c, m)-space so that $Domain(D_o) \subseteq R_{ul}$.*   □

It is observed from the construction of m_l, c_u, m_u, c_l bounds that given any D_o we may be able to compute them numerically. That is, if D_o is indeed a *DSLS* then the bounds m_l, c_u, m_u, c_l really exist. However, if D_o is any arbitrary set of 2-D digital points and not really a *DSLS*, $Domain(D_o)$ does not exist.

In view of the above corollary, we immediately get an algorithm from iterative refinement to test whether a given digital set is a valid digital straight line or not.

Corollary 3.3. *$Domain(D_o)$ is empty if and only if $m_l \geq m_u$ or $c_l \geq c_u$ or both. In other words, the (m, c) bounds are consistent if and only if the digital set D_o satisfies the chord property.*   □

We illustrate the above equivalence through a trivial yet interesting example.

Example 3.2. *Let us consider a generic digital set $D = \{y_i : y_i = i, 0 \leq i \leq a; y_i = a, a \leq i \leq a + b; a, b \text{ integer} \geq 1 \text{ and } a + b = n\}$. This corresponds to a chain code where b 0s follow a 1s.*

Consider m_u and c_l for this set. Assuming the consistency of these bounds, we can show that
$c_l^k = a(1 - m_u^{k-1})$ and $m_u^k = (a + 1 - c_l^{k-1})/(a + b)$.
Solving this mutual recurrence we get,

$$c_l^k = a(1 - 1/b)(1 - (a/(a + b))^r), r = \lfloor k/2 \rfloor$$

$$m_u^k = (1/b) + (1 - 1/b)(a/(a + b))^s, s = \lceil k/2 \rceil, \ k \geq 0, \ a, b \geq 1.$$

Hence, $c_l = a(1 - 1/b)$ and $m_u = 1/b$.

Again from consistency $c_l < c_u^0 = y_0 + 1 = 1$. That is, $1/a + 1/b > 1$ or $min(a, b) = 1$, since a and b are integers.

Similarly, $c_u = 1$ and $m_l = 1 - 1/a$, if $min(a, b) = 1$, whereas for $min(a, b) > 1$.

$$c_u^k = a(1 - 1/b) - a(1 - 1/a - 1/b)(1 + b/a)^r, r = \lfloor k/2 \rfloor,$$

$$m_l^k = 1/b + (1 - 1/a - 1/b)(1 + b/a)^s, s = \lfloor (k-1)/2 \rfloor, k \geq 0.$$

Therefore, $c_l < c_u$ and $m_l < m_u$ if and only if $min(a, b) = 1$. Now it is easy to see that the chord property for the particular kind of digital set reduces to $1/a + 1/b > 1$ or $min(a, b) = 1$. Hence, in this case, consistency and the chord property are equivalent. It may be noted that in the absence of the chord property, $c_u \to -\infty$ and $m_l \to +\infty$ violate consistency.

Next we illustrate the domain of the *DSLS* of Example 3.1.

Example 3.3. $n = 4, D_o = \{0, 1, 1, 1, 2\}. m_o = 0.4$ and $c_o = 0.6$. From Example 3.1, $m_l = 1/4; m_u = 1/2, c_l = 1/2$ and $c_u = 1$. Diagrammatically, we show the domain in Fig. 3.7.

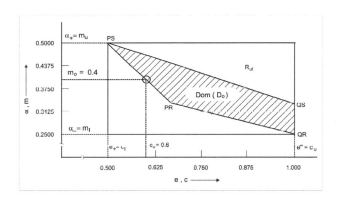

FIGURE 3.7: The domain $Dom(D_o)$ and the bounding rectangle R_{ul} of D_o for $m_o = 0.4$ and $c_o = 0.6$ (Example 3.3). The domain has been computed using Theorem 3.10. The original parameter value has been shown by an '*.'

It can be shown that for any valid *DSLS* D_o, $Domain(D_o)$ is a quadrilateral or a triangle formed in the (c, m)-plane. First, we prove the following lemma.

Lemma 3.5. *The points* (c_l, m_u), $(c, m_l^*(c))$, *and* (c_{PR}, m_{PR}) *are collinear for all* $c, c_l \leq c \leq c_{PR}$. $\qquad\square$

Similarly, we can show that $\forall c, c_{PR} \leq c < c_u$, and the points (c_{PR}, m_{PR}), $(c, m_l^*(c))$, and (c_u, m_l) are collinear.

Thus, the lower boundary of the region $Domain(D_o)$ is a pair of straight lines in the (c, m)-plane defined by the three points (c_l, m_u), (c_{PR}, m_{PR}), and (c_u, m_l).

Similarly, the upper boundary of $Domain(D_o)$ can also be formulated as a pair of straight lines defined by (c_l, m_u), (c_{QS}, m_{QS}), and (c_u, m_l), where $Q = (h, y_h+1)$, $S = (h', y_{h'}+1)$ (Fig. 3.7) , $c_{QS} = (h'y_h - hy_{h'} + h' - h)/(h' - h)$ and $m_{QS} = (y_{h'} - y_h)/(h' - h)$.

Note that $m_{PR} = m_{QS}$. In Example 3.1, the domain is a quadrilateral with vertices $(c_l, m_u) = (1/2, 1/2)$, $(c_{PR}, m_{PR}) = (2/3, 1/3)$, $(c_u, m_l) = (1, 1/4)$, and $(c_{QS}, m_{QS}) = (1, 1/3)$ (Fig. 3.7).

In special cases, where Q and S (or P and R) coincide, the region $Dom(D_o)$ becomes triangular because either $(c_{QS}, m_{QS}) = (c_l, m_u)$ or $(c_{PR}, m_{PR}) = (c_u, m_l)$. This leads us to the following theorem.

Theorem 3.11. *The domain $Domain(D_o)$ for a DSLS D_o is a quadrilateral, in general determined by the following vertices: (c_l, m_u), (c_{PR}, m_{PR}) and (c_u, m_l), and (c_{QS}, m_{QS})* □

Finally, we would like to highlight that there is a direct correspondence between the iterative refinement (I_R) algorithm presented here and the analysis formulated in other relevant works [78, 4, 134]. The crux of the analysis of digital straight lines lies in the identification of these four points: P, Q, R and S. We get them from Theorem 3.11 as $P = (g', y_{g'})$, $Q = (h, y_h + 1)$, $R = (g, y_g)$, and $S = (h', y_{h'} + 1)$. It follows, then, that $PS = (c_l, m_u)$, $QR = (c_u, m_l)$, and $m_u = \alpha_+$, $m_l = \alpha_-$, $c_l = e_+$ and $c_u = e_-$ in Dorst's notation [78]. Thus, there exists a direct correspondence between the iterative refinement scheme and the (n, q, p, s)-characterization.

3.2.4 Speed and Convergence of Iterative Refinement

The complexity of the I_R algorithm to compute c_l, m_u, c_u, and m_l depends heavily on the speed of convergence of the proposed iterative scheme. The worst case time complexity of the iterative algorithm has not been proven theoretically. However, with some modifications the complexity is reduced drastically. In the following we state the modified algorithms, and present experimental evidence of reduction of complexity. The proof is left as an exercise.

In the algorithm presented in Theorem 3.6, in every iteration the values of $m_l^k, m_u^k, c_l^k, c_u^k$ are computed from the values of $m_l^{k-1}, m_u^{k-1}, c_l^{k-1}, c_u^{k-1}$. Let us call this *parallel I_R algorithm*. The first improvement in this algorithm is to use the most recent estimate to compute the bounds of m and c instead of computing all bounds in parallel for the same k. The resulting algorithm is given in Algorithm 4.

The correctness of Algorithm 4 follows from the I_R algorithm. It is also noted that this sequential version improves the speed of the earlier algorithm by approximately a factor of two!

Another heuristic may be suggested to achieve more speed. In this case, when we compute m_l^{k+1} and c_u^{k+2}, we also tabulate i, j, where $m_l^{k+1} = (y_i - c_u^k)/i$; and $c_u^{k+2} = (y_j + 1 - m_l^{k+1}j)$.

Then we calculate the gradient m and the slope c of the line joining the

---

Algorithm 4: Sequential Updates for Iterative Refinement

Algorithm Sequential I_R

$c_l = y_0$; { Initialize c_l^0 }

repeat

$\qquad m_u = \min_i (y_i + 1 - c_l)/i$; { compute m_u^{k+1} from c_l^k }

$\qquad c_l = \max_i (y_i - m_u i)$ { compute c_l^{k+2} from m_u^{k+1} }

until changes in m_u and c_l become negligible.

$c_u = y_0 + 1$;

repeat

$\qquad m_l = \max_i (y_i - c_u)/i$; { compute m_l^{k+1} from c_u^k }

$\qquad c_u = \min_i (y_i + 1 - m_l i)$ { compute c_u^{k+2} from m_l^{k+1} }

until changes in m_l and c_u become negligible.

end (Algorithm Sequential I_R)

---

points (i, y_i) and $(j, y_j + 1)$. We show that $m_l^{k+1} \leq m \leq m_l$ and $c_u^{k+2} \geq c \geq c_u$. Let us call this version the *sequential I_R with simultaneous solution*.

In reality, often the m and c values obtained by solving such simultaneous equations are better approximations of m_l and c_u than m_l^{k+1} and c_u^{k+1}. Since our iteration scheme is strictly monotone, the use of m and c as new initial values can only hasten the convergence. See Fig. 3.8 for a comparison of the speed of convergence of these three versions of the I_R algorithms.

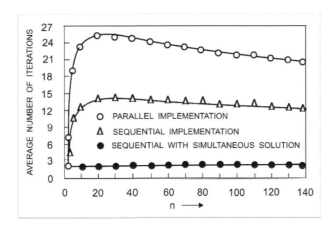

FIGURE 3.8: Convergence speed of I_R algorithms.

3.2.5 Length Estimators

Consider that the length of a *CSLS* l is to be estimated given its digitization D. D is the *DSLS* and may be represented by a chain code C. But the chain code C is not typically used to estimate the properties of the pre-digitized *CSLS*. Most of the digitized image (in this case the chain code C) is characterized by a fixed number of parameters, also called a tuple t. For example, we discuss in Section 3.1.2 that (n, q, p, s) provides a 4-tuple charac-terization given any chain code C of a *DSLS*. The length estimator is expressed in terms of t, say $g(t)$. Thus, the length of all lines whose chain code could have generated a tuple t would be estimated by the same $g(t)$. Let the length of a line l be $f(l)$ and let l be digitized to generate a chain code C that is char-acterized by a tuple t. It would be important to minimize the error between $f(l)$ and $g(t)$ for all lines l whose chain code after characterization becomes t. Two types of estimators are considered here.

The best linear unbiased estimator (BLUE) [79] minimizes the mean square error (MSE) between $f(l)$ and $g(t)$ over all l, which are in $Domain(t)$, the domain of t. This is equivalent to

$$g_{BLUE}(t) = \int_{Domain(t)} f(l)p(l)dl$$

where $p(l)$ is the probability density of lines.

The most probable estimator of the length is the most probable original value of length given t is obtained as

$$g_{MPO}(t) = f(argmax\{p(l)|l \in Domain(t)\})$$

where $argmax\{p(l)\}$ indicates the value of l maximizing $p(l)$ in that range. The estimators that are provided in this section are found in [76, 79].

Before we discuss the estimators for different characterizations of the chain code, let us formulate the length of a line segment and the probability density function of random lines in the 2-D plane.

The length of a line with m slope, and c intercept between $x = 0$ and $x = n$ is given by

$$f(c, m, n) = n\sqrt{1 + m^2}.$$

The probability density function of random lines in the 2-D plane is taken as uniform in polar parameters and is given by

$$p(c, m, n) = \sqrt{2}(1 + m^2)^{-3/2}.$$

As a measure of the accuracy of the length estimator g we use $RDEV(g, n)$, which is defined by

$$RDEV(g(t), n) = (1/n)\sqrt{\sum_t \int_{Domain(t)} (g(t) - f(l))^2 p(l)dl}.$$

3.2.5.1 (n)-characterization [76, 79]

This is the simplest characterization of a chain code and is given by the number of elements in it.

$$g_{MPO}(n) = n$$

$$g_{BLUE}(n) = (\pi/4)\sqrt{2}n = 1.111n$$

For both estimators, RDEV tends to be a constant as $n \to \infty$. This means that the accuracy of these estimators cannot be increased beyond a certain point, even by increasing the sampling density. $RDEV(g_{MPO}(n)) = 16\%$ and $RDEV(g_{BLUE}(n)) = 11\%$.

3.2.5.2 (n_e, n_o)-characterization [76, 79]

In (n_e, n_o)-characterization, the number of 0s in a chain code is computed as n_e and the number of 1s in a chain code is computed as n_o. This is also known as *odd–even* characterization.

$$g_{MPO}(n_e, n_o) = \begin{cases} \frac{1}{2}\sqrt{3} & \text{if } (n_e, n_o) = (0, 1) \\ \sqrt{(n_e + n_o)^2 + n_o^2} & \text{elsewhere} \end{cases}$$

$$\begin{aligned}
g_{BLUE}(n_e, n_o) =\ & \frac{\sqrt{2}}{n}\{\frac{m+1}{n}tan^{-1}\frac{m+1}{n} - 2\frac{m}{n}tan^{-1}\frac{m}{n} + \frac{m-1}{n}tan^{-1}\frac{m-1}{n} \\
& - \frac{1}{2}log(1 + (\frac{m+1}{n})^2) + log(1 + (\frac{m}{n})^2) - \frac{1}{2}log(1 + (\frac{m-1}{n})^2)\}
\end{aligned}$$

RDEV for both these estimators are $O(n^{-1})$.

3.2.5.3 (n, q, p, s)-characterization [76, 79]

The MPO estimator for this characterization is given by

$$g_{MPO}(n, q, p, s) = n\sqrt{1 + (\frac{p}{q})^2}$$

As (p/q) is a better estimate of the slope of a line, this estimator is the best of all. The asymptomatic error for this estimator is $O(n^{-3/2})$.

3.3 Three-Dimensional Digital Straight Line Segments

In this section, we review a few preliminary notions of three-dimensional geometry and introduce digitization schemes in 3-D. Digitization of a 3-D line yields a set of discrete points in 3-D. These points are usually denoted by a chain code string. We restrict ourselves to the *standard situation* (to be defined later) with commonly used 26-connected chain code strings on the cubic grid of discrete points. This is followed by a discussion on proper geometric or algebraic characterization of a 3-D *DSLS*. This characterization enables us to define an algorithm to detect whether a given set of 3-D grid points can be the digitization of a 3-D *CSLS*.

3.3.1   Geometric Preliminaries, Digitization, and Characterization

A straight line in 3-D may be specified in several possible ways. For example, coordinates of two points in 3-D uniquely determine a line. Similarly, the slope of the straight line together with the Cartesian coordinates of a point lying on the line also represents it uniquely. The slope, however, can be stated (Fig. 3.9) either in terms of direction cosines commonly denoted by l, m, n, or it can be given by θ and ϕ, where ϕ is the angle made by the projection of the line on the XY plane with the X-axis and θ is the angle between the line and its projection on the XY plane.

The quantities l, m, n, and θ, ϕ are related by the following equations:

$$l = \cos\theta\cos\phi$$
$$m = \cos\theta\sin\phi$$
$$n = \sin\theta$$

and it is well known that

$$l^2 + m^2 + n^2 = 1.$$

Thus, a 3-D line segment can be specified by the initial point (x_1, y_1, z_1), the slope of the line given by ϕ, and θ and x_2, the X-coordinate of the final point.

Lemma 3.6. *Any straight line segment, L, in 3-D (not parallel to Y or Z-axis) can be uniquely identified by its projections on XY and XZ planes and vice versa [39].*

Proof: Without any loss of generality, we may assume that line segment L lies in the first octant and is contained within the planes $x = 0$ and $x = x_2$. Let the initial point of L be $(0, y_1, z_1)$. If θ and ϕ together denote the slope of L, then the equation of L_Y, the projection of L on XZ plane, is given by $z = x\sec\phi\tan\theta + z_1$. Similarly, the equation for L_Z, the projection on XY plane, is $y = x\tan\theta + y_1$. Since the 4-tuple (θ, ϕ, y_1, z_1) is fixed for L, the above equations are unique. Also, L_Y, L_Z are line segments on the $y = 0$ and $z = 0$ planes, respectively, enclosed within the lines $x = 0$ and $x = x_2$.

Conversely, given the equations of the projection segments $y = px + y_1$ and $z = qx + z_1$ between $x = 0$ and $x = x_2$, the slope of the original line in 3-D

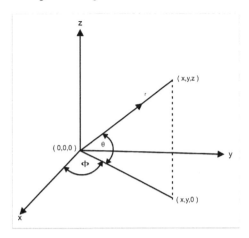

FIGURE 3.9: The slopes of a 3-D line segment given by θ and ϕ. r shows the length of the line.

can be written as the pair $\phi = \tan^{-1}(p)$ and $\theta = \tan^{-1}(q/\sqrt{1+p^2})$. The first point of the 3-D line segment is $(0, y_1, z_1)$ and the final point is at $x = x_2$. $\square$

Thus, any straight line segment L in 3-D can be denoted by a pair of line segments L_Z, L_Y in 2-D representing the projections of L on the XY and XZ planes, respectively. Next, we state that without any loss in generality, for any straight line, there exists a coordinate transformation such that ϕ lies within 0 and $\pi/4$ and θ lies between 0 and $\tan^{-1}(\sin\phi)$.

Lemma 3.7. *Without any loss of generality, for any 3-D line segment, $0 \leq \phi \leq \pi/4$, $0 \leq \theta \leq \tan^{-1}\sin(\phi)$ for a proper choice of axes [39].* $\square$

From a different perspective, the parameters that characterize a 3-D *CSLS* L may as well be its direction cosines l, m, n, the initial point $(0, y, z)$, and x_2, the X-coordinate of the final point of L. It is no restriction to assume that L lies on the first octant, which means all of l, m, n are non-negative. Moreover, in accordance with the relationship $l^2 + m^2 + n^2 = 1$, these three quantities cannot be zero simultaneously. Thus, we may select the axes properly to ensure the $l \geq m \geq n \geq 0$ and $l + m + n > 0$. Actually these conditions and the ones stated in the previous lemma are satisfied simultaneously with the same choice of axes.

Lemma 3.8. *For all lines in the first octant, the pair of conditions $l \geq m \geq n \geq 0$ and $l + m + n > 0$ is equivalent to the pair of conditions $\tan\phi \leq 1$, $\tan\theta \leq \sin\phi$ for the same choice of axes [39].* $\square$

When a *CSLS* L in the first octant starting from $(0, y_1, z_1)$ satisfies these sets of inequalities, L is said to be in the *standard situation*. In the following, unless otherwise stated, a 3-D *CSLS* L is always assumed to be in the *standard situation*.

3.3.1.1   Digitization of a 3-D Line Segment

In the following, we define different digitization schemes used in the case of 3-D *CSLS*.

Grid Intersection Quantization (for 3-D lines) [4]: Whenever a 3-D line crosses a coordinate plane the nearest digital point on that plane to the point of crossing becomes a point of $D(l)$, the digital representation of l. This may also be termed as Grid Intersection Quantization in (GIQ) 3-D.

Digitization scheme (adapted from [201]): Let the crossing point of the line (segment) L and the hyperplane $x_1 = j$ be $P_j : (j, p_2, p_3)$. Then we say that the digital point $P'_j : (j, r_2, r_3)$, where $p_m - 1/2 < r_m \leq p_m + 1/2$, for $m = 2$ and 3, is the nearest digital point to P_j and that P'_j is the digital image of P_j. The set consisting of all points P'_j is said to be the digital image of the line (segment) L.

Object Boundary Quantization (OBQ): The OBQ digitization for 3-D CSLS is defined in the following.

Definition 3.6. *Let L be a 3-D line and $D(L)$ denote its digitization. For integral values of i from 0 to n, the point $(i, \lfloor y(i) \rfloor, \lfloor z(i) \rfloor)$ belongs to $D(L)$ if and only if $(i, y(i), z(i))$ is a point on L. In other words, whenever a 3-D line segment L crosses a coordinate plane $x = i$ at the point $(i, y(i), z(i))$, the grid point $(i, \lfloor y(i) \rfloor, \lfloor z(i) \rfloor)$ becomes an element of the digital image $D(L)$ of L for $i = 0$ to n.* $\square$

The following theorem is restated from [4, 201], which holds for a line segment in the *standard situation*.

Theorem 3.12. *Let L be a 3 -D DSLS and l be a 3-D CSLS. L is the digital image of l if and only if the projections of L on the $y = 0$ and $z = 0$ planes are the GIQ images of the projections of l on the same two coordinate planes.* $\square$

Next, we consider characterization of *DSLS* using OBQ digitization as discussed above. As $y(i) = i \tan \phi + y_1$ and $z(i) = i \sec \phi \tan \theta + z_1$, therefore the grid points of $D(L)$ can be written as (x_i, y_i, z_i) where

$$
\begin{aligned}
x_i &= i, \\
y_i &= \lfloor i \tan \phi + y_1 \rfloor, \text{ and} \\
z_i &= \lfloor i \sec \phi \tan \theta + z_1 \rfloor \text{ for all } i, 0 \leq i \leq n.
\end{aligned}
$$

We have already seen that the equation of L_Z is $y = x \tan \phi + y_1$ and that of L_Y is $z = x \sec \phi \tan \theta + z_1$. Since the slopes of these lines are not greater than unity, $D(L)_Z$ is the OBQ image of L_Z and $D(L)_Y$ is the OBQ image of L_Y. The converse is also true. This leads us to the following theorem.

Theorem 3.13. *$D(L)$ is the digital image of L if and only if $D(L)_Z$, $D(L)_Y$ are the OBQ images of L_Z and L_Y, respectively.* $\square$

The (0-1) chain code (C) representation of L is a sequence of 2-tuples where the i-th tuple, C_i, is given by,

$$C_i = (C_{i,1} = (y_{i+1} - y_i), C_{i,2} = (z_{i+1} - z_i)) \text{ for } 0 \leq i \leq n-1.$$

It is interesting to note that if $y_1 = \lfloor y_1 \rfloor + y_1'$ and $z_1 = \lfloor z_1 \rfloor + z_1'$, then the chain code representation of the 3-D line $y = px + y_1$ and $z = qx + z_1$ is the same as that of $y = px + y_1'$ and $z = px + z_1'$.

Before concluding this section, we observe the relationship between the digitization procedure described and the one proposed by Kim [4] which was later refined by [201]. In fact we see that the scheme discussed so far and the schemes from [4, 201] are very close to each other and the difference between them is the same as that between OBQ and GIQ in 2-D.

3.3.1.2 Characterization of a 3-D DSLS

Theorem 3.12 enables us to decompose the discrete image of a line segment in its projections without losing any information. This also motivates us to characterize and analyze the discrete line segment $D(L)$ in terms of its projections. Using Theorem 3.12, we can easily define an algorithm (see Algorithm 5) to determine the straightness of a 3-D digital arc that runs in $O(n)$ time.

Algorithm 5: To Check a DSLS in 3-D (adapted from [201])

Algorithm *Check_3D_DSLS*

Input: The Set S of n points $p_i : (x_i, y_i, z_i)$ for $1 \leq i \leq n$.

Output: true/false if S is/is not a 3-D *DSLS*.

1: Compute S_Y and S_Z, the projections of S onto XZ and XY planes respectively.

2: Determine for each projection S, whether S_Y and S_Z are 2D-*DSLS*s in the two-dimensional planes XZ and XY planes respectively. If any test fails, then return false. Otherwise, return true.

End Check_3D_DSLS

Step 1 of Algorithm 5 can be clearly performed in $O(3n)$ time, given a set of n points in the plane. There also exists an $O(n)$ algorithm to determine whether or not the set is a 2-D *DSLS*. This test is applied 2 times, leading to $O(2n)$ time for Step 2.

3.3.2 Characterization of Chain Codes of 3-D DSLS

In this section, we have used the OBQ digitization scheme for a 3-D *CSLS*. From the 3-D *DSLS*, we compute the chain codes of its projections on two principal coordinate planes as discussed before. The chain codes are then represented by a fixed number of elements. These representations of chain code are called *tuples*. Four different characterization schemes of the chain code of a 3-D *DSLS* are discussed in this section. These characterizations are used in constructing the domain of the *DSLS*. For details, interested readers may refer to [39].

3.3.2.1 *n*-characterization

This is the most rudimentary characterization scheme where a chain code string C is characterized by the total number of 2-tuples it contains. This is very similar to the (n)-characterization in the case of a 2-D *DSLS*.

3.3.2.2 (n, n_{o1}, n_{o2})-characterization

This is also an extension of odd–even characterization of a tuple obtained from a 3-D *DSLS*. Here, n_{0j}, $j = 1$ or 2, is the total number of 1s in the j-th component of the tuples in C. Subsequently, (n, n_{o1}) and (n, n_{o2}) are the odd–even characterizations of the projections of the 3-D line on XY and XZ planes, respectively.

3.3.2.3 $(n, n_{o1}, n_{c1}, n_{o2}, n_{c2})$-characterization

This is yet another improvement over the last characterization scheme. Here, n_{cj} denotes the total number of occurrences of the consecutive pair of unequal elements in the j-th component of the chain code C. That is, n_{cj} is the cardinality of the set $\{i|$ such that $C_{i,j} \neq C_{i+1,j}\}$.

3.3.2.4 $(n, q_1, p_1, s_1, q_2, p_2, s_2)$-characterization

This is similar to and an extension of the faithful characterization of 2-D *DSLS*s.

Let C_Z (C_Y) be the chain code containing only the first (second) elements of C_i for all i in the original string C. q_1, p_1, and s_1 (or q_2, p_2, s_2) represent the period, number of ones in one period, and the phase shift equivalent to the intercept of the string C_Z (C_Y).

Combining Theorem 3.13 and the results in [78] and observing that C_Z (C_Y) denote the chain code for projection of the line on the XY (XZ) plane, we can get a faithful characterization of C in 3-D.

3.3.3 Length Estimators for Different Characterizations

Let $L(t)$ serve as the length attributed to all digitizations with characterization t. For a string of n elements, the total length of the part of the continuous contour between $x = 0$ and $x = n$ is given by

$$f(n, \theta, \phi) = n \sec \phi \sec \theta.$$

Naturally, this correct length is independent of y_1 and z_1. Now, the continuous line segments leading to a particular tuple t are all found within the region $Domain(t)$. As a measure of the error for the length estimator $L(t)$, we use a quantity dubbed $RDEV(L(t), n)$ from [79]. It is the root mean square difference between the original length and the estimated length, averaged over all strings of n elements and divided by n. The last division is performed to render the measure scale invariant and assumes a sampling density of n^3 points per unit cube.

$$RDEV(L(t), n) =$$

$$(1/n) \sqrt{\sum_{\forall t \in Domain(t)} \int \int \int \int (L(t) - f(n, \phi, \theta)^2 p(\phi, y_1, \theta, z_1) \, d\theta \, dz_1 \, d\phi \, dy_1}$$

where $p(\phi, y_1, \theta, z_1)$ is the probability density function of the lines in (ϕ, y_1, θ, z_1) space. We assume that y_1 and z_1 are uniformly distributed in the range $[0, 1)$, and θ and ϕ are uniformly distributed in the range $[0, \pi/4)$ and $[0, \tan^{-1}(\sin \phi))$, respectively. The ranges are selected in accordance with our discussion in the previous sections. Hence we get,

$$p(\theta, \phi, y_1, z_1) = (1/(\pi/4 - 0)) * (1/(\tan^{-1} \sin(\phi) - 0))$$
$$= 4/(\pi * \tan^{-1}(\sin \phi)).$$

For notational convenience, $I(L(t); e, f; g, h; a, b; c, d)$ denotes the following integration:

$$\int_a^b \int_c^d \int_e^f \int_g^h (L(t) - n \sec \phi \sec \theta)^2 . (4/\pi) / \tan^1(\sin \phi)) \, dz_1 \, d\theta \, dy_1 \, d\phi.$$

3.3.3.1 (n)-characterization [39]

A simple estimator for the n-characterization has the form $L(n) = \alpha n$. For the choice of α to be 1 we have $L(n) = n$. Now, the RDEV for this estimator can be found:

$$X = I(n; 0, \tan^{-1}(\sin \phi); 0, 1; 0, \pi/4; 0, 1).$$
$$RDEV(L(n), n) = \sqrt{X}/n = 21.77\%.$$

Of course, this RDEV is true asymptotically.

To compute the unbiased linear estimator using this characterization, we have to find an α that minimizes the RDEV. Thus, we have to minimize $RDEV(\alpha n, n)$ with respect to α. Setting $d(RDEV)/d\alpha$ to 0 and solving for α, we get optimum $\alpha = 1.1307$ and the corresponding error becomes 15.26%.

3.3.3.2 (n, n_{o1}, n_{o2})-characterization [39]

Here, we use (n, n_{o1}) and (n, n_{o2}) separately to estimate the length of the projection on the XY and XZ planes. Let these estimates be denoted by $|L_Z|$ and $|L_Y|$, respectively, and the estimator for the 3-D line be $|L| = \sqrt{(|L_Y|^2 + |L_Z|^2 - n^2)}$. $|L_Y|$ and $|L_Z|$ may be evaluated using several methods.

To design a very good estimator for the length of L, $|L_Y|$ and $|L_Z|$ are to be chosen properly so that the *RDEV* for L is minimized. This global minimization seems to be very difficult. Instead, we discuss a simple method of estimation. We select good estimators for L_Y locally, and for simplicity, we use a similar functional form for L_Z. It is true that the estimators so obtained are not optimal but the performance of these estimators is seen to be satisfactory both theoretically and practically.

Depending on different choices of $|L_Y|$ and $|L_Z|$, we may use different length estimators. In one method, $|L_Y|$ may be $(n - n_{o1}) + \sqrt{2}n_{o1}$ and $|L_Z|$ may be $(n - n_{o2}) + \sqrt{2}n_{o2}$. The 3-D estimator so formed is called L_1. In the other method, the 2-D estimators are computed using the formula $L_k(n, n_o) = 0.948(n - n_o) + 1.343n_o$ and the resultant 3-D estimator is named L_2.

Another interesting estimator is the Euclidean length between the first and the last digital point. This turns out to be the length between $(0, 0, 0)$ and (n, n_{o1}, n_{o2}). Thus, the estimator is $L_0 = \sqrt{(n^2 + n_{o1}^2 + n_{o2}^2)}$. As this characterization is not a faithful one, the image of the line joining the above mentioned points is not the given digital set. However, this line is *close* to one probable line producing the same set of discrete points. This *closeness* arises as a consequence of the famous *chord property*, which holds for the projections on 2-D. This additional property of this estimator has made it worth investigating.

3.3.3.3 $(n, n_{o1}, n_{c1}, n_{o2}, n_{c2})$-characterization [39]

In this case also, we combine two 2-D estimators to obtain the desired 3-D estimator. The 2-D estimators are calculated using the formula $L_c(n, n_o, n_c) = 0.980(n - n_o) + 1.406n_o - 0.091n_c$.

3.3.3.4 $(n, q_1, p_1, s_1, q_2, p_2, s_2)$-characterization

Since this is the faithful characterization, in most cases, the line defined by the slopes $\tan\phi = (p_1/q_1)$ and $\sec\phi\tan\theta = (p_2/q_2)$ reproduces the same digital set on quantization. Therefore, in most cases, this line turns out to be one probable pre-image of the image data. Thus, another three-dimensional estimator of interest is given by

$$L_0 = n\sqrt{(1 + (p_1/q_1)^2 + (p_2/q_2)^2)}.$$

For different values of n, the average error of each length estimator has been reported in [39].

3.4 Digital Plane Segments

Planes come next after straight lines in geometry. Naturally, digitization and characterization for digital planes have already been the focus of many researchers [110, 85]. The digital plane segment (*DPS*) is the digitization of a plane segment in continuous space. Thus, a *DPS* is a set of discrete points in 3-D that could have been resulted by the digitization of a plane segment.

Definition 3.7. *A support face or (support) of a DPS is a plane* **P** *that goes through one or more points of the DPS so that all other points of the DPS lie on one side of* **P**. □

Forchhammer has mapped the problem of finding the domain of a digital plane segment (*DPS*) to the problem of computing the intersection of n half-spaces in 3-D in [85]. Though Forchhammer's algorithm runs in optimal time, it is considerably complex for implementation. On the other hand, a simple algorithm from Kim [110] constructs a support face that, by definition, is a reconstruction of a digital plane segment. It is not true that a support face always exists for a *DPS*. However, most often such a support face is likely to exist. In this section we outline an optimal algorithm to compute a support face of a DPS, if it exists, exploiting ideas from both Kim and Forchhammer. We also present other important characterizations of a digital plane segment.

Another equally fundamental problem regarding a *DPS* is the estimation of the area of the original predigitized plane segment. The problem of estimating the surface area of 3-D objects has been discussed in [81] for grey images. We discuss here a net code representation of a digital plane segment [42], and based on this representation, a number of area estimators for discrete planes are described.

3.4.1 Digitization and Netcode Representation

A plane, $P(a, b, c)$ is expressed by the equation $z = ax + by + c$ where a, b, and c are any real number. By suitable axes transformation and translation of the origin, any plane may be equivalently given by $z = ax + by + c$, $0 \leq b \leq a \leq 1$. In the latter case, the plane is said to be in the *standard situation*. In subsequent discussion, all planes are assumed to be in the standard situation.

We define a digital plane with respect to a model of digitization as defined below.

Definition 3.8. *Digitization $D(p)$ of a plane segment P ($P : z = ax + by + c$, $0 \leq b \leq a \leq 1$), in $0 \leq x, y \leq n$, is defined as follows: $D(P) = \{(i, j, k) | i, j, k$ integers, $0 \leq i, j \leq n$ and $k = \lfloor ai + bj + c \rfloor\}$* □

Clearly, $D(p)$ contains $(n + 1)^2$ points, which are represented by an $(n + 1) \times (n + 1)$ matrix v where $v(i, j) = k$. Therefore, every row v_i (or column

v_j) of v is an n-*DSLS*. The i-th row *DSLS* v_i consists of the following digital points:

$$v_i = (\lfloor bj + (ai + c)\rfloor \text{ where } 0 \le j \le n).$$

Similarly, the j-th column *DSLS* v_j is defined as follows:

$$v_j = (\lfloor ai + (bj + c)\rfloor \text{ where } 0 \le i \le n).$$

Definition 3.9. *v is an n-DPS if there exists a plane p such that $D(p) = v$ in $0 \le x, y \le n$.* □

Let P and P' be two planes given by the equations $z = ax + by + c$ and $z = a'x + b'y + c'$, respectively. Let v be $D(P)$ and v' be $D(P')$. Then the two n-*DPS*s v and v' are identical if and only if all the row and the column n-*DSLS*s are the same for them. That is, we have this lemma:

Lemma 3.9. *$\forall i, j\ v(i, j) = v\prime(i, j)$ if and only if $\forall i(\forall j(v_i(j) = v_i'(j)))$ and $\forall j(\forall i(v_j(i) = v_j'(i)))$.*

Proof: The proof follows from simple rearrangement. □

Similar to the chain code description of a digital line segment we describe a *net code* representation of a *DPS* in 3-D. If p is a plane and v is its digitization $D(P)$, then v consists of only those grid points that are nearest and below p on the vertical grid lines. Now consider a point $p : (i, j, k)$ of v on the vertical line $x = i$, $y = j$. Looking forward from P we encounter two points p_1 and p_2 on vertical lines $x = i+1$, $y = j$ and $x = i$, $y = j+1$. Depending on the difference between the Z-coordinates of p and p_1 we associate a '0' or a '1' to p. Similarly another '0' or '1' is attached to p by considering the Z-coordinates of p and p_2. Associating a netcode element to every point $v_{i,j}$ of v, $0 \le i, j \le n-1$, in this fashion we construct a forward net of the n-DPS v. A backward net of v is analogously defined.

Definition 3.10. *Let $v_{i,j} = (i, j, z_{i,j}), 0 \le i, j \le n-1$ and $P : z = ax + by + c$. If the netcode at $v_{i,j}$ is $N_{i,j}$, then $N_{i,j} =< z_{i+1,j} - z_{i,j}, z_{i,j+1} - z_{i,j} >$. We also write $N_{i,j}(1) = $ the first component $= z_{i+1,j} - z_{i,j}$ and $N_{i,j}(2) = $ the second component $= z_{i,j+1} - z_{i,j}$.* □

It may seem at a first glance that four types of net code elements are required to describe the forward net but we prove in the next lemma that one out of the four possible codes is not needed.

Lemma 3.10. *In the forward net of an n-DPS v, only three types of netcode elements viz. 00, 10, 11 can occur.*

Proof: Let P be given by $z = ax + by + c$ and $a \ge b$. In this case we claim that the code 01 cannot appear.

Let us assume that there exists some i, j such that $N_{i,j} = 01$. So $z_{i,j+1} = \lfloor ai + b(j + 1) + c\rfloor = z_{i,j+1}$.

Now, $z_{i+1,j} = \lfloor a(i+1) + bj + c \rfloor \geq \lfloor ai + b(j+1) + c \rfloor = z_{i,j+1}$.

Therefore, if the second component of the net code is a 1 then the first component is bound to be a 1, and hence the code 01 can never occur for $a \geq b$. Similarly, if $b > a$, then the code 10 will not appear. Clearly, only three codes may appear in the netcode. □

3.4.2 Geometric Characterization

Similar to the chord property to characterize a 2-D *DSLS*, Kim and Rosenfeld defined a chordal triangle property to characterize a *DPS*. The chordal triangle property is defined in the following.

Definition 3.11. *[111] S, A set of digital points in 3-D, is said to have the chordal triangle property iff for any p_1, p_2, $p_3 \in S$, every point of the triangle $p_1 p_2 p_3$ is at $L_\infty - distance < 1$ from some point of S.*

$L_\infty - distance$ between two points $p : (x_1, y_1, z_1)$ and $q : (x_2, y_2, z_2)$ is defined as $max\{|x_1 - x_2|, |y_1 - y_2|, |z_1 - z_2|\}$. □

Kim and Rosenfeld [111] also proved the following theorem.

Theorem 3.14. *A simple digital surface is a digital plane iff it has the chordal triangle property.* □

Definition 3.12. *Vertical distance $V_d(\mathbf{P}, \mathbf{S})$ between a plane $\mathbf{P}$ and a set of points S is defined as $max_{\mathbf{p} \in \mathbf{S}}$ { Vertical distance between $\mathbf{P}$ and $\mathbf{p}$ }.* □

Definition 3.13. *$\mathbf{P}$ is a Nearest Supporting Plane (SP) of a set of points S if and only if the points in S lie completely on one side of $\mathbf{P}$ and $V_d(\mathbf{P}, \mathbf{S}) < 1$.*

Further, if S lies below $\mathbf{P}$ then $\mathbf{P}$ is a nearest upper supporting plane (NUSP); otherwise, P is a lower nearest supporting plane (NLSP). □

The characterization of a *DPS* S is obtained following the same approach as described in [110]. The following lemma is an adaptation of Theorem 15 of [110].

Lemma 3.11. *A set of points S is an n-DPS if and only if there is an nearest (Upper or Lower) Support Plane $\mathbf{P}$ of S.* □

Now, let us define a support face of a *DPS* **P**.

Definition 3.14. *A face F of the convex hull of a DPS $\mathbf{P}$ is an Upper Support Face (USF) of an n-DPS $\mathbf{P}$ if and only if the plane $\mathbf{Q}$ containing F is an NUSP of $\mathbf{P}$. The Lower Support Face (LSF) is defined analogously.* □

Kim claimed that (Theorem 16, [110]) **P** is an *n-DPS* if and only if there exists a support face of **P**. It is important to note that there are counter examples of this claim.

To find the support face, Kim constructed the convex hull $CH(\mathbf{P})$ and examined whether some face of this hull is a Nearest SP of **P**. As there are n^2 points in **P**, the construction of $CH(\mathbf{P})$ takes $O(n^2 log n)$ time [110].

It can be shown that it is sufficient to construct the convex hull of a subset of points of a *DPS* **P** to obtain the *USF F*. In that case, a *USF* of **P** can be found in $O(n^2)$ time if it exists, where $(n + 1)^2$ is the number of points in **P**. We observe that the *LSF* may also be obtained in $O(n^2)$ time similarly.

Recall that if D is a *DSLS*, then it can be represented by a four tuple (n, p, q, s) and the following line is a pre-image of D (refer to Section 3.1.2).

$$L : y = \frac{p}{q}(x - s) + \lceil \frac{sp}{q} \rceil$$

We note that L is a *Nearest Upper Support Line* that must pass through at least one point of D. Let the rightmost and leftmost point of L be called *limiting points* of L.

Definition 3.15. *The limiting point set* $Lim(\mathbf{P})$ *of a DPS* **P** *is defined as:* $Lim(\mathbf{P}) = \{\mathbf{p} | \mathbf{p} \in \mathbf{P}$ *and* **p** *is a limiting point of some column DSLS* $\mathbf{P_j}, 0 \leq \mathbf{j} \leq \mathbf{n} \}$. ☐

The following theorem is central to the algorithm to compute a USF of a DPS.

Theorem 3.15. *If there is a USF F of an n-DPS* **P***, then F must be a face of the convex hull of the set* $Lim(\mathbf{P})$ *[42].* ☐

Thus to search for the Upper Supporting Face F of a *DPS* **P** it is sufficient to construct the convex hull of the points belonging to the set $Lim(\mathbf{P})$ only. Next we present the **Algorithm Find_Upper_Support_Face** to compute the *USF* (see Algorithm 6).

In the next theorem we analyze the time complexity of the algorithm.

Theorem 3.16. *The algorithm runs in* $O(n^2)$ *time where* $(n + 1)^2$ *is the number of points in the DPS* **P***.* ☐

3.4.3   Characterization by Convex Hull Separability

Stojmenovic in [201] presented an accurate characterization by using Convex Hull Separability.

Let S be a set of 3-D points and $S_{z+1} = \{(i, j, k + 1) : (i, j, k) \in S\}$. A plane γ in the Euclidean space separates the sets S_1, S_2 in 3-D digital space iff S_1 and S_2 are in opposite open half-spaces defined by γ. The following theorem from to [201] provides a characterization of a digital plane.

Theorem 3.17. *A set S in 3-D digital space is a subset of a digital plane iff there exists a plane that separates S from* S_{z+1}. ☐

Algorithm 6: To Compute the Upper Support Face of a DPS

Algorithm Find_Upper_Support_Face

Input: The n-DPS v.

Output: An Upper Support Face of v.

1: For each column j {

Compute the limiting points of v_j and include them in the set $Lim(v)$.

Compute the domain of v_j using $Dorst's$ [76] algorithm.}

2: For each row i, compute the domain of v_i from [76].

3: Construct the convex hull H of the points in $Lim(v)$.

4: For each face F of H {

flag $\leftarrow$ true;

Let $p : z = ax + by + c$ be the plane of F.

For each row i {if $((a.i + c), b) \notin$ domain of v_i then flag $\leftarrow$ false; }

For each column j {if $((b * j + c), a) \notin$ domain of v_j then flag $\leftarrow$ false; }

If $flag = true$ then return$(true, a, b, c, F)$.

}

End Find_Upper_Support_Face

Arithmetic geometry, as briefly introduced in [85] and developed in [172], provides a uniform approach to the study of digitized hyperplanes in n dimensions. Basic definitions follow the general idea of specifying lower and upper supporting planes. We discuss here the three-dimensional case. Let a, b, c, μ and $\omega > 0$ be integers.

Definition 3.16. $D_{(a,b,c,\mu,\omega)} = \{(i, j, k) \ in \ 3 - D \ digital \ space \ : \ \mu < ai + bj + ck < \mu + \omega\}$ *is called an arithmetic plane with normal* $n = (a, b, c)^T$, *intercept* μ, *and arithmetic thickness* ω. □

An arithmetic plane is a generalization of an arithmetic line $D_{(a,b,\mu,\omega)} = \{(i, j) \ in \ 2 - D \ digital \ space \ : \ \mu < ai + bj < \mu + \omega\}$. From Reveilles' theorem on arithmetic lines [172] we know that naive lines with $\omega = max\{|a|, |b|\}$ are the same as digital lines. If $\omega = max\{|a|, |b|, |c|\}$, then the arithmetic plane $D_{(a,b,c,\mu,\omega)}$ is called a naive plane. They have shown that a finite DPS γ in the grid-point model is characterized by the property that it is between two supporting planes

$$ai + bj + ck = \mu \ \text{and} \ \ ai + bj + ck = \mu + c.$$

The upper supporting plane is a translation of the lower supporting plane (by translation vector $(0, 0, 1)$). The main diagonal direction of both (under the assumption $0 < a < b < c$) is $(-1, -1, +1)$, and the main diagonal distance between both planes is less than or equal to $\sqrt{3}$. In [46], Coeurjolly et al. also presented a theorem to compute all the pre-images of a set of digital points representing a 3-D digital plane segment.

3.4.4 Area Estimators

This subsection introduces various estimators for the actual area of a pre-digitized planar segment corresponding to a given netcode of an *n-DPS*. As in the case of 3-D line segments we shall characterize the netcode by a tuple t that consists of a fixed number of parameters extracted from the digital image. The parameters of these tuples are then combined in several ways to formulate various estimators $A(t)$.

In order to measure the performance quantitatively, we define an error function associated with an estimator $A(t)$, which we call the Relative Deviation $RDEV(A(t), N)$ where N is a netcode. It is the square root of the normalized (with respect to n^2) mean square error (MSE) of the estimator $A(t))$ in the area measurement averaged over all plane segments producing the tuple t. Formally, if $A = n^2 \sqrt{(1 + a^2 + b^2)}$ is the area of the original plane segment (we use $m = n$ here), then

$$RDEV(A(t), N) = \sqrt{(\sum \int \int \int_{Domain(t)} (A(t) - A)^2) p(a, b, c) \, da \, db \, dc) / n^2}$$

where $Domain(t)$ is the set of continuous plane segments (i.e., a, b, c triplet) having the same characterization t of their digitizations.

3.4.4.1 n^2-characterization

The simplest way to characterize a netcode N is to count the number of netcode elements. Obviously, it is n^2. This is also the crudest estimator of the actual area.

$$A_0 = n^2$$

Clearly, A_0 is a biased estimator and therefore the unbiased estimator is of the form

$$A_1 = \mu n^2, \text{where } \mu \text{ is a constant.}$$

3.4.4.2 (n_1, n_2, n_3) -characterization

From Lemma 3.10, we can say that there are three types of codes in a netcode N. Using the number of each of the three codes, we provide a new characterization for N as follows:

Definition 3.17. *A netcode N is characterized by (n_1, n_2, n_3) if $n_1 = |((i, j) : N_{i,j} = 00)|, n_2 = |((i, j) : N_{i,j} = 10)|$ and $n_3 = |((i, j) : N_{i,j} = 11)|$.* □

This is a direct extension of the (n_e, n_o)- characterization for straight lines. Also, $(n_1 + n_2 + n_3) = n^2$.

The exact relationship between (n_1, n_2, n_3) and a, b, c is not yet reported. However, we provide here a relation between them for large n.

Lemma 3.12. *For an n-DPS , $n_3 = bn^2$, $n_2 = (a - b)n^2$ and $n_1 = (1 - a)n^2$ when $n \to \infty$ [42].* □

We can also characterize the backward net by three similar parameters, namely, n_1^b, n_2^b and n_3^b. As a corollary to the previous lemma, we conclude that for large n, $n_i \approx n_i^b$, where $i = 1, 2, 3$. From the description of v in terms of forward and backward nets, we may visualize the digitization as a triangular tessellation in 3-D. The areas of these triangles are better estimators for the original area. Three types of triangles δ_1, δ_2, and δ_3, may arise in our context depending on the three types of netcode present. The vertices of these triangles are:

$$\delta_1 \ : \ \{(i, j, k), (i + 1, j, k), (i, j + 1, k)\}$$
$$\delta_2 \ : \ \{(i, j, k), (i + 1, j, k + 1), (i, j + 1, k)\}$$
$$\delta_3 \ : \ \{(i, j, k), (i + 1, j, k + 1), (i, j + 1, k + 1)\}$$

The areas of these triangles are $|\delta_1| = 1/2$, $|\delta_2| = 1/\sqrt{2}$, and $|\delta_3| = \sqrt{3}/2$. Thus, the corresponding estimator is given by

$$(n_1 + n_1^b)|\delta_1| + (n_2 + n_2^b)|\delta_2| + \sqrt{3}(n_3 + n_3^b)|\delta_3|,$$

which asymptotically approximates to

$$A_2 = n_1 + \sqrt{2}n_2 + \sqrt{3}n_3$$

for large n.

Though A_2 is expectedly better than A_0 or A_1, this is again a biased estimator usually overestimating the original. Actually all these estimators A_0, A_1, A_2 are expressed as linear combinations of n_1, n_2, and n_3. Depending on the choice of the coefficients we can think of a class of simple estimators, which are of the form $\mu_1 n_1 + \mu_2 n_2 + \mu_3 n_3$, where μ_1, μ_2, μ_3 are constants. Important among them is the one for which the RDEV is the least. There is an effort [42] to obtain better estimates by minimizing RDEV with respect to the coefficients.

3.4.4.3 Non-linear Estimators

A major common drawback of all simple estimators discussed so far is that the RDEV of them does not decrease beyond a certain value (i.e., the asymptotic bound obtained theoretically), even if we go on increasing the sampling density. To offer a solution to this problem, two non-linear estimators are used [42].

An example of these estimators is given below:

$$
\begin{aligned}
A_4 &= n^2\sqrt{(1 + (n_3/n^2)^2 + ((n_2 + n_3)/n^2)^2)} \\
&= \sqrt{((n_1 + n_2 + n_3)^2 + (n_2 + n_3)^2 + n_3^2)}.
\end{aligned}
$$

The RDEV of A_4 is seen to be zero as n tends to infinity.

The RDEV and bias of the different area estimators and their bias are pictorially presented in Fig. 3.10 [42]. They establish that the performance of all these estimators except A_4 cannot be improved beyond a certain limit by increasing the sampling density alone.

3.5 Summary

In this chapter, we have dealt with digitized straight line segments in 2 and 3 dimensions and digitized plane segments. In each case, we discussed digitization schemes and various characterizations of the digitized figures. We also discussed about several property estimators for these discrete shapes.

For 2-D and 3-D digital straight line segments, we outlined algorithms to compute one probable continuous figure that could be the pre-image of a given digital set of points. In fact, we described algorithms to compute the domain of a digitized 2-D straight line segment, which is the set of all continuous

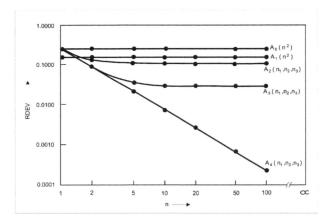

FIGURE 3.10: Plot of RDEV vs. n for various area estimator. (the first set).

Reprinted from the proceedings of Conference on Vision Geometry, 1832(1993), S. Chattopadhyay et al., Digital Plane Segments, 150–161, Copyright (1993), with permission from SPIE.

line segments that are the pre-images of the given discrete straight line segment (*DSLS*). We also presented in detail an iterative refinement algorithm to analyze a *DSLS* and compute its domain.

We also described several length estimators of digitized straight lines in two and three dimensions. The accuracies of these estimators were also discussed. A study of the performance of the estimators shows the importance of characterizing the chain code in a much better way. It reveals that a richer characterization leads to better estimators, and naturally the faithful characterization yields the best length estimator for 2-D and 3-D *DSLS*.

A netcode (like a chain code for lines) is described as a representation of a digital plane segment. Some of its properties were examined and many area estimators were introduced in terms of the number of different elements in the netcode.

Exercises

1. Suppose that a point p in 3-D is denoted by (p_x, p_y, p_z), and let $p_{z=0}$ denote the point $(p_x, p_y, 0)$.

   A set of digital points in 3-D S is called even iff its projection onto the xy-plane is one-to-one, and for every quadruple (p, q, r, s) of points in S such that $p_{z=0} - q_{z=0} = r_{z=0} - s_{z=0}$, we have $(p_z - q_z) - (r_z - s_z) \le 1$.

   Prove that a simple digital surface is a digital plane iff it has the evenness property.

2. A corridor is a closed region between two parallel straight lines. If a corridor contains k points including the points on the boundary, then the corridor is called a k-dense corridor (KDC). Of all $KDCs$, the one with the smallest width is called the narrowest KDC ($NKDC$).

   Suppose D is a $DSLS$ with n points and (n, q, p, s) is Dorst's characterization of D. Prove that the slope of $NnDC$ through D is (p/q).

3. If the digitization scheme of Stojmenovic is assumed for 3-D straight lines, then prove that each of the planes $x = i$, $y = j$ or $z = k$ contains exactly one digital point of the line (segment) L.

4. The r, s noise cover of $L : y = mx + c$, denoted by $N_{r,s}(L)$, is given by $N_{r,s}(L) = \{(i, j) | 0 \le i \le n, mi + c - s - 1 < j < mi + c + r, i, j \text{ integers }\}$.

   Suppose the real image of a line is I, where for every i, there would be multiple y_i around $\lfloor mi + c \rfloor$. Modify the iterative refinement technique to obtain r, s, m, c so that $N_{r,s}(L)$ includes I.

5. Consider the algorithm *Algorithm Sequential LR*. If for some k, $m_l^{k+1} = (y_i - c_u^k)/i$ and $c_u^{k+2} = (y_j + 1 - m_l^{k+1} j)$ then prove that the new estimates from the straight line $L : y = mx + c$ passing through (i, y_i) and $(j, y_j + 1)$ satisfy $m_l \ge m \ge m_l^{k+1}$ and $c_u \le c \le c_u^{k+2}$.

6. Provide a counter-example where **P** is an n-*DPS*, but there does not exist a support face.

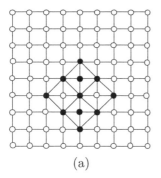 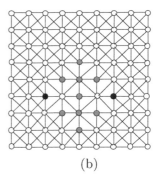

(a)   (b)

FIGURE 1.3: Topological configurations on the same set of foreground pixels with (a) (8,4), and (b) (4,8).

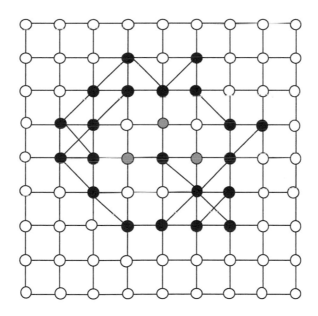

FIGURE 1.6: A connected component of foreground pixels surrounds orange pixels in background in $(8, 4)$ digital grid.

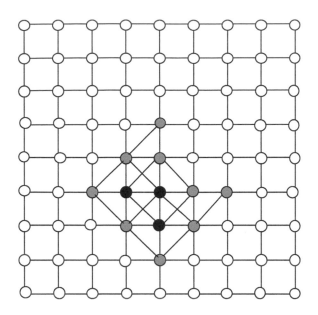

FIGURE 1.7: The border (orange pixels) and interior (black pixels) of a connected component of foreground pixels in $(8, 4)$ digital grid.

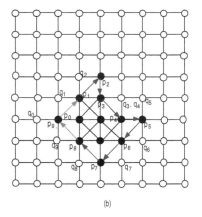

FIGURE 1.8: (a) The order of searching a foreground pixel in the neighborhood of a border pixel at p with a background neighbor at q in an $(8, 4)$ digital grid. The order follows clockwise movement starting from q. (b) The sequence of pairs of border pixels (p_i, q_i), where p_i belongs to foreground, and q_i belongs to background, respectively, for the point set as shown in Fig. 1.7.

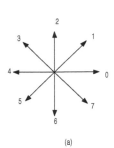

(a)

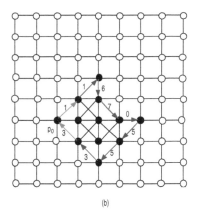

(b)

FIGURE 1.9: (a) Codes of discrete Orientations in $(8, 4)$ 2-D grid, and (b) Chain code of a contour starting from $p_0 \equiv 116705533$.

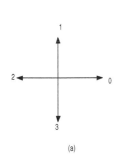

(a)

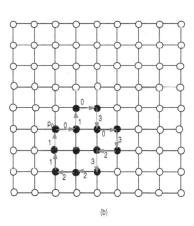

(b)

FIGURE 1.10: (a) Codes of discrete Orientations in $(4, 8)$ 2-D grid, and (b) Chain code of a contour starting from $p_0 \equiv 010303232211$.

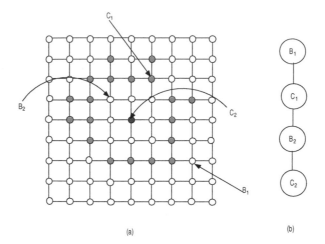

(a)

(b)

FIGURE 1.19: (a) Connected components of foreground points (differently colored) and background points (white) of the 2D point set in a (8,4) grid, and (b) Corresponding adjacency tree.

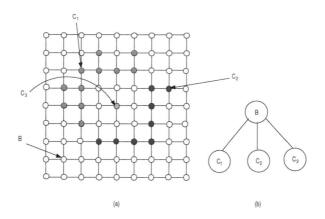

(a)

(b)

FIGURE 1.20: (a) Connected components of foreground points (differently colored) and background points (white) of the 2D point set in a (4,8) grid, and (b) Corresponding adjacency tree.

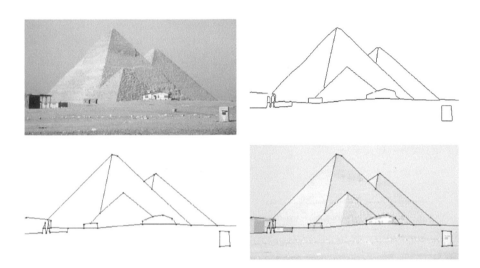

FIGURE 4.1: Polygonal approximation of the image "pyramid". **Top-left:** 8-bit gray-scale image of "pyramid". **Top-right:** The edge map of "pyramid" is considered to be a real-world digital curve. Note that the edge map is subject to the parameter(s) specified in the edge extraction algorithm. **Bottom-left:** Polygonal approximation of the edge map with the vertices shown in red color and the edges in blue color. **Bottom-right:** Polygonal approximation superimposed on the original gray-scale image shows how well the algorithm can approximate a real-world image.

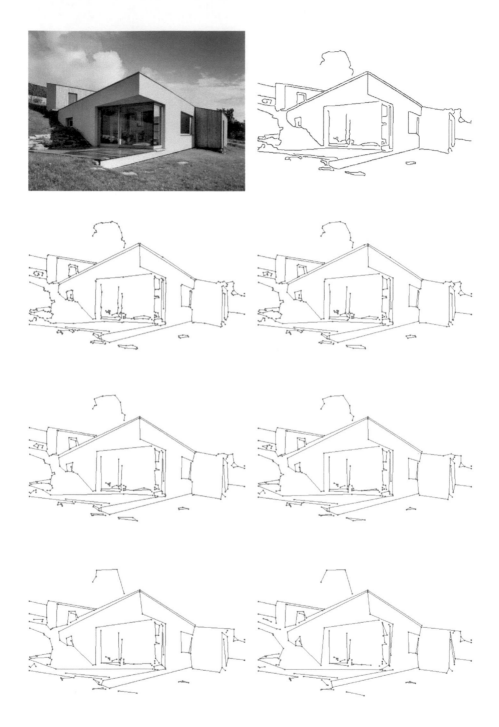

FIGURE 4.8: Result of polygonal approximation on the (thinned) edge map of a real-world image.

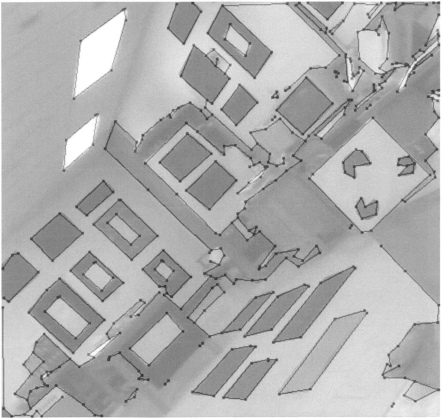

FIGURE 4.9: Result of straight edge detection (bottom) on 'lab' image (top) with $\tau = 4$. (See color insert.)

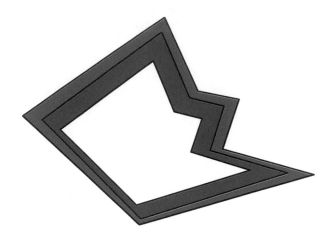

FIGURE 6.1: Computation of DT using iterative scan from boundary points.

FIGURE 7.1: Some typical examples of 3D models (triangular faces shown randomly colored). From top-left to bottom-right: `icosahedron` (20 identical equilateral triangular faces), `teapot`, `turbine`, `cow`, `pickup-van`.

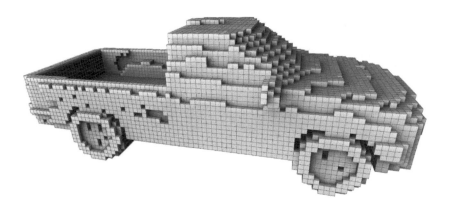

FIGURE 7.3: Voxelation of `pickup-van` shown in Fig. 7.1.

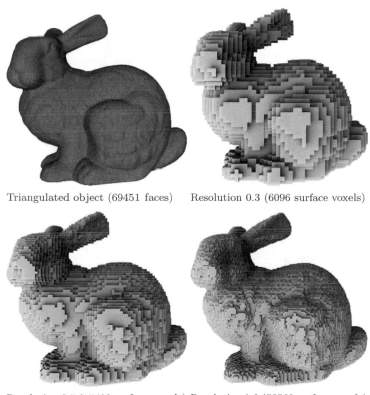

Triangulated object (69451 faces)   Resolution 0.3 (6096 surface voxels)

Resolution 0.5 (15488 surface voxels) Resolution 1.0 (58560 surface voxels)

FIGURE 7.4: Voxelation of `bunny` at different resolutions.

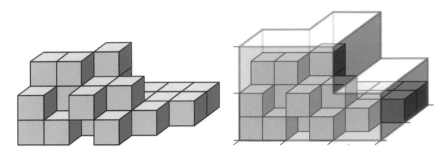

FIGURE 7.5: An object A in $\mathbb{Z}^3$ (left) and its outer isothetic cover for $g = 2$ (right).

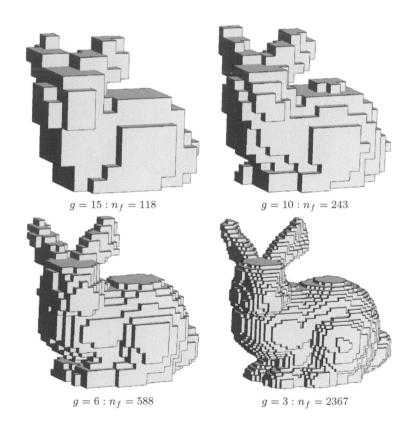

$g = 15 : n_f = 118$ 　　　　 $g = 10 : n_f = 243$

$g = 6 : n_f = 588$ 　　　　 $g = 3 : n_f = 2367$

FIGURE 7.6: Isothetic covers of Stanford Bunny for different grid sizes; n_f = number of cover faces defined as isothetic polygons. (See color insert.)

Reprinted from *Proc. 14th International Workshop on Combinatorial Image Analysis: IWCIA11*, LNCS **6636**: 70–83, N. Karmakar *et al.*, Copyright 2011, with permission from Springer.

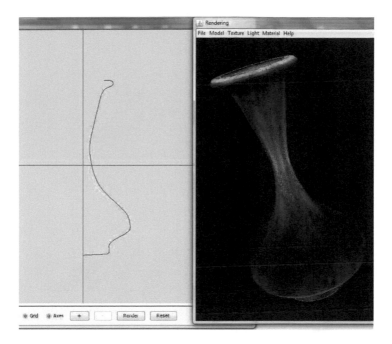

FIGURE 7.7: A snapshot of a part of the algorithm in action: The digital generatrix (shown in the left pane) and the corresponding digital surface resembling a flowerpot generated in the right pane.

Reprinted from *International Journal of Arts and Technology*, **4**: 196–215, G. Kumar *et al.*, Copyright 2011, with permission from Inderscience Publishers.

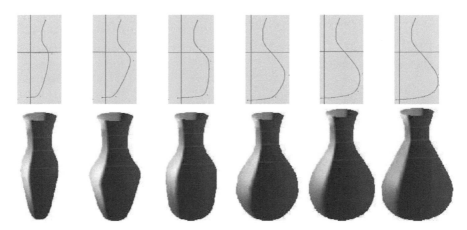

FIGURE 7.8: Simulating the local effect of a "potter's hand" by inserting the control points (from left to right). Other operations (e.g., deletion and repositioning of control points) are also incorporated in the algorithm.

Reprinted from *International Journal of Arts and Technology*, **4**: 196–215, G. Kumar *et al.*, Copyright 2011, with permission from Inderscience Publishers.

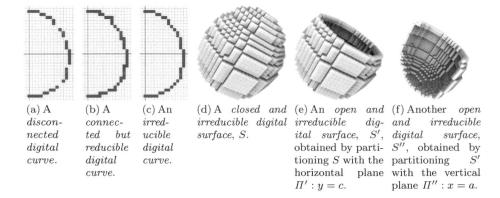

(a) A disconnected digital curve.

(b) A connected but reducible digital curve.

(c) An irreducible digital curve.

(d) A *closed and irreducible digital surface, S.*

(e) An *open and irreducible digital surface, S′,* obtained by partitioning S with the horizontal plane $\Pi' : y = c$.

(f) Another *open and irreducible digital surface, S″,* obtained by partitioning S' with the vertical plane $\Pi'' : x = a$.

FIGURE 7.9: An illustration of connectivity and irreducibility of digital curves and surfaces.

Reprinted from *International Journal of Arts and Technology*, **4**: 196–215, G. Kumar *et al.*, Copyright 2011, with permission from Inderscience Publishers.

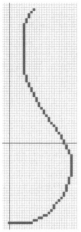

(a) A *digital generatrix* in the form of an *irreducible digital curve*.

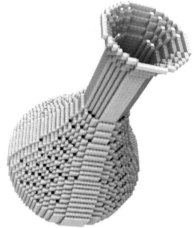

(b) A *disconnected surface* of revolution created due to *missing voxels*.

(c) The set of *missing voxels*.

(d) The *connected and irreducible surface* of revolution when missing voxels are taken care of.

FIGURE 7.10: How a *connected and irreducible surface* of revolution is created by fixing the missing voxels, failing which a disconnected surface would be produced.

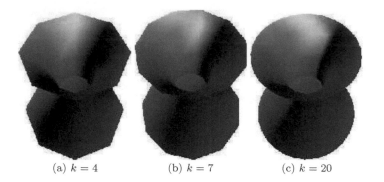

(a) $k = 4$ (b) $k = 7$ (c) $k = 20$

FIGURE 7.11: Results on quad-decomposition for a digital jug: Approximating generating digital circles by regular $2k$-gons.

Reprinted from *International Journal of Arts and Technology*, **4**: 196–215, G. Kumar *et al.*, Copyright 2011, with permission from Inderscience Publishers.

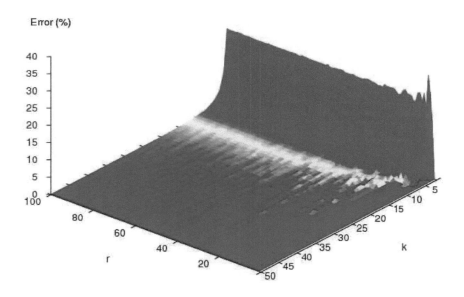

FIGURE 7.13: 3D plot of error versus k and radius r of a digital circle. Note that, with higher values of k, the resultant $2k$-gon covering the digital circle has lesser error. (See color insert.)

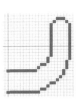

(a) A *double layered irreducible digital curve.*

(b) A *thick-walled disconnected digital surface of revolution.*

(c) Missing voxels.

(d) A *thick-walled irreducible digital surface of revolution.*

(e) A part of the irreducible digital surface.

FIGURE 7.15: A thick- and hollow-walled irreducible digital surface of revolution.

Reprinted from *International Journal of Arts and Technology*, **4**: 196–215, G. Kumar *et al.*, Copyright 2011, with permission from Inderscience Publishers

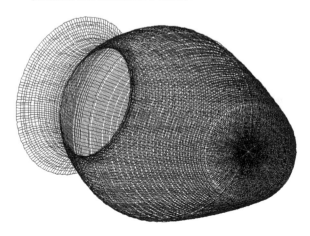

FIGURE 7.16: Quad decomposition of a wheel-thrown digital vase for texturing.

Reprinted from *International Journal of Arts and Technology*, **4**: 196–215, G. Kumar *et al.*, Copyright 2011, with permission from Inderscience Publishers.

FIGURE 7.17: A digital wheel-thrown uni-voxel thick "bowl" created by an irreducible digital curve segment as the digital generatrix. The surface is then decomposed into quads for texture mapping with suitable illumination and shadow formation.

Reprinted from *International Journal of Arts and Technology*, **4**: 196–215, G. Kumar *et al.*, Copyright 2011, with permission from Inderscience Publishers.

FIGURE 7.18: Potteries with thick walls.

Reprinted from *International Journal of Arts and Technology*, **4**: 196–215, G. Kumar *et al.*, Copyright 2011, with permission from Inderscience Publishers.

Chapter 4

Digital Straightness and Polygonal Approximation

Digital straight segments (DSS) started gaining special attention since the 1960s from the viewpoint of their theoretical formulation [28, 86, 87, 166, 180]. Many interesting properties of DSS have been discovered in later periods by various researchers, which are mostly related to the *theory of words and numbers* [7, 32, 117] and *continued fractions* [113, 115, 119, 143, 209]. With the proliferation of digitization and vectorization of graphical objects and visual imageries, uses of these properties have been investigated by different researchers for different application-specific problems related with computer graphics and image analysis. The most significant among these is to determine whether or not a given digital curve segment S is a DSS, and its algorithmic solutions for several defining criteria have been reported in the literature in the 1980s and

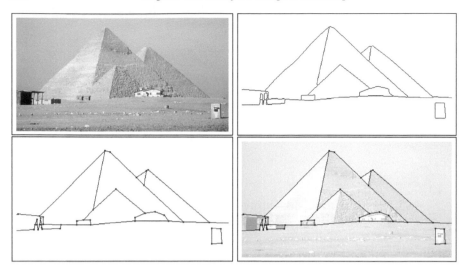

FIGURE 4.1: Polygonal approximation of the image 'pyramid'. **Top-left:** 8-bit gray-scale image of 'pyramid'. **Top-right:** The edge map of 'pyramid' is considered to be a real-world digital curve. Note that the edge map is subject to the parameter(s) specified in the edge extraction algorithm. **Bottom-left:** Polygonal approximation of the edge map with the vertices shown in red color and the edges in blue color. **Bottom-right:** Polygonal approximation superimposed on the original gray-scale image shows how well the algorithm can approximate a real-world image. (See color insert.)

1990s [50, 73, 74, 122, 142, 195]. In the previous chapter, characterization of digital straightness property of a set of points in a 2-D grid by determining the domain of a candidate DSLS was briefly discussed. In this chapter, we present a more direct approach for determination of the straightness of a digital arc and demonstrate its application in polygonal approximation of boundaries of objects in an image.

A typical example of this application is shown in Fig. 4.1. It depicts a polygonal approximation of a real-world digital curve defining the edge map of the 'pyramid' image.[1] The merits of this approach lie in its superior performance on computational speed and quality of results compared to other methods of polygonization of edge points. The polygonal approximation is achieved (refer to Fig. 4.1) only in a few milliseconds from the digital curve, since it uses only a few primitive operations in the integer domain only. This shows the potential strength of a digital-geometric technique compared to the others, which are mostly based on parametric approaches, such as distance criteria, usage of masks, eigenvalue analysis, Hough transform, etc.

---

[1]Source: The Berkeley Segmentation Dataset and Benchmark, http://www.eecs.berkeley.edu/Research/Projects/CS/vision/bsds/.

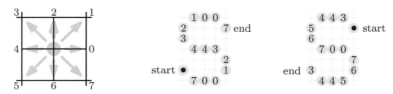

FIGURE 4.2: Chain codes and their enumeration in defining a (irreducible) digital curve segment.

4.1 Digital Straightness

In Euclidean space, a *straight line* consists of points lying evenly on itself [27]. A *digital straight line* (DSL) is a sequence of digital points satisfying certain straightness properties in an appropriate sense. With a similarity lying in their constitution by points, there arise fundamental differences in their very definitions and related properties compared to those of a Euclidean straight line. Some of these are stated below.

Similar to a Euclidean straight line, the property of evenness of points lying on a line, which was stated as the definition of a real straight line by Euclid, was reiterated by both Freeman and Rosenfeld [87, 115, 176] in the 1960s–70s. The difference is that Euclid stated the *evenness* as a definition, whereas the concept of *evenness* for DSL was formalized and proved in [176]. Clearly, this indicates *evenness* as a strong necessary condition in order that a digital curve segment is *digitally straight*. A thorough discussion of the formalization of evenness of points constituting a digital straight segment (DSS) or a DSL is provided later in Sec. 4.1.2.[2] A DSL is infinitely long and a DSS is a finite segment of a DSL. A DSS is an irreducible digital curve segment that is *digitally straight*. An *irreducible digital curve segment* (referred to as a discrete arc in Section 3.1 of Chapter 3) is a sequence of digital points having two distinct endpoints, each with one neighbor (in 8-neighborhood), and each other point having two neighbors from the curve segment (Fig. 4.2).

Another interesting aspect that distinguishes a DSS (DSL) from a real/continuous straight line segment (straight line) is the *cutting syndrome*, which is as follows. If a real line segment is cut into two (or more) segments, then each segment remains straight. This is also true in digital geometry; if a DSS is cut into two parts, then each of them would still be digitally straight. However, from the two subsequences of digital points representing the cut-off parts, the correspondence is not straightforward. The reason is as follows. Let p and q be two digital points; pq denote their connecting real line segment, and $DSS(p, q)$ denotes the sequence of digital points obtained by the digitization of pq. See Sec. 4.1.1, Eq. 4.1 in particular, for digitization of a straight line

---

[2]A DSS is also referred to as DSLS, as in Chapter 3.

segment. If we cut $DSS(p, q)$ at any intermediate point $r \in DSS(p, q)$, then the sequence of digital points in $DSS(p, q)$ from p to r may not be the same as $DSS(p, r)$; similarly, this difference may be observed between the sequence of digital points in $DSS(p, q)$ from r to q and $DSS(r, q)$.

4.1.1   Slopes and Continued Fractions

In Sec. 3.1.2, a number-theoretic characterization of a DSS (or a DSLS) is presented. In this section, we elaborate it further by elucidating the link between the rational slope of a straight line and the chain code of its digital image. We illustrate this relationship by using Euclid's famous algorithm for expressing a rational number in the form of continued fractions.

We consider that the digitization process of a straight line is by the method of *nearest integral coordinates* (NIC), as discussed in Sec. 3.1. This means that given a real line l, its corresponding DSL is the sequence of digital points such that the isothetic distance of each digital point p from l is at most $\frac{1}{2}$. The *isothetic distance* of $p(i, j)$ from l is given by $\min\{|x - i|, |y - j|\}$, where (j, x) and (i, y) are the respective (real) points of intersection of l with the horizontal and the vertical lines passing through p. As mentioned earlier, a DSS is a finite segment of a DSL, and hence it may be obtained by cutting the DSL at two arbitrary digital points, namely p and q, and the cut-off DSS is very likely to differ from $DSS(p, q)$ obtained by digitization of the real line segment, pq. Hence, given a digital curve segment as a sequence of digital points, S, we cannot decide whether S is digitally straight simply by verifying whether S is identical with the DSS formed by joining the endpoints (i.e., first and last points) of S. Following the NIC digitization process, we formally define a DSS as follows:

Definition 4.1. *If $p = (i_p, j_p)$ and $q = (i_q, j_q)$, and without loss of generality, if $i_p < i_q$ and the slope of pq lies between 0 and 1, then for each vertical line $x = i$ with i lying in the integer interval $[i_p, i_q]$, we have a unique pixel (i, j) in $DSS(p, q)$ where j is rounded off from the y-coordinate of the intersection point of the real line $\overline{pq}$ with $x = i$. Hence, the set of points defining the DSS from p to q is given by*

$$DSS(p, q) = \left\{ (i, j) \in \mathbb{Z}^2 \mid i_p \leqslant i \leqslant j_p, j = round(y), \frac{y - j_q}{j_q - j_p} = \frac{i - i_q}{i_q - i_p} \right\}.$$
(4.1)

$\square$

Clearly, given two digital points p and q, the subsequent $DSS(p, q)$ involves a mapping from $\mathbb{R}^2$ to $\mathbb{Z}^2$. On the contrary, the problem related to digital straightness is the reverse and involves the mapping from $\mathbb{Z}^2$ to $\mathbb{R}^2$. For, given the sequence of digital points constituting a digital curve segment S, its digital straightness should be verified not only with respect to the real line segment joining the terminal points of S, but also with respect to other line segments

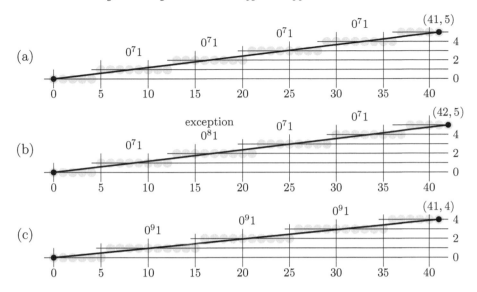

FIGURE 4.3: Periodicity in chain-code composition of DSS with integer endpoints.

lying very close to this real line segment. In other words, we have to verify whether there exists any real line segment—whose endpoints need not be digital points—whose digitization produces S. In particular, we have to verify whether S is a portion of some DSL, which is again the digitization of some real line (of infinite length). Solutions for this problem and some related ones may be seen in [50, 73, 74, 122, 142, 166, 195].

Several interesting works have revealed that DSL and DSS have a close relationship with continued fractions [119, 143, 209]. Fig. 4.3 shows a few examples of DSS whose two endpoints are digital points. For each of these, we have a real line segment whose two endpoints coincide with the endpoints of the concerned DSS. Thus, there exists a real line l with rational slope corresponding to this DSS such that the digitization of l contains an infinite concatenation of the DSS. In essence, given the real line l (with rational slope), its DSL has a periodicity in its constitution, which can be obtained by an analysis of continued fractions. For the DSS shown in Fig. 4.3(a), the slope of the concerned real line / DSL is $\frac{5}{41}$, which can be expressed as

$$\frac{5}{41} = \cfrac{1}{8 + \cfrac{1}{5}} = [8, 5].$$

Note that, in continued fraction, $[q_1, q_2, \ldots, q_n]$ implies

$$\cfrac{1}{q_1 + \cfrac{1}{q_2 + \cfrac{1}{\ddots \cfrac{1}{q_{n-1} + \cfrac{1}{q_n}}}}}.$$

As explained in the coming section, we get the corresponding chain-code (Fig. 4.2; see also Chapter 1) representation from this as $0^8(0^71)^4$ (here, k consecutive 0s are written as 0^k for brevity), which defines the period of the DSL. It indicates that there are four consecutive runs of identical composition, i.e., 0^71, following (and preceding) a single run of 0^8. A *run* is given by the maximum sequence of contiguous digital points lying on the same horizontal or vertical line. The first and the last runs of the DSS should be ignored, since they are partial runs with respect to the concerned DSL. In fact, if we produce the real line segment in both directions, then the partial runs would grow into complete runs; and for the DSS of Fig. 4.3(a), both the left and the right runs would become 0^8.

In Fig. 4.3(b), the DSL has slope

$$\frac{5}{42} = \cfrac{1}{8 + \cfrac{2}{5}} = \cfrac{1}{8 + \cfrac{1}{2 + \cfrac{1}{2}}} = [8, 2, 2],$$

wherefore its period becomes $0^810^710^81(0^71)^2$, a portion of this being contained in the DSS. Notice that, for this DSS, we get 0^81 as an *exceptional run* among the runs of 0^71, which is the *predominant run*. The exceptional run for the DSL in Fig. 4.3(a) was also 0^81, which was not present in the DSS as 0^81 was less frequent. The occurrence of such exceptional runs have been addressed with utmost importance to conceptualize digital straightness from the perspective of word theory and number theory. The predominant run and the exceptional run may change if there is a slight change of slope, as exemplified in Fig. 4.3(c). Here the DSL has slope

$$\frac{4}{41} = \cfrac{1}{10 + \cfrac{1}{4}} = [10, 4],$$

which yields the period $0^{10}1(0^91)^3$, thus giving $0^{10}1$ and 0^91 as the respective exceptional and predominant runs.

$$[q_1, q_2, \ldots, q_n] = \cfrac{1}{q_1 + \cfrac{1}{q_2 + \cfrac{1}{\ddots \cfrac{1}{q_{n-1} + \cfrac{1}{q_n}}}}}.$$

4.1.1.1 Analyzing a Continued Fraction

A brief overview of the procedure for getting the period of a DSL (with rational slope) is given here; for details, see [34, 115, 116, 209]. Let the slope of a DSS $= a/b$ ($1 < a < b; a, b \in \mathbb{Z}$). Using the Euclidean algorithm, it can be expressed as

$$\frac{a}{b} = \cfrac{1}{q_1 + \cfrac{1}{q_2 + \cfrac{1}{\ddots \cfrac{1}{q_{n-1} + \cfrac{1}{q_n}}}}}$$

$$= [q_1, q_2, \ldots, q_n] \tag{4.2}$$

where $q_1, q_2, \ldots, q_n$ are positive integers

$$= \frac{\alpha_n q_n + \beta_n}{\gamma_n q_n + \delta_n}, \text{ where } \alpha_n, \beta_n, \gamma_n, \delta_n \text{ are defined by } q_1, q_2, \ldots, q_{n-1}$$

$$= \frac{(\alpha_{n-1} q_{n-1} + \beta_{n-1}) q_n + \alpha_{n-1}}{(\gamma_{n-1} q_{n-1} + \delta_{n-1}) q_n + \gamma_{n-1}} \tag{4.3}$$

$$= \frac{(\alpha_{n-1} q_{n-1} + \beta_{n-1})(q_n - 1) + \alpha_{n-1}(q_{n-1} + 1) + \beta_{n-1}}{(\gamma_{n-1} q_{n-1} + \delta_{n-1})(q_n - 1) + \gamma_{n-1}(q_{n-1} + 1) + \delta_{n-1}}. \tag{4.4}$$

Eq. 4.3 is obtained from Eq. 4.2 based on the observation that $[q_1, q_2, \ldots, q_n]$ (n elements) and $[q_1, q_2, \ldots, q_{n-1} + \frac{1}{q_n}]$ ($n - 1$ elements) are equivalent, and

$$\left[q_1, q_2, \ldots, q_{n-1} + \frac{1}{q_n} \right] = \frac{\alpha_{n-1} \left(q_{n-1} + \frac{1}{q_n} \right) + \beta_{n-1}}{\gamma_{n-1} \left(q_{n-1} + \frac{1}{q_n} \right) + \delta_{n-1}}.$$

Eq. 4.4 is obtained from Eq. 4.3 simply by manipulating a few terms in order to perform *concatenation* of two slopes (fractions), which is defined below.

Definition 4.2. *The concatenation of two fractions a_1/b_1 and a_2/b_2 is given by $(a_1/b_1) \otimes (a_2/b_2) = a/b$, where $a = (a_1 + a_2)/c$ and $b = (b_1 + b_2)/c$ for an integer c such that $\gcd(a, b) = 1$.* □

We can use the *splitting formula* to recursively split and concatenate the slope $a/b = [q_1, q_2, \ldots, q_n]$ into *atomic slopes* ($[q] = 1/q$ with chain code $= 0^{q-1}1$) as follows.

$$[q_1, q_2, \ldots, q_n] = \begin{cases} [q_1, q_2, \ldots, q_{n-1} + 1] \otimes (q_n - 1)[q_1, q_2, \ldots, q_{n-1}] \\ \qquad\qquad \text{if } n \text{ is even;} \\[2mm] (q_n - 1)[q_1, q_2, \ldots, q_{n-1}] \otimes [q_1, q_2, \ldots, q_{n-1} + 1]. \\ \qquad\qquad \text{if } n \text{ is odd.} \end{cases} \tag{4.5}$$

Note that Eq. 4.5 can be obtained from Eq. 4.4 by using the following fact for $n \geqslant 1$.

$$\alpha_n \delta_n - \beta_n \gamma_n = (-1)^n. \tag{4.6}$$

Example: On finding the period of a DSL with slope $= \frac{38}{87}$, using Eq. 4.5.

$$\frac{38}{87} = \cfrac{1}{2 + \cfrac{1}{3 + \cfrac{1}{2 + \frac{1}{5}}}}$$

$$= [2, 3, 2, 5] \Rightarrow n = 4 \text{ is even}$$

$$= [2, 3, 3] \otimes 4 \cdot [2, 3, 2] \Rightarrow n = 3 \text{ is odd}$$

$$= 2 \cdot [2, 3] \otimes [2, 4] \otimes 4([2, 3] \otimes [2, 4]) \Rightarrow n = 2 \text{ is even}$$

$$= 2([3] \otimes 2 \cdot [2]) \otimes ([3] \otimes 3 \cdot [2]) \otimes 4([3] \otimes 2 \cdot [2] \otimes [3] \otimes 3 \cdot [2])$$
$$\Rightarrow \text{ atomic slopes}$$

and hence the period

$$= (001(01)^2)^2 \, 001(01)^3 \, (001(01)^2 \, 001(01)^3)^4$$

$$= (001)(0101)(001)(0101)(001)(010101)(001)(0101)(001)(010101)(001)$$
$$(0101)(001)(010101)(001)(0101)(001)(010101)(001)(0101)(001)(010101).$$

See the following example that demonstrates the operation of concatenation. We have seen that $\frac{38}{87} = [2, 3, 2, 5] = [2, 3, 3] \otimes 4 \cdot [2, 3, 2]$. Now,

$$[2, 3, 3] = \cfrac{1}{2 + \cfrac{1}{3 + \frac{1}{3}}} = \frac{10}{23} \quad \text{and} \quad [2, 3, 2] = \cfrac{1}{2 + \cfrac{1}{3 + \frac{1}{2}}} = \frac{7}{16}.$$

Thus,

$$[2, 3, 3] \otimes 4 \cdot [2, 3, 2] = \frac{10}{23} \otimes 4 \cdot \frac{7}{16} = \frac{10 + 4 \times 7}{23 + 4 \times 16} = \frac{38}{87}.$$

4.1.2 Periodicity

As mentioned and exemplified in the previous section, a DSL can be represented as a *word* w of infinite length built from the chain-code alphabet $A = \{0, 1, \ldots, 7\}$. The *infinite word* w *is periodic* if $w = v^\omega$, for some *basic segment* $v \in A^* \smallsetminus \{\epsilon\}$. For example, $w = \ldots 0001000100010001 \ldots$ is periodic with $v = 0001$; we can set also 0010 as the period for w as it is an infinite word, but that is not important; what matters here is that w is periodic and its *period* is $|v| = 4$. Studies on such periodicity related to digital straightness date back to 1970s. The following theorem is important in this context.

Theorem 4.1 (Brons [34]). *Rational digital rays are periodic and irrational digital rays are aperiodic.* □

Note that a *rational (irrational) digital ray* implies the DSL corresponding to a real ray with rational (irrational) slope. As evident from Theorem 4.1, the problem of periodicity or aperiodicity becomes crucial while determining the digital straightness of a digital curve segment. In case a DSS is a part of DSL with irrational slope, the period can never be found; and even if the slope is rational, the DSS may not be sufficiently long in comparison with the period so as to accommodate a basic segment of the DSL. Further, the characterization of the basic segment is also quite important. The following theorem provides some clue to address the aforesaid problems using the properties of self-similarity in a word of finite or infinite length.

Theorem 4.2 (Freeman [87]). *The chain-code sequence of a DSS or DSL should have the following three properties:*

F1: At most two types of elements (chain codes) can be present, and these can differ only by unity, modulo 8;

F2: One of the two element values always occurs singly;

F3: Successive occurrences of the element occurring singly are as uniformly spaced as possible.

□

The properties in Theorem 4.2 were illustrated by examples and based on heuristic insights. Further, Property F3 is not precise enough for a formal proof [160]. Nevertheless, it explicates how the composition of a straight line in the digital space resembles that in the Euclidean space as far as the evenness (Section 4.1) in distribution of its constituting points is concerned. A few examples are given below to clarify the idea.

1. 101012100 is not a DSS, since F1 fails (three elements are present).

2. 01101000 is not a DSS, since F2 fails (none of 0 and 1 occurs singly).

3. 010100010 is not a DSS, since F3 fails (singular element 1 is not uniformly spaced).

4. 010010010 is a DSS, since F1–F3 are true.

It may be noted that, for a chain code sequence 010100100, the question of *uniform spacing* (as stated in F3) of the singular element 1 remains unanswered, because there is one 0 between the first two consecutive 1s, and there are two 0s between the next two consecutive 1s. The first formal characterization of DSS, which also brought in a further specification of Property F3, was however provided a few years later in [176], as stated in the following properties of DSS.

Theorem 4.3 (Rosenfeld [176]). *Necessary conditions for a DSS (DSL) are as follows.*

R1:   *The runs have at most two directions, differing by 45°, and for one of these directions, the run length must be 1.*

R2:   *The runs can have only two lengths, which are consecutive integers.*

R3:   *One of the run lengths can occur only once at a time.*

R4:   *For the run length that occurs in runs, these runs can themselves have only two lengths, which are consecutive integers; and so on.*

$\square$

It may be noted that the above four properties, R1–R4, still do not allow a formulation of sufficient conditions for the characterization of a DSS, but they specify F3 by a recursive argument on run lengths. Thus, 010100100 qualifies as a DSS by R1–R4, since the intermediate run-lengths (i.e., 1 and 2) of 0s are consecutive.

4.2   Approximate Straightness

In a digital image containing one or more objects with fairly straight edges, the set of (approximate) digital straight line segments carries strong geometric information of the underlying objects. Hence, the concept of approximate digital straight segments (ADSS) has been proposed in [13]. In the ADSS, some of the most fundamental properties of DSS are preserved and some are relaxed (see Fig. 4.4). The number of ADSS extracted from a set of digital curve segments S in a real-world scenario is usually fewer than that of DSS cover, since many visually straight segments may fail to satisfy all stringent

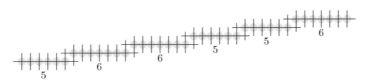

S_1 with chain code $0^5 10^6 10^6 10^5 10^5 10^6$ (from left to right) ($p = 5, q = 6, l = 5, r = 6$) that does not satisfy R3, since both the run lengths 5 and 6 have non-singular occurrences in the code of run lengths: 566556. Thus, S_1 is not a DSS but an ADSS.

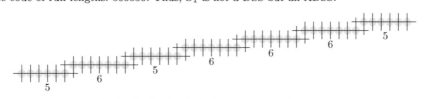

S_2 with chain code $0^5 10^6 10^5 10^6 10^6 10^6 10^5$ ($p = 5, q = 6, l = 5, r = 5$) does not satisfy R4, since in the run length code 5656665, the runs of 6 have lengths 1 and 3 that are not consecutive. Here also, S_2 is not a DSS but an ADSS.

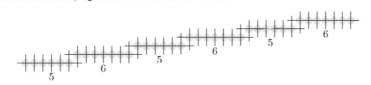

S_3 with chain code $0^5 10^6 10^5 10^6 10^5 10^6$ ($p = 5, q = 6, l = 5, r = 6$) that satisfies (R1–R4) and (c1, c2). So, S_3 is an ADSS as well as a DSS.

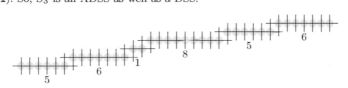

S_4 with chain code $0^5 10^6 1010^8 10^5 10^6$ ($p = 1, q = 8, l = 5, r = 6$) that does not satisfy R2 and conditions (c1, c2). Thus, S_4 is neither a DSS nor an ADSS.

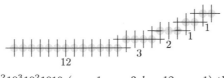

S_5 with chain code $0^{12} 10^3 10^2 1010$ ($p = 1, q = 3, l = 12, r = 1$) that violates R2 and is, therefore, not a DSS. Further, as $q - p (= 2) \not\leq d (= 1)$ and $l - p (= 11) \not\leq e (= 1)$, it fails to satisfy (c1) and (c2); hence, it is not even an ADSS.

FIGURE 4.4: Instances of digital curve segments showing the significance of properties and conditions related to DSS and ADSS recognition.

properties (R3 and R4, Theorem 4.3) of a DSS. For example, in Fig. 4.4, the DSS are all *exactly straight*, whereas the ADSS are *visually straight*.

The concept of ADSS can also be used for an efficient polygonal approximation. Since the set of ADSS provides an elegant and compact representation of digital curve segments, it is very effective in producing approximate polygons (or, polychains) using a single parameter. The whole process consists of two stages: extraction of ADSS and polygonal approximation. The major features are as follows:

- The detection of ADSS is based on chain-code properties; only primitive integer operations, such as comparison, increment, shift, and addition (subtraction) are required.

- Does not use any recursion, and thus saves execution time.

- To obtain the polygonal approximation, only the endpoints of ADSS are required with a few integer multiplications.

- The actual approximation of a digital curve segment never oversteps the worst-case approximation for a given value of a control parameter.

Several other methods [43, 44, 95, 220] have been proposed recently for (approximate) line detection. Most of the conventional parametric approaches are based on certain distance criteria, usage of masks, eigenvalue analysis, Hough transform, etc. In contrast, the ADSS-based method relies on utilizing some of the basic properties of DSS for extraction of ADSS. Earlier algorithms for approximating a given digital curve segment may be seen in [2, 8, 103]. Several variants of this technique were proposed later [16, 17, 163, 188, 191]. The class of polygonal approximation algorithms, in general, can be broadly classified into two categories: one in which the number of vertices of the approximate polygon(s) is specified, and the other where a distortion criterion (e.g., maximum Euclidian distance) is used.

Most of the existing polygonal approximation algorithms, excepting a few, have super-linear time complexities, for example, $O(N)$ in [210], $O(MN^2)$ in [163], $O(N^2)$ in [188] and [191], and $O(N^3)$ in [173], where M denotes the number of segments, and N the total number of points representing the input set of digital curve segments. A comparative study of these algorithms can be found in [13, 223]. Further, in order to analyze curvature, most of them require intensive floating point operations [3, 83, 89, 205, 219]. For other details, the reader may look at [10, 80, 161, 210, 217, 183, 205, 224, 225]. The ADSS-based polygonal approximation proposed in [13] uses only integer operations, and yields a suboptimal polygonal approximation with linear time complexity (see [13]).

4.2.1   Extraction of ADSS

In the algorithm DETECT-ADSS, designed for extraction of ADSS from a digital curve segment S, we have used R1 along with certain modifications in R2. However, we have dropped R3 and R4, since they impose very tight restrictions on S to be recognized as a DSS. Such a policy has been done in order to successfully extract the ADSS, and some of the advantages are as follows:

- avoiding tight DSS constraints, especially while representing the gross pattern of a real-world image with digital aberrations/imperfections;

- enabling extraction of ADSS from a curve segment, thereby straightening a part of it when the concerned part is not exactly *digitally straight*;

- reducing the number of extracted segments, thereby decreasing storage requirements and run-time in subsequent applications;

- reducing the CPU time of ADSS extraction; and

- usage of integer operations only (e.g., to compute $\lfloor (p+3)/4 \rfloor$, 3 is added with p, followed by two successive right shifts).

Since the chain code of a curve segment is taken in a one-dimensional list, S, the ADSS may be characterized by the following sets of parameters:

- *Orientations parameters*: n (non-singular element), s (singular element), l (length of leftmost run of n), and r (length of rightmost run of n). They play decisive roles on the orientation (and the digital composition, thereof) of the concerned ADSS. For example, in Fig. 4.4, the curve S_1 has n = 0, s = 1, and chain code $0^5 10^6 10^6 10^5 10^5 10^6$ having $l = 5$ and $r = 6$.

- *Run-length interval parameters*: p and q, where $[p, q]$ is the range of possible lengths (excepting l and r) of n in S that determines the level of approximation of the ADSS, subject to the following two conditions:

$$(c1) \quad q - p \leqslant d = round(p/2). \tag{4.7}$$

$$(c2) \quad (l - p), (r - p) \leqslant e = round(p/2). \tag{4.8}$$

While implementing DETECT-ADSS, we strictly adhere to R1, as it is directly related to the overall straightness of S. However, we have modified the stricture in R2 by considering that the run lengths of n can vary by more than unity, depending on the minimum run length of n. The rationale of modifying R2 to Condition c1 is that, while approximating the extracted line segments from S, an allowance of approximation (d) specified by c1 is permitted. Given a value of p, the amount d by which q is in excess of p indicates the deviation of the ADSS from the actual/real line, since ideally (for a DSS) q can exceed from p by at most unity (the significance of d in characterizing an ADSS is detailed out in [13]).

Apart from d, the other parameter, namely e, is incorporated in c2, which, along with c1, ensures that the extracted ADSS is not badly approximated owing to some unexpected values of l and r. The DSS properties R1–R4, however, do not give any idea about the possible values of l and r (depending on n). Further, in the algorithm for DSS recognition [50], l and r are not taken into account for adjudging the DSS characteristics of a curve segment. However, we impose some bounds on the possible values of l and r, in order to ensure a reasonable amount of straightness at either end of an extracted ADSS. The values of d and e are heuristically chosen so that they become computable with integer operations only. Some other values, like $d = \lfloor (p+3)/4 \rfloor$ and $e = \lfloor (p+1)/2 \rfloor$, or so, may also be chosen provided the computation is realizable in the integer domain and does not produce any undesirable ADSS. For example, in Fig. 4.4, the curve S_5 has $p = 1, q = 3, l = 12, r = 1$. In our case (Eqns. 4.7 and 4.8), therefore, we get $d = 1$ and $e = 1$ resulting in a violation of (c2) by l; thus S_5 will not be accepted as an ADSS.

To justify the rationale of c1 and c2, we consider a few digital curves, S_1–S_5, as shown in Fig. 4.4. It is interesting to observe that, although each of S_1 and S_2 has the appearance of a digital line segment, they fail to hold all the four properties of DSS simultaneously, as shown in their respective figures. The curve S_1 violates R3, and the curve S_2 violates R4. However, they satisfy R1, c1, and c2, and therefore each of them is declared as an ADSS. Similarly, the curve S_3 satisfies R1–R4 and (c1, c2); it is both an ADSS and a DSS. However, none of the curves S_4 and S_5 can be identified as a DSS or an ADSS because of the violation of R2, c1, and c2.

4.2.2 Algorithm DETECT-ADSS

Algorithm 7 shows the algorithm DETECT-ADSS for extracting ADSS from the chain code of each digital curve segment, say S_k, stored in the list S. This requires n_k repetitions from Step 2 through Step 23, where n_k is the number of ADSS in S_k. Let the ith repetition on S_k produce the ADSS $\mathbf{L}_i^{(k)}$. Recognition of $\mathbf{L}_i^{(k)}$ is prompted by finding its corresponding parameters $(\mathsf{n}, \mathsf{s}, l)$ using the ADSS-PARAMS procedure in Step 2 of DETECT-ADSS. This is followed by checking/validating the following:

(i) Property R1: Step 4 and Step 10;

(ii) Condition (c1): **while** loop check at Step 9;

(iii) Condition (c2): on the leftmost run length l in Step 8 and Step 11, and on the rightmost run length r in Step 14.

Proof of correctness: For each ADSS, $\mathbf{L}_i^{(k)}$, we show that property R1 and conditions (c1) and (c2) are simultaneously satisfied. We also show that $\mathbf{L}_i^{(k)}$ is maximal in length in S_k in the sense that inclusion of the character (n or s or any other in $\{0, 1, \ldots, 7\}$) (or a substring of characters) that immediately precedes or follows the part of the digital curve segment corresponding to $\mathbf{L}_i^{(k)}$ in S_k does not satisfy the ADSS property/conditions.

Algorithm 7: DETECT-ADSS to find out the ordered list $\mathcal{A}$ of end points of ADSS in the input curve S that contains the chain code for each connected component.

Algorithm DETECT-ADSS (S)

1. $\mathcal{A} \leftarrow \{1\}, u \leftarrow 1$
2. ADSS-PARAMS (S, u)
3. $c \leftarrow l$
4. **if** $s - n \pmod 8 \neq 1$
   then go to Step 20
5. $p \leftarrow q \leftarrow$ next run-length of n
6. $d \leftarrow e \leftarrow round(p/2)$
7. $c \leftarrow c + 1 + p$
8. **if** $l - p > e$
   then go to Step 20
9. **while** $q - p \not\geqslant d$
10. $\quad k \leftarrow$ next run-length of n
11. $\quad$ **if** $l - k > e$ **then**
12. $\qquad c \leftarrow c + 1 + k$
13. $\qquad$ **break**
14. $\quad$ **if** $k - p \leqslant e$
    $\quad$ **then** $c \leftarrow c + 1 + k$
15. $\quad$ **else** $c \leftarrow c + 1 + p + e$
16. $\quad$ **if** $k < p$ **then**
17. $\qquad p \leftarrow k$
18. $\qquad d \leftarrow e \leftarrow round(p/2)$
19. $\quad$ **if** $k > q$ **then** $q \leftarrow k$
20. $u \leftarrow u + c$
21. $\mathcal{A} \leftarrow \mathcal{A} \cup \{u\}$
22. $u \leftarrow u + 1$
23. repeat from Step 2
    until S is finished

Procedure ADSS-PARAMS (S, u)

1. $i \leftarrow u, l \leftarrow 1$
2. **if** $S[i] = S[i + 1]$ **then**
3. $\quad n \leftarrow S[i]$
4. $\quad$ find $s(\neq n)$ after $S[i + 1]$
5. $\quad l \leftarrow$ leftmost run length of n
6. **else**
7. $\quad (n, s) \leftarrow (S[i], S[i + 1])$
8. $\quad i \leftarrow i + 1$
9. $\quad$ **while** $S[i + 1] \in \{n, s\}$
10. $\qquad$ **if** $S[i] = S[i + 1]$ **then**
11. $\qquad\quad$ **if** $S[i] = s$ **then**
12. $\qquad\qquad$ swap n and s
13. $\qquad\quad l \leftarrow 0$
14. $\qquad\quad$ **return**
15. $\qquad$ **else**
16. $\qquad\quad i \leftarrow i + 1$
17. **return**

While checking R1 in Step 4 or Step 10, if an expected n or s is not found at the desired place in S_k, then the current ADSS, $\mathbf{L}_i^{(k)}$ ends with the previously checked valid characters. This is explicit in Step 4 and implicit in Step 10. Thus $\mathbf{L}_i^{(k)}$ satisfies R1, and is maximal from its starting point and the finishing end, since either it is the first ADSS in S_k, or the previous ADSS, $\mathbf{L}_{i-1}^{(k)}$, was maximal.

Now, for each new run (of n), (c1) is verified in Step 9—excepting the leftmost run, l, which is not required since p (and q) does not exist for a single run—after appropriately updating p and q in Step 17 and Step 19, respectively, whenever necessary. In Step 9, if it is found that q is unacceptably large (i.e.,

$q \not\leqslant p + d$), then the **while** loop (steps 10–19) is not executed, and the current ADSS, $\mathbf{L}_i^{(k)}$, ends with the truncated part of that run (truncated maximally, i.e., up to length $p + e$, in Step 15 of the previous iteration) as its rightmost run, r.

For checking (c2), however, we have to be more careful. For the second run (i.e., the run immediately following l) of the current ADSS, (c2) is checked (with respect to l) in Step 8. It may be noted that, if $l - p > e$, then (c2) is not satisfied, and so the first two runs (l and its successor) trivially constitute an ADSS by Step 7; because for two runs, we get only l and r (and no p or q), and no relation is imposed between l and r to define an ADSS.

For the third and the subsequent run(s), if any, the corresponding run length is stored in k (Step 10). If some (small enough) k violates (c2), then that k is treated as r (steps 11–13), and the current ADSS ends with that run as the rightmost run (of run length k), whereby the maximality criterion of the ADSS is fulfilled. Otherwise, if k does not exceed the maximum possible length of the rightmost run (checked in Step 14), then we consider k as a valid run of the current ADSS (Step 14), else we truncate it to the maximum permissible length ($p + e$) as the rightmost run (Step 15). Note that, if $k > p + e$, then $k > q$ (for $p + e \geqslant p + d \geqslant q$), and Step 19 updates q to k, whence (c1) will be false in Step 9 in the next iteration, and so, the ADSS will end here with the (maximally) truncated part ($p + e$) as its rightmost run. □

Time Complexity: Determination of the parameters $(\mathbf{n}, \mathbf{s}, l)$ in ADSS-PARAMS consists of two cases—the first one (Steps 2–5) being easier than the second (Steps 6–16). In either of these two cases, the procedure searches linearly in S for two distinct (but not necessarily consecutive) chain code values and determines the parameters accordingly. As evident from the loop in either case, the three parameters are obtained using only a few integer comparisons. The number of comparisons is $l + 1$ for the first case, and that for the second case is the number of characters in S until two consecutive non-singular characters are found.

The parameters $\mathbf{n}, \mathbf{s}, l$ obtained in ADSS-PARAMS are successively passed through a number of checkpoints, as mentioned earlier, which take constant time, as evident in Steps 3–8 of DETECT-ADSS. In Step 5 of DETECT-ADSS, the first run length of $\mathbf{n}$ is measured immediately after the leftmost run length of $\mathbf{n}$, if any, and it starts from the first non-singular character out of the two consecutive characters detected in ADSS-PARAMS. In Step 10 of DETECT-ADSS, we have another simple (and silent) loop that determines in linear time each valid run of $\mathbf{n}$ in S, the validity criteria being verified and updated in Steps 9–19, each of these steps taking constant time. Hence, for the ADSS, $\mathbf{L}_i^{(k)}$, the algorithm DETECT-ADSS, together with the procedure ADSS-PARAMS, takes linear time; wherefore the time complexity for extraction of all ADSS in S is strictly linear on the number of points in S.

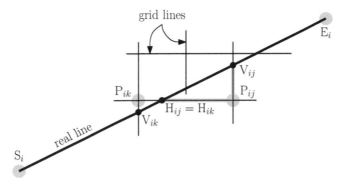

FIGURE 4.5: Isothetic distance of P_{ij} (from $\overline{S_i E_i}$) is $\overline{P_{ij} V_{ij}}$, which is greater than $\frac{1}{2}$, thereby making P_{ij} an error point, whereas, P_{ik} is not an error point, since the isothetic distance of P_{ik} is $\overline{P_{ik} V_{ik}}$, which is less than $\frac{1}{2}$.

4.2.3 Error Points

An ADSS extracted from an input digital curve segment may not be a perfect DSS. There may occur erroneous points. An erroneous point or error point is one whose isothetic distance (i.e., minimum of the vertical distance and the horizontal distance) from the real straight line corresponding to the concerned ADSS is greater than $1/2$. To check whether a point is an error point or not, we use Eq. 4.9, stated as follows.

Let S_i and E_i be the start point and the end point of the ith ADSS, and let P_{ij} be the jth digital point on the ith ADSS, as shown in Fig. 4.5. Let $\overline{S_i E_i}$ denote the real line segment joining S_i and E_i. Then it can be shown that $P_{ij} = (x_p, y_p)$ is an error point corresponding to the line $\overline{S_i E_i}$, if and only if

$$2 \left| \mathbf{x}_{\mathrm{ES}} \mathbf{y}_{\mathrm{EP}} - \mathbf{x}_{\mathrm{EP}} \mathbf{y}_{\mathrm{ES}} \right| - \max \left\{ \left| \mathbf{x}_{\mathrm{ES}} \right|, \left| \mathbf{y}_{\mathrm{ES}} \right| \right\} > 0, \qquad (4.9)$$

where, $\mathbf{x}_{\mathrm{ES}} = x_c - x_s$, $\mathbf{y}_{\mathrm{ES}} = y_e - y_s$, etc. Note that (x_s, y_s) and (x_e, y_e) denote the respective coordinates of the start point and the end point of the ADSS under consideration. Although Eq. 4.9 is not required at any stage in our algorithm, it enables us to check whether or not P_{ij} is an error point without using any floating point arithmetic.

4.3 Polygonal Approximation

Extraction of the ADSS for each curve $\mathcal{C}_k$ in the given set (binary image) $\mathcal{I} = \{\mathcal{C}_k\}_{k=1}^K$ of digital curves generates an ordered set of ADSS, namely, $\mathcal{A}_k = \langle \mathbf{L}_i^{(k)} \rangle_{i=1}^{n_k}$, corresponding to $\mathcal{C}_k$. In each such set $\mathcal{A}_k$, several consecutive

ADSS may occur, which are approximately collinear, and therefore may be combined together to form a single segment.

Let $\langle \mathbf{L}^{(k)} \rangle_{j_1}^{j_2}$ be the maximal (ordered) subset of the ADSS starting from $\mathbf{L}_{j_1}^{(k)}$ that conforms to some approximation criterion. Then these $j_2 - j_1 + 1$ segments in A_k are combined together to form a single straight line segment starting from the start point of $\mathbf{L}_{j_1}^{(k)}$ and ending at the end point of $\mathbf{L}_{j_2}^{(k)}$. This procedure is repeated for all such maximal subsets of A_k in succession to obtain the polygonal approximation (in case C_k is a closed curve) or polychain approximation (in case C_k is open), namely $\mathcal{P}_k$, corresponding to C_k.

In this algorithm, depending on the approximation criterion, we have used a greedy method of approximating the concerned curve C_k starting from the very first ADSS in A_k. Determination of a minimal set of DSS (and a minimal set A_k of ADSS, thereof) corresponding to a given curve C_k is known to be computationally intensive [115, 180]; thus for real-time applications, a near-optimal but speedy solution is often preferred rather than the optimal one.

4.3.1   Approximation Criterion

There are several variants of approximation criteria available in the literature [183]. These algorithms were tested with two variants of the approximation measures using area deviation by Wall and Danielsson [210], which are realizable in a purely integer domain subject to few primitive operations only. The approximation criterion is defined with respect to the *approximation parameter* or *error tolerance*, denoted by τ, as follows.

4.3.1.1   Cumulative Error

Let $\langle \mathbf{L}^{(k)} \rangle_{j_1}^{j_2}$, be an ordered subset of A_k as discussed above. Then under the cumulative error criterion $C_{\sum}$, the ADSS ($j_2 - j_1 + 1$ in number) in A_k are replaced by a single straight line segment starting from the start point of $\mathbf{L}_{j_1}^{(k)}$ and finishing at the end point of $\mathbf{L}_{j_2}^{(k)}$, if:

$$\sum_{j=j_1}^{j_2-1} \left| \triangle \left(s(\mathbf{L}_{j_1}^{(k)}), e(\mathbf{L}_j^{(k)}), e(\mathbf{L}_{j_2}^{(k)}) \right) \right| \leqslant \tau d_\top \left(s(\mathbf{L}_{j_1}^{(k)}), e(\mathbf{L}_{j_2}^{(k)}) \right) \qquad (4.10)$$

where, $s(\mathbf{L}_j^{(k)})$ and $e(\mathbf{L}_j^{(k)})$ represent the respective start point and the end point of the ADSS $\mathbf{L}_j^{(k)}$, etc. The start point of $\mathbf{L}_j^{(k)}$ coincides with the end point of the preceding ADSS, if any, in A^k, and the end point of $\mathbf{L}_j^{(k)}$ coincides with the succeeding one, if any. In Eq. 4.10, $|\triangle (p, q, r)|$ denotes twice the magnitude of area of the triangle with vertices $p = (x_p, y_p)$, $q = (x_q, y_q)$, and $r = (x_r, y_r)$, and $d_\top (p, q)$ the maximum isothetic distance between two points p and q. Since all these points are in two-dimensional digital space, the above measures are computable in the integer domain as shown in the following equations.

$$d_\top (p, q) = \max \left\{ |x_p - x_q|, |y_p - y_q| \right\} \qquad (4.11)$$

$$\triangle\,(p,q,r) = \begin{vmatrix} 1 & 1 & 1 \\ x_p & x_q & x_r \\ y_p & y_q & y_r \end{vmatrix} \tag{4.12}$$

From Eq. 4.12, it is evident that $\triangle\,(p,q,r)$ is a determinant that gives twice the signed area of the triangle with vertices p, q, and r. Hence, the ADSS in the given subset are merged to form a single straight line segment, say $\widetilde{\mathbf{L}}$, provided the cumulative area of the triangles ($j_2 - j_1$ in number), having $\widetilde{\mathbf{L}}$ as base and the third vertices being the end points of the ADSS (excepting the last one) in the subset $\langle\mathbf{L}^{(k)}\rangle_{j_1}^{j_2}$, does not exceed the area of the triangle with base $\widetilde{\mathbf{L}}$ (isothetic length) and height τ.

4.3.1.2 Maximum Error

With similar notations as mentioned above, using the maximum error criterion $C_{\max}$, the ADSS in $\langle\mathbf{L}^{(k)}\rangle_{j_1}^{j_2}$ would be replaced by a single piece, provided the following condition is satisfied.

$$\max_{j_1 \leqslant j \leqslant j_2-1} \left| \triangle\left(s(\mathbf{L}_{j_1}^{(k)}), e(\mathbf{L}_{j}^{(k)}), e(\mathbf{L}_{j_2}^{(k)}) \right) \right| \leqslant \tau d_\top\left(s(\mathbf{L}_{j_1}^{(k)}), e(\mathbf{L}_{j_2}^{(k)}) \right) \tag{4.13}$$

The rationale of considering two such criteria are as follows. Since we would be replacing a number of ADSS, which are almost straight, and more importantly, are not ordinary digital curves of arbitrary patterns and arbitrary curvatures, the end point of each ADSS makes a triangle with the replacing segment, namely $\widetilde{\mathbf{L}}$; wherefore the sum of the areas of triangles formed by the end points of these ADSS in combination with the replacing line $\widetilde{\mathbf{L}}$ gives a measure of error due to approximation of *all ADSS in* $\langle\mathbf{L}^{(k)}\rangle_{j_1}^{j_2}$ by $\widetilde{\mathbf{L}}$. Alternatively, if we are guided by the worst-case approximation, that is, if the mostly digressing ADSS is considered to estimate the error, then the maximum of the areas of these triangles should be considered as the error measure for approximation of *worst ADSS in* $\langle\mathbf{L}^{(k)}\rangle_{j_1}^{j_2}$ by $\widetilde{\mathbf{L}}$.

Empirical observations as reported in [13], reveal that the above two criteria are essentially similar in the sense that they produce almost identical polygons for different digital curves for different values of the error tolerance (i.e., τ). This is quite expected as far as the output is concerned.

As mentioned earlier in Sec. 4.3, to construct a polygonal approximation we consider the start point of the first ADSS (i.e., $\mathbf{L}_{j_1}^{(k)}$), and the end point of the last ADSS (i.e., $\mathbf{L}_{j_2}^{(k)}$). This can by justified as follows.

Fact 1. The sum (for criterion C_Σ) or the maximum (for criterion $C_{\max}$) of the isothetic distances of the end points of each ADSS from the replacing line $\widetilde{\mathbf{L}}$ never exceeds the specified error tolerance τ. This follows easily on expansion of the left hand-side of the corresponding Eqns. 4.10 and 4.13, and from the fact that the term $d_\top\left(s(\mathbf{L}_{j_1}^{(k)}), e(\mathbf{L}_{j_2}^{(k)}) \right)$ represents the isothetic length of $\widetilde{\mathbf{L}}$.

Fact 2. Since each ADSS $\mathbf{L}_j^{(k)}$ is approximately a DSS, we consider $\#p \in \mathbf{L}_j^{(k)}$,

---

Algorithm 8: Algorithm MERGE-ADSS for polygonal approximation of a sequence of ADSS in $\mathcal{A}$ using criterion $\mathrm{C_{max}}$.

Algorithm MERGE-ADSS($\mathcal{A}, n, \tau$)
1. **for** $m \leftarrow 1$ to n
2.   **for** $S \leftarrow 0, i \leftarrow 1$ to $(n - m - 1)$
3. $S \leftarrow S + \triangle(\mathcal{A}[m], \mathcal{A}[m + i], \mathcal{A}[m + i + 1])$
4. $dx \leftarrow |\mathcal{A}[m].x - \mathcal{A}[m + i + 1].x|$
5. $dy \leftarrow |\mathcal{A}[m].y - \mathcal{A}[m + i + 1].y|$
6. $d \leftarrow \max\{dx, dy\}$
7.   **if** $S \leqslant d\tau$
8.   delete $\mathcal{A}[m+i]$ from $\mathcal{A}$
9.   **else**
10.   **break**
11. $m \leftarrow m + i - 1$

---

such that the isothetic distance of p from DSL passing through the end points of $\mathbf{L}_j^{(k)}$ exceeds unity (as testified in our experiments). Although for sufficiently long ADSS, this may not hold for the underlying conditions (c1) and (c2) as stated in Sec. 4.2.1; however, in our experiments with real-world images, this was found to hold. In the case of any violation, some heuristics may be employed to find the error points and to find smaller ADSS to resolve the problem.

4.3.2 Algorithm for Polygonal Approximation

The algorithm for polygonal approximation of a sequence of ADSS in the set $\mathcal{A}$, using the approximation criterion of Eq. 4.10, is described in Algorithm 8. To take care of the criterion $\mathrm{C_{max}}$ of Eq. 4.13, a similar procedure may be written.

Final Time Complexity: As explained in Sec. 4.2.2, the time complexity for extracting the ADSS in a set of digital curves, $\mathcal{I} = \{\mathcal{C}_k\}_{k=1}^K$, is given by $\Theta(N_1) + \Theta(N_2) + \ldots + \Theta(N_K) = \Theta(N)$, where $N(= N_1 + N_2 + \ldots + N_K)$ is the total number of points representing $\mathcal{I}$. Now, in the algorithm MERGE-ADSS, we have considered only the ordered set of vertices of the ADSS corresponding to the curves, so that the worst-case time complexity in this stage is linear in N. Hence, the overall time complexity is given by $\Theta(N) + \mathcal{O}(N) = \Theta(N)$, regardless of the error of approximation τ.

4.3.3 Quality of Approximation

The goodness of an algorithm for polygonal approximation is quantified, in general, by the amount of discrepancy between the approximate polygon(s) (or polychain(s)) and the original set of digital curves. There are several measures to assess the approximation of a curve $\mathcal{C}_k$, such as

(i) compression ratio CR $= N_k/M_k$, where N_k is the number of points in $\mathcal{C}_k$ and M_k is the number of vertices in the approximate polygon $\mathcal{P}_k$;

(ii) the integral square error (ISE) between $\mathcal{C}_k$ and $\mathcal{P}_k$.

Since there is always a trade-off between CR and ISE, other measures may also be used [97], [184], [187]. These measures, however, may not always be suitable for some intricate approximation criterion. For example, the figure of merit [187], given by FOM $=$ CR/ISE, may not be suitable for comparing approximations for some common cases, as shown in [183]. In [208], the percentage relative difference, given by $((E_{approx} - E_{opt})/E_{opt}) \times 100$, has been used, where E_{approx} is the error incurred by a suboptimal algorithm under consideration, and E_{opt} the error incurred by the optimal algorithm, under the assumption that same number of vertices are produced by both the algorithms. Similarly, one may use two components, namely fidelity and efficiency, given by $(E_{opt}/E_{approx}) \times 100$ and $(M_{opt}/M_{approx}) \times 100$, respectively, where M_{approx} is the number of vertices in the approximating polygon produced by the suboptimal algorithm and M_{opt} is the same produced by the optimal algorithm subject to same E_{approx} as the suboptimal one [183].

The algorithm presented here is not constrained by the number of vertices M_k of the output polygon $\mathcal{P}_k$, and therefore, the measures of approximation where M_k acts as an invariant, are not applicable. Instead, we have considered the error of approximation, namely τ, as the sole parameter in our algorithm, depending on which, the number of vertices M_k corresponding to $\mathcal{P}_k$ will change. A high value of τ indicates a loose or slacked approximation, whence the number of vertices M_k decreases automatically, whereas a low value of τ implies a tight approximation, thereby increasing the number of vertices in the approximate polygon. Hence, in accordance with the usage of τ in both the methods, one based on criterion $C_{\sum}$ and the other on C_{max}, the total number of vertices $M = M_1 + M_2 + \ldots + M_K$ in a set of approximate polygons $\{\mathcal{P}_k\}_{k=1}^{K}$ corresponding to the input set of digital curves, namely $\mathcal{I} = \{\mathcal{C}_k\}_{k=1}^{K}$, versus τ, provides the necessary quality of approximation. Since the total number of points lying on all the points in $\mathcal{I}$ characterizes (to some extent) the complexity of $\mathcal{I}$, we consider the compression ratio (CR) as a possible measure of approximation.

Another measure of approximation is given by how much a particular point $(x, y) \in \mathcal{C}_k \in \mathcal{I}$ has deviated in the corresponding polygon $\mathcal{P}_k$. If $\widetilde{p} = (\widetilde{x}, \widetilde{y})$ is the point in $\mathcal{P}_k$ corresponding to $p = (x, y)$ in $\mathcal{I}$, then for all points in $\mathcal{I}$, this measure is captured by the variation of the number of points with isothetic deviation $d_\perp$ with respect to $d_\perp$, where the *(isothetic) deviation from p to $\widetilde{p}$* is given by

$$dev_\perp(p \to \widetilde{p}) = \min\{|x - \widetilde{x}|, |y - \widetilde{y}|\}. \tag{4.14}$$

Further, since $dev_\perp(p \to \widetilde{p})$ depends on the chosen value of τ in our algorithm, the fraction of the number of points in $\mathcal{I}$ with deviation $d_\perp$ varies plausibly with τ. So, the *isothetic error frequency* (IEF) (or, simply *error frequency*), given by

$$f(\tau, d_\perp) = \frac{1}{N} \left| \{ p \in \mathcal{I} : dev_\perp(p \to \widetilde{p}) = d_\perp \} \right|, \qquad (4.15)$$

versus τ and $d_\perp$, acts as the second measure that provides the error distribution for the polygonal approximation of $\mathcal{I}$.

4.4 Approximation on Gray-Scale Images

Edges carry meaningful gray-level discontinuities and hence define the boundary of an object present in a real-world image. A geometric economy of representation of these edges is furthered to a great extent if straight edges can be derived directly from a gray-scale image. A sequence of straight edges results in an efficient polygonal representation of a digital object, which, in turn, can be used for subsequent applications, such as volume rendering and multi-resolution modeling [137, 194], image/video retrieval [144], shape coding [159, 112, 191], etc.

Conventional polygonal (or poly-chain) approximation of a (thinned) digital curve [13, 183, 205, 225] of an object in a gray-scale image requires the following steps:

1. Find the edge map (usually multi-pixel thick).

2. Thin the edge map by a suitable algorithm.

3. Find the polygonal form by edge tracking.

The entire procedure based on the above steps is, therefore, not only susceptible to pitfalls of the adopted edge extraction algorithm and subsequent thinning, but also affected by inter-stage dependence and high runtime. Instead of this, we can extract the (apparently) straight edges present in a gray-scale image without resorting to a conventional edge-detection algorithm (which usually produces thick and rough edges) and without using any thinning. The salient features of the algorithm, as explained in [168], are as follows:

i) *Exponential averaging of Prewitt response [92].* An edge is always one-pixel thick, after being detected.

ii) *Checking the straightness.* An edge is detected as a sequence of piecewise linear components.

iii) *Speedy execution.* Simple integer operations are required while checking the straightness of (a part of) an edge.

4.4.1 Commencing a Straight Edge

We visit each point of the image in row-major order, and compute its Prewitt response [92]. If the (Prewitt) response at a point p exceeds the threshold value, T ($= 100$ in our experiments), and the response is a local maximum in $N_8(p)$ (i.e., the response of each 8-neighbor of p does not exceed that of p), then p is a start point, namely $p_s(i_s, j_s)$. The next point on the edge commencing from p_s is obtained from the responses in $N_8(p_s)$. The direction d_s from p_s is the chain code from p_s to its neighbor having the maximum response. In case of multiple maxima (which indicates multiple edges incident at p_s), we consider each of them, one by one, to find the straight edges from p_s.

To get the (straight-)edge point next to the current point p, we need not apply the convolution at each neighbor of p with the Prewitt operator (in order to get their responses, and the maximum/maxima, thereof). Instead, in our algorithm, checking the Prewitt responses at three neighbors corresponding to three directions suffices: d, $(d+1)(\mathrm{mod}\ 8)$, and $(d+7)(\mathrm{mod}\ 8)$, where d is the chain code of p. It may be noted that, by dint of Property R1, no other neighbor can be the next point on the current edge. Hence, out of the three possible neighbors, we select the one having the maximum response as the next point, provided its response is greater than T (in case of multiple maxima, we consider each, as explained earlier). Further, to capture the previous information as a part of the predicted direction, we have taken the exponential average of Prewitt responses, as explained next.

4.4.2 Exponential Averaging

It is a quick and effective method, as shown in [168], to estimate the edge strength at an edge point using its own response and the weighted contribution of responses at the previous edge points. In order to compute the exponential average of the response at a point p, we need to consider the responses that have already been computed and stored—at the previous points up the straight edge. In an 8-bit image, since the maximum possible (Prewitt) response at a point p is 3×256, and since an exponentially decreasing factor of $\frac{1}{2}$ for the contributing points up the edge is taken, eight previous responses are enough to compute the exponential average at p. Hence, the average response at p is given as follows:

$$R_E = \frac{R_P + \frac{1}{2}R_E^{(-1)} + \frac{1}{2^2}R_E^{(-2)} + \ldots + \frac{1}{2^8}R_E^{(-8)}}{1 + \frac{1}{2} + \frac{1}{2^2} + \ldots + \frac{1}{2^8}} \qquad (4.16)$$

where $R_E^{(-i)}$ ($1 \leqslant i \leqslant 8$) denotes the exponentially averaged response at the ith previous point, R_P denotes the (Prewitt-operator-based) response at the current point p, and R_E is the exponentially averaged response at p. Such a strategy gives a boost to p to be an edge point when it shows a weak response R_P (possibly due to noise) but lies on the edge. If the point p that lies on the edge but has weak response is not taken into consideration, then a straight edge may halt at this point. Consideration of the exponentially averaged response, therefore, increases the chance of including a point with weak response in the current straight edge.

Eq. 4.16 has a notable advantage in its implementation: its denominator can be expressed as

$$\sum_{i=0}^{\infty} \frac{1}{2^i} - \frac{1}{2^9} \sum_{i=0}^{\infty} \frac{1}{2^i} = \left(1 - \frac{1}{2^9}\right) \sum_{i=0}^{\infty} \frac{1}{2^i} = \left(1 - \frac{1}{2^9}\right) \cdot 2 \approx 2.$$

Thus, computation of R_E at the current point p reduces to

$$R_E = \frac{1}{2}\left(R_P + \frac{1}{2}R_E^{(-1)} + \frac{1}{2^2}R_E^{(-2)} + \ldots + \frac{1}{2^8}R_E^{(-8)}\right) \qquad (4.17)$$

Further, since the responses $\{R_E^{(-i)} : i = 1, 2, \ldots, 8\}$ at the ith previous points are already computed before reaching the current point p, these responses are used to compute R_E as per Eq. 4.17 in a simpler way. Only one right shift and one additive operation in the integer domain are required for this. We store $\frac{1}{2}(R_E^{(-1)} + \frac{1}{2}R_E^{(-2)} + \ldots + \frac{1}{2^8}R_E^{(-9)})$ at the previous step—computed from the previous responses—in a variable E. At p, we first add R_P with E to compute $R_P + E$ and then apply a right shift to $R_P + E$ to compute the current value of $R_E(= \frac{1}{2}(R_P + E))$, which is reassigned to E.[3] This updated value of E is carried forward to the next step to compute $R_E^{(+1)}$ in a similar way.

For multiple maximum responses exceeding T, the next direction is chosen in a way so that the properties R1 and R2 are ensured. For example, if p lies on a straight edge having 1 and 0 as the respective singular and non-singular codes, and there are two maxima with chain codes 0 and 7, then we choose 0 as it satisfies R1. If no response at any direction exceeds T, then we stop at p and search for a new start point.

4.4.3   Checking the Straightness

To check the straightness of an edge, we verify whether the properties R1 and R2 are true for the chain code of every new entry. Further, since R2 is inherently stringent on deciding the straightness of a curve (run-lengths of the non-singular code can differ at most by unity), we allow a relaxation on R2 by considering an integer interval $[p, q]$ as the range of run-length of the non-singular code n in a straight edge. The limits of the run-length interval are

$$\begin{aligned} p &= \max\big(1, (1 - \alpha)l\big) \\ q &= (1 + \alpha)l \end{aligned} \qquad (4.18)$$

---

[3]Since $E = \frac{1}{2}R_E^{(-1)} + \frac{1}{2^2}R_E^{(-2)} + \ldots + \frac{1}{2^9}R_E^{(-9)}$, we get $R_E = \frac{1}{2}(R_P + E)$ on neglecting the smallest term $\frac{1}{2} \cdot \frac{1}{2^9}R_E^{(-9)} = \frac{1}{2^{10}}R_E^{(-9)}$ in $\frac{1}{2}E$, and so the updated value of E for the next step is $\frac{1}{2}(R_P + E)$.

where l is length of the leftmost run of n in the current straight edge, and $\alpha(> 0)$ is the factor of relaxation. Clearly, on increasing the value of α, $(q - p)$ increases, giving more relaxation to the property R2. Thus, if α is high, then fewer straight edges would be produced at the cost of an error of approximation; whereas, if α is low, then we obtain better information about the straight edges at the cost of increasing their count.

4.4.4 Finishing the Edge Sequence

The boundary of an object in a gray-scale image is represented as a sequence of straight edges, namely $\langle e_1, e_2, \ldots, e_m \rangle$, with their vertices $\langle v_1, v_2, \ldots, v_m \rangle$ being stored in order in a stack S. Note that the end point of e_i coincides with the start point of e_{i+1} for $i = 1, 2, \ldots, m - 1$. Finally, if the end point of e_m coincides with the start point of e_1, then the edges $\langle e_1, e_2, \ldots, e_m \rangle$ form a closed polygon; otherwise, the extracted straight edges make a poly-chain.

The vertices are popped from the stack S, one by one, in order to reduce the number of straight edges defining the boundary of an object. If a sequence of vertices, thus popped from S, are *almost collinear*, then they are combined together to form a single edge. In effect, if $\langle e_i, e_{i+1}, \ldots, e_j \rangle$ be the maximal subset of straight edges that are almost collinear, then these $j - i + 1$ edges are combined to a single edge. The process is repeated for all such maximal subsets in succession to obtain a reduced set of (almost) straight edges corresponding to the object boundary.

We have used the technique based on area deviation [210]. If $p(x_p, y_p)$, $q(x_q, y_q)$, and $r(x_r, y_r)$ are three consecutive vertices popped from S, then the magnitude of area of the triangle pqr is

$$\Delta = \frac{1}{2} \left| (x_q y_r - y_q x_r) + (x_r y_p - y_r x_p) + (x_p y_q - y_p x_q) \right|.$$

Now, for a tolerance of approximation, τ (= 2 as used in [168]),[4] if Δ does not exceed $\tau \cdot d(p, r)$, where $d(p, r)$ is the distance between p and r, then the two edges, $\langle p, q \rangle$ and $\langle q, r \rangle$, are merged into a single edge, $\langle p, r \rangle$.

4.5 Examples

Some typical examples of DSS and ADSS have been shown in Fig. 4.4. It is evident from these examples that an ADSS is not only reasonably straight, but also fewer in count while covering a digital curve segment compared to

---

[4]$\tau = 2$ implies that each of the dropped vertices is at most at a distance 2 from the combined edge [210].

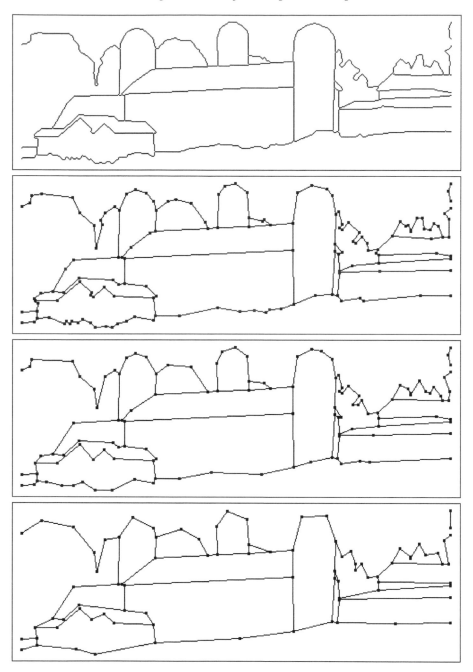

FIGURE 4.6: Results on the 'factory' image. From top to bottom: input and polygonal approximations for $\tau = 2, 3, 6$.

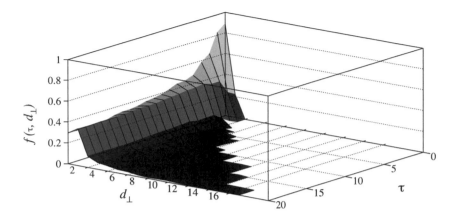

FIGURE 4.7: Quality of approximation for the 'factory' image shown in Fig. 4.6. The plot of IEF $= f(\tau, d_\perp)$ versus the error tolerance τ and the isothetic distance $d_\perp$ corresponding to criterion C_{max} has been shown, and that corresponding to criterion C_Σ is quite similar. Note that here $\tau = 0$ corresponds to the number of vertices without any polygonal approximation (i.e., with ADSS only).

the count of DSS in covering the same segment. This fact is used to expedite the subsequent algorithm for polygonal approximation. Further, the ADSS-recognition algorithm is much faster than a DSS extraction algorithm because of the recursive nature of the latter.

Fig. 4.6 demonstrates the polygonal approximation with the criterion C_{max} for the 'factory' image[5] for a few values of τ. In Fig. 4.7, the plot on IEF using criterion C_{max} only has been shown, since the plot with criterion C_Σ is quite similar. It may be observed from this plot that, for higher values of τ, the amount of maximum isothetic deviation $(d_{\perp(max)})$ falls quite short of the permissible limit, i.e., τ; in particular, for a given value of τ, the number of points $(N_{d_\perp})$ with deviation $d_\perp$ decreases almost monotonically with $d_\perp$. In Fig. 4.8, approximate polygons for another image with few values of τ, corresponding to the approximation criterion C_{max}, have been shown.

Fig. 4.9 shows the result of using the algorithm directly on a gray-scale image, as explained in Sec. 4.4. To compute the Prewitt responses, T is taken as 100. The chain code properties R1 and R2 are used with $\alpha = 2$ (Eq. 4.18) to detect the straight edges. To reduce the number of edges by merging the *almostcollinear* edges (Sec. 4.4.4), τ is taken as 2. Both the Sobel and the Prewitt operators can be used to find the straight edges in a gray-scale im-

---

[5]Source: The   Berkeley   Segmentation   Dataset   and   Benchmark, http://www.eecs.berkeley.edu/Research/Projects/CS/vision/bsds/.

age, by specifying appropriate values of the parameters. It is observed that the algorithm based on the concepts of exponential averaging and straightness checking produces marginally better results when the Prewitt operator is chosen instead of the Sobel operator. For checking the robustness to image rotation, the algorithm is tested on all the 360 images rotated from 0^0 to 359^0, and their outputs are recorded. It is noticed from each such output that the sequence of straight edges describing an object in the rotated image remains the same or nearly the same as that of the original one; occasionally, there occurs a variation due to changes in gray-value distribution around the edges owing to the subject rotation.

4.6   Summary

In this chapter we discuss the characterization of DSS and discuss its application in polygonal approximation by relaxing some of the criteria for declaring a digital arc as a DSS. It is evident from the discussion and the algorithms that a set of ADSS extracted from a set of digital curves is significantly smaller in size than that of a DSS extracted from the same, although each ADSS can be treated as sufficiently straight for various practical applications. The CPU time needed for ADSS extraction is remarkably less than that for DSS extraction. For polygonal approximation, the set of ADSS serves well to determine a suboptimal solution from an arbitrary set of digital curves. It is observed from the experimental results and analysis that the polygon vertices are densely located in and around the regions with high curvature, and sparsely in the regions with low curvature, owing to the fact that the length of an ADSS (alternatively, a DSS) is small in the former region but high in the latter.

Optimizing the set of ADSS to cover a given digital curve segment is a promising area of research, since the output set of the algorithm DETECT-ADSS depends on the start point and the direction of the traversal. For a gray-scale image, the straight edges can also be extracted using the properties of digital straightness so as to preserve the geometric information of the image.

The algorithm EXTRACT-ADSS is a typical example of the application of digital geometry for image analysis using the concept of digital straightness. In subsequent chapters we will deal with the geometry of curved lines and surfaces in digital space.

Exercises

1. Given two continued fractions $[p_1, p_2, \ldots, p_m]$ and $[q_1, q_2, \ldots, q_n]$, suggest an algorithm to find the smaller of the two. Explain its time complexity.

2. Derive Eq. 4.5 from Eq. 4.4 and Eq. 4.6.

3. Show that the continued fraction of $\frac{46}{87}$ is $[1, 1, 8, 5]$; then show that the period of a DSL with slope $\frac{46}{87}$ is $(01)^8(011)((01)^7(011))^4$.

4. Use the concept of Algorithm DETECT-ADSS to design an algorithm to verify whether a digital curve segment is digitally straight.

5. A digital line segment op is drawn from $o(0, 0)$ to $p(35, 455)$. Obtain the chain code sequence that uniquely represents op.

6. A DSS consists of n_0 0s and n_1 1s. Given that $n_0 \gg n_1$, suggest an efficient way of representing such a DSS so as to reduce the storage requirement.

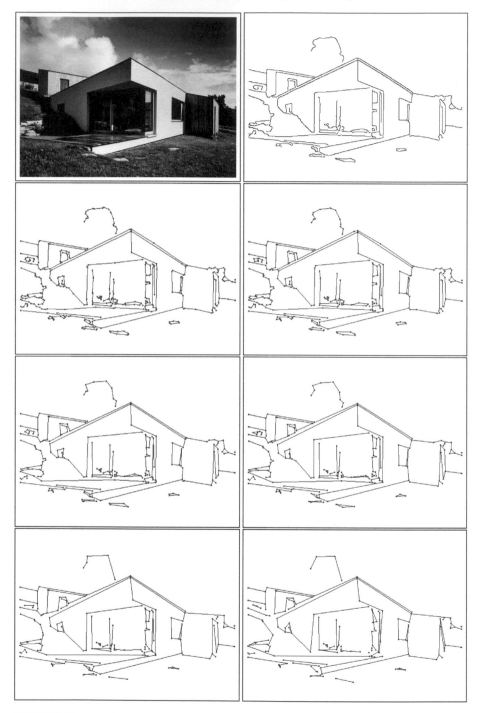

FIGURE 4.8: Result of polygonal approximation on the (thinned) edge map of a real-world image. (See color insert.)

FIGURE 4.9: Result of straight edge detection (bottom) on 'lab' image (top) with $\tau = 4$. (See color insert.)

Chapter 5

Parametric Curve Estimation and Reconstruction

Reconstruction of the original continuous curve from a given set of digital points representing its digitization is an important problem in digital image analysis. It is understood that digitization is a lossy transformation. Therefore, reconstruction of the original curve from its digitization is, in general, impossible. This loss of information opens up a number of questions, the most important of them being: *Is it possible to obtain the given digital image by digitizing a particular kind of curve?* In Chapter 3, we considered this problem for straight lines. In this chapter we investigate the problem under a more general context in 2-D.

To pose the question formally, consider the class of planar curves algebraically by an equation $f(x, y, P) = 0$ where x and y are spatial variables, $P = (p_1, p_2, ..., p_k)$ is a set of control parameters and f is a function relating x, y, and P. For example, $f(x, y, m, c)$ may be $y - mx - c$ where $P = (m, c)$ and $k = 2$. Let D_o be the given set of digital points. Then the earlier question can be restated as: *Does there exist a vector $P_1 \in R^k$ (R denotes the set of real numbers) so that the digitization of $f(x, y, P_1) = 0$ yields the given set D_o?*

This problem is known as a reconstruction problem of a digitized image

159

D_o with respect to a given function f. A solution to this problem computes at least one P such that digitization of $f(x, y, P) = 0$ produces D_o.

As a generalization of the reconstruction problem, we may want to find out the set of all $P \in R^k$ so that the digitization of $f(x, y, P) = 0$ produces D_o. The solution to this problem, known as a domain construction problem, is a region S in k-dimensional space such that

$$\forall P \in S, D(f(x, y, P)) = D_o$$

where D represents the digitization function.

We may view the digitization procedure D as a transformation of a curve $f(x, y, P_o) = 0$ to the corresponding digitized data D_o. Then the domain construction problem is one of finding an inverse mapping from digitized data to the specification of the curve. Formally, if $D(f(x, y, P_o)) = D_o$, then the domain of digitization is defined as

$$Domain(D_o, f) = \{P | P \in R^k \text{ and } D(f(x, y, P)) = D_o\}.$$

Note that due to the intrinsic nature of this *one-to-many* inverse mappings, we cannot compute the exact original of a digital image and have to be content with specifying one or all of its possible original(s). It is in this view we say that a solution to the domain construction problem involves the *estimation* of the actual parameters (though the term *estimation* occurs in statistics with a different connotation).

The problems of reconstruction and domain construction for a straight line segment have been discussed in detail in Chapter 3. Conics are next to straight lines in order of complexity. In this chapter, we discuss algorithms to solve the reconstruction and domain construction problem of canonical digital conics. We go further and delve into designing algorithms to solve the domain construction problem for a class of digitized curves including digital circles.

5.1   Digital Conics in Canonical Form

A general conic is represented by the equation $ax^2 + by^2 + cxy + dx + fy + g = 0$. Therefore, six parameters are needed to characterize such conics. As the number of parameters is large, we restrict our attention to an important subset of general conics, which are in *canonical form*.

Definition 5.1. *A conic is said to be in canonical form if its center lies on the origin of the coordinate system and if its axes are parallel to the coordinate axes.* □

In canonical form, circles and parabolas are represented by one unknown parameter, while ellipses and hyperbolas are characterized by two parameters.

From the symmetry of such conics, it is sufficient to concentrate on the first quadrant only, though the results can be extended easily over other quadrants too. In the following discussion, all conics are assumed to be in canonical form unless otherwise stated.

The OBQ scheme as discussed in Section 3.1 of Chapter 3 has been chosen for digitizing conics in canonical form. For the sake of illustration, let us consider an ellipse in canonical form $E(a, b)$ where a and b denote the semi-major and semi-minor axes, respectively. In Fig. 5.1, we have shown the OBQ contour of an ellipse whose $a = 12.5$ and $b = 10.5$.

Let D^* denote the OBQ image of any curve for the ellipse with $a = 12.5$ and $b = 10.5$.

$D^* = \{(0, 10), (1, 10), (2, 10), (3, 10), (4, 9), (5, 9), (6, 9), (7, 8), (8, 8), (9, 7),$
$(10, 6), (10, 5), (11, 4), (11, 3), (12, 2), (12, 1), (12, 0)\}$.

Definition 5.2. *We define the Y-digitization $D_Y = D_Y(E)$ of an ellipse as follows (in the first quadrant only).*

$D_Y = \{(i, y_i) : i, y_i \in \mathbb{Z} \text{ integer, } 0 \leqslant i \leqslant \lfloor a \rfloor \text{ and } y_i = \max\{y | y \in \mathbb{Z} \text{ and } i^2/a^2 + y^2/b^2 \leqslant 1\}\}$. *That is,* $y_i = \lfloor b\sqrt{(1 - i^2/a^2)} \rfloor, 0 \leqslant i \leqslant \lfloor a \rfloor$. $\quad\square$

Since for every i, y_i is unique, an alternative notion for D_Y lists $(\lfloor a \rfloor + 1)$ y_i values only, as $D_Y = (y_0, y_1, y_2, ..., y_{\lfloor a \rfloor})$. We shall use both these notations interchangeably. We can similarly quantize E along the Y-axis and define the X- digitization D_X. Finally, the total digitization is denoted by $D = D(E) = D_X \cup D_Y$. Note that D_X and D_Y may not be disjoint.

Example 5.1. *In Fig. 5.1,*

$$
\begin{aligned}
D_X \;=\; & (12, 12, 12, 11, 11, 10, 10, 9, 8, 6, 3) \text{ or} \\
& \{(12, 0), (12, 1), (12, 2), (11, 3), (11, 4), (10, 5), (10, 6), \\
& (9, 7), (8, 8), (6, 9), (3, 10)\} \\
D_Y \;=\; & (10, 10, 10, 10, 9, 9, 9, 8, 8, 7, 6, 4, 2) \text{ or} \\
& \{(0, 10), (1, 10), (2, 10), (3, 10), (4, 9), (5, 9), (6, 9), (7, 8), \\
& (8, 8), (9, 7), (10, 6), (11, 4), (12, 2)\} \\
D \;=\; & \{(0, 10), (1, 10), (2, 10), (3, 10), (4, 9), (5, 9), (6, 9), (7, 8), \\
& (8, 8), (9, 7), (10, 6), (10, 5), (11, 4), (11, 3), (12, 2), \\
& (12, 1), (12, 0)\}.
\end{aligned}
$$

In digitization of an ellipse, we may note that both D_X and D_Y are required. This is so because digitization results in grid points having multiple x-coordinate values for the same y-coordinate and multiple y-coordinate values for the same x-coordinate.

It is easy to see in this example that $D = D^*$. This is not a coincidence and holds in general for every ellipse. The formal discussions leading to the *proof* is presented later in Theorem 5.12.

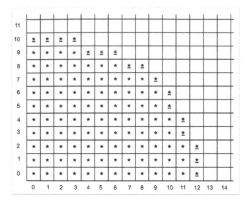

FIGURE 5.1: Digitization of an ellipse with $a_o = 12.5$ and $b_o = 10.5$. Grid points inside the ellipse have been marked with '*'. The boundary (that is the OBQ image) has been highlighted by underlining. The chaincode of the boundary (from left to right, i.e., clockwise) is '0007007077676766'.

Reprinted from *CVGIP: Graphical Models and Image Processing*, 54(5)(1992), S. Chattopadhyay et al., Parameter Estimation and Reconstruction of Digital Conics in Normal Positions, 385–395, Copyright(1992), with permission from Elsevier.

The above result is significant because D can be directly constructed from the image, whereas the formal characterization can be carried out in $D = D_X \cup D_Y$. In the rest of the section we view the OBQ simply as a pair of sets D_X and D_Y. We have another interesting property of D or D^*.

Lemma 5.1. *For any a, b denoting semi-major and minor axes of an ellipse in canonical form, D is 8-connected.*   □

Hence, D can be represented by a chain code such as 0007007077676766.

The above discussion, though reasoned for an ellipse, holds equally well for any convex closed curve, like a circle. In case of open, unbounded convex curves like parabolas or hyperbolas we assume it to be truncated by a horizontal or a vertical line, say $y = n$ or $x = n$, where n is the maximum range up to which the digitization is performed. Then the above analysis can again be applied.

We consistently use the sub-(super-)script 'o' to represent the parameters and quantities related to the original conic and use symbols without postscripts for estimated ones. For example, a_o and b_o are the semi-axes of the original ellipse E_o with digitization $D_o(= D(E(a_o, b_o)))$, whereas a and b stand to mean the estimates of a_o and b_o, respectively. We define the domain $Domain(D_o)$ of D_o as follows.

Definition 5.3. *$Domain(D_o)$ of an ellipse given a set of digital points, D_o, is the set of all possible values of a and b of an ellipse that allow for reconstruction, that is,*

$$Domain(D_o) = \{(a, b) | D(E(a, b)) = D_o\}.$$

$\square$

Domains and digitizations of other conics are defined analogously.

5.2   Circles and Parabolas in Canonical Form

In this section, we study digitized images of circles and parabolas in canonical form. We elucidate the method to construct the domain of digital circles while a similar method for domain construction of digitized parabolas is easy to get. Unless otherwise stated, circles and parabolas are always assumed to be in canonical form.

The restrictions of a conic in canonical form allows us to characterize a circle, say $C(r)$, or a parabola, say $P(a)$, by one parameter only, where r is the radius of the circle and a is the semi-latus rectum of the parabola. Given a digitization D_o of a circle, we present a simple algorithm to determine the domain $Domain(D_o)$ of the possible radius values of all continuous circles, which can be quantized to D_o.

Given D_o, it is easy to separate D_X^o and D_Y^o. Mathematically,

$$D_X^o = \{(x_i, i) : 0 \leqslant i \leqslant \lfloor r_o \rfloor \text{ and } x_i = \lfloor \sqrt{(r_o^2 - i^2)} \rfloor\}, \text{ and}$$
$$D_Y^o = \{(i, y_i) : 0 \leqslant i \leqslant \lfloor r_o \rfloor \text{ and } y_i = \lfloor \sqrt{(r_o^2 - i^2)} \rfloor\}.$$

From D_X^o, $\sqrt{(r_o^2 - i^2)} - 1 < x_i \leqslant \sqrt{(r_o^2 - i^2)}, 0 \leqslant i \leqslant \lfloor r_o \rfloor$.
Rearranging terms, $\sqrt{(x_i^2 + i^2)} \leqslant r_o < \sqrt{((x_i + 1)^2 + i^2)}$.
Similarly, $\sqrt{(y_i^2 + i^2)} \leqslant r_o < \sqrt{((y_i + 1)^2 + i^2)}, 0 \leqslant i \leqslant \lfloor r_o \rfloor$.
Combining the above inequalities we get two bounds on r_o, namely r_l and r_u, as follows.

$$r_l = \max_i(\max(\sqrt{(x_i^2 + i^2)}, \sqrt{(y_i^2 + i^2)})), \text{ and}$$
$$r_u = \min_i(\min(\sqrt{((x_i + 1)^2 + i^2)}, \sqrt{((y_i + 1)^2 + i^2)})).$$

From the symmetry of the circle, $D_X^o = D_Y^o$ in the list notation. Hence,

$$r_l = \max_i(\sqrt{(x_i^2 + i^2)}) = \max_i(\sqrt{(y_i^2 + i^2)}), \text{ and}$$
$$r_u = \min_i(\sqrt{((x_i + 1)^2 + i^2)}) = \min_i(\sqrt{((y_i + 1)^2 + i^2)}).$$

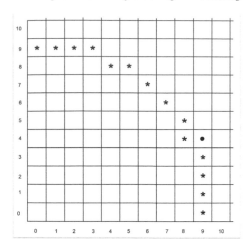

FIGURE 5.2: Digitization of a circle with $r_o = 9.7$, where (9,3) defines r_l and (9,4) defines r_u (Example 5.2).

Reprinted from *CVGIP: Graphical Models and Image Processing*, 54(5)(1992), S. Chattopadhyay et al., Parameter Estimation and Reconstruction of Digital Conics in Normal Positions, 385–395, Copyright (1992), with permission from Elsevier.

It is immediately evident that $r_l \leqslant r_o < r_u$.

Theorem 5.1. $r_l \leqslant r < r_u$ *if and only if* $D(C(r)) = D_o$. *In other words,* $Domain(D_o) = [r_l, r_u)$.

Proof: The proof follows from the definitions of r_l and r_u. □

Example 5.2. *Let* $r_o = 9.7$. *So,* $D_X^o = D_Y^o = \{9, 9, 9, 9, 8, 8, 7, 6, 5, 3\}$.

Clearly, $r_l = \sqrt{(9^2 + 3^2)} = \sqrt{90} = 9.4868$ and $r_u = \sqrt{(9^2 + 4^2)} = \sqrt{97} = 9.8489$, and any $C(r)$ with r lying between 9.4868 and 9.8489 will have the same digitization. The situation is shown in Fig. 5.2.

r_l and r_u can be interpreted geometrically too. Let A be the farthest point from origin O in D_o. Then $r_l = OA$. A circle with radius r_l has the same digitization as D_o. Any circle with $r < r_l$ misses the point A from $D(C(r))$. Note that r_l^2 is expressible as a sum of two perfect squares. Now to find r_u search for the least integer greater than r_l^2 that can be expressed as the sum of the squares of two integers i and j. Set the sum to $r_u^2 = i^2 + j^2$. So any circle with $r \geqslant r_u$ includes the point (i, j) extra in $H(C(r))$ with respect to $H(C(r_o))$ where $H(C(r))$ denotes the set of grid points contained in a canonical circle of radius r. As r_u^2 is the least such number, a continuous dilation of the circle $C(r)$ from $r = r_l$ first hits a grid point (i, j). Hence, for all r, $r_l \leqslant r < r_u$ we have $D = D_o$. The situation is depicted in Fig. 5.2, for example, with the point (i, j) marked with a dot. Note that none of 91, 92, 93, 94, 95, 96 can be expressed as a sum of two squares.

The analysis for the parabola $y^2 = 4a_o x$ can be carried out in an analogous manner. Here, $x_i = \lceil i^2/4a_o \rceil, 0 \leqslant i \leqslant y_n$ and $y_i = \lfloor \sqrt{(4a_o i)} \rfloor, 0 \leqslant i \leqslant n$, where the parabola is truncated by $x = n$. The bounds on a_o can be formed as in the case of a circle:

$$a_l = \max(\max_i(y_i^2/4i), \max_i(i^2/4x_i)) \text{ and}$$

$$a_u = \min(\min_i((y_i + 1)^2/4i), \min_i(i^2/4(x_i - 1))) \text{ and } i \geqslant 1.$$

Clearly then, Theorem 5.2 follows.

Theorem 5.2. $a_l \leqslant a < a_u$ *if and only if* $D(P(a)) = D_o$ *and* $Domain(D_o) = [a_l, a_u)$. □

Example 5.3. *(Fig. 5.3) Let* $a_o = 4.3$ *and* $n = 10$. *So,* $D_X^o = (0, 1, 1, 1, 1, 2, 3, 3, 4, 5, 6, 8, 9, 10)$ *and* $D_Y^o = (0, 4, 5, 7, 8, 9, 10, 10, 11, 12, 13)$. *We get* $a_l = 4.225000$ *and* $a_u = 4.321429$. *The grid point defining* a_l *is* $(10, 13)$ *and that defining* a_u *is* $(7, 11)$.

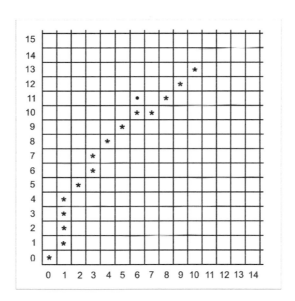

FIGURE 5.3: Digitization of a parabola with $a_o = 4.3$ (Example 5.3).

Reprinted from *CVGIP: Graphical Models and Image Processing*, 54(5)(1992), S. Chattopadhyay et al., Parameter Estimation and Reconstruction of Digital Conics in Normal Positions, 385–395, Copyright (1992), with permission from Elsevier.

5.3 Estimation of Major and Minor Axes of an Ellipse in Canonical Form

In contrast to the one-parameter characterization of circles and parabolas we need two parameters, namely a and b, to specify an ellipse or hyperbola in canonical form. A straightforward analysis of the discrete point inequalities cannot determine the possible domain of a and b values in this case. However, we show that the *iterative refinement* scheme introduced in Chapter 3 (refer to Section 3.2) can be appropriately modified to first find a tight rectangular bound for the domain and then to compute the domain itself. In the ensuing discussions, we first present the iterative refinement scheme for the bounds, then prove their various properties, discuss the reconstruction algorithm, and finally present the domain theorem [41].

Let us highlight a general property of two intersecting ellipses E_1 and E_2.

Lemma 5.2. *Let $E_1 = E(a_1, b_1)$ and $E_2 = E(a_2, b_2)$ be two ellipses such that $a_1 > a_2$ and $b_1 < b_2$. If $E_1 \cap E_2 = (x', y')$, then*

> (A)   *At every $x, x \geqslant x' (\leqslant x')$ E_1 lies above (below) E_2 and*
>
> (B)   *At every $y, y \geqslant y' (\leqslant y')$ E_1 lies to the left (right) of E_2*

where $\cap$ denotes the point of intersection of two continuous ellipses in the first quadrant.

Proof: The proof is immediate from the equation of the ellipse.   □

5.3.1 Reconstruction of Ellipse

We present the main result in the next theorem [41].

Theorem 5.3. *Suppose that the upper and lower bounds of a_o and b_o are defined by the following iterative algorithm where $k \geqslant 0$:*

$$b_l^0 = \lfloor b_o \rfloor = y_0, b_l^{k+1} = \max_i(y_i/\sqrt{(1 - i^2/(a_u^k)^2)}),$$

$$b_u^0 = \lfloor b_o \rfloor + 1 = y_0 + 1, b_u^{k+1} = \min_i((y_i + 1)/\sqrt{(1 - i^2/(a_l^k)^2)}),$$

$$a_l^0 = \lfloor a_o \rfloor = x_0, a_l^{k+1} = \max_i(x_i/\sqrt{(1 - i^2/(b_u^k)^2)}),$$

$$a_u^0 = \lfloor a_o \rfloor + 1 = x_0 + 1, a_u^{k+1} = \min_i((x_i + 1)/\sqrt{(1 - i^2/(b_l^k)^2)}),$$

then there exist b_l, b_u, a_l and a_u such that

> (A) $\lim_{k \to \infty} b_l^k = b_l, \lim_{k \to \infty} b_u^k = b_u, \lim_{k \to \infty} a_l^k = a_l, \lim_{k \to \infty} a_u^k = a_u,$ *and*

$$(B) \quad b_l < b_o < b_u \text{ and } a_l < a_o < a_u.$$

Proof: First, we establish Part (A).

By induction on k, we can easily prove that

$$(i) a_u^{k-1} \; \leqslant \; a_u^{k-2} \text{ implies } b_l^k \geqslant b_l^{k-1}, k \geqslant 2, \text{ and}$$
$$(ii) b_l^{k-1} \; \geqslant \; b_l^{k-2} \text{ implies } a_u^k \leqslant a_u^{k-1}, k \geqslant 2.$$

Applying (i) and (ii) alternately we get,

$$a_u^0 \; \geqslant \; a_u^1 \geqslant a_u^2 \geqslant ... \geqslant a_u^k \geqslant a_u^{k+1} \geqslant ..... > a_o, \text{ and}$$
$$b_l^0 \; \leqslant \; b_l^1 \leqslant b_l^2 \leqslant ... \leqslant b_l^k \leqslant b_l^{k+1} \leqslant ..... < b_o.$$

Hence either there exists a $k = k'$ such that $a_u^k = a_u^{k'}, k \geqslant k'$, or there exists an infinite sequence of a_u^k satisfying $a_u^k < a_u^{k-1}, k \geqslant 1$. It implies that $a_u^k, k \geqslant 0$ is a monotone decreasing sequence. Again, it is found that $a_u^k > a_o$. Hence, $\lim_{k \to \infty} a_u^k = a_u$. Numerically, we can estimate a_u by selecting a small quantity $\epsilon > 0$ and terminating with $|a_u^{k-1} - a_u^k| < \epsilon$.

Other cases can be derived likewise. Part (B) follows from Part (A). $\quad\square$

From the last theorem it is clear that there exist p and q, $0 \leqslant p \leqslant x_0$ and $0 \leqslant q \leqslant y_0$ such that $(p, y_p) \in D_Y^o, (x_q, q) \in D_X^o$, and

$$a_u = (x_q + 1)/\sqrt{(1 - q^2/b_l^2)} \text{ and } b_l = y_p/\sqrt{(1 - p^2/a_u^2)}. \qquad (5.1)$$

So $E_{ul} = E(a_u, b_l)$ passes through at least two grid points, namely (p, y_p) and (x_{q+1}, q). Intuitively, the iteration stops by hitting the *extreme* grid points (p, y_p) and $(x_q + 1, q)$. This can be compared with the case of a circle where $r_u = \sqrt{((x_q + 1)^2 + q^2)}$ gives an extreme of the radius value. Similar results also hold for a_l, b_u and E_{lu}

Let us consider an example now.

Example 5.4. *Take $a_o = 12.5$, $b_o = 10.5$, and $\epsilon = 10^{-6}$. $D_X^o = (12, 12, 12, 11, 11, 10, 10, 9, 8, 6, 3)$ and $D_Y^o = (10, 10, 10, 10, 9, 9, 9, 8, 8, 7, 6, 4, 2)$. The convergence can be easily established. We get $a_l = 12.220202$, $a_u = 12.533591 (q = 3$ and $x_q = 11)$, $b_l = 10.392305$ $(p = 8$ and $y_p = 8)$, $b_u = 10.583005$.*

We state various properties of these bounds leading to the reconstruction algorithm.

Lemma 5.3. *For any D_o, $q < y_p$ and $p < x_q + 1$, where p, q, x_q, y_p are as defined in Eq. 5.1 [41].*

Proof: Since $(p, y_p) \in D_Y^o$, we have $b_o \sqrt{(1 - p^2/a_o^2)} \geqslant y_p$. Replacing p by $a_u \sqrt{(1 - y_p^2/b_l^2)}$ (from b_l) and rearranging terms we get,

$$y_p^2 \geqslant (a_u^2 - a_o^2)/(a_u^2/b_l^2 - a_o^2/b_o^2).$$

Again, from $(x_q, q) \in D_X^o$ and $x_q + 1 = a_u\sqrt{(1 - q^2/b_l^2)}$, we get

$$q^2 < (a_u^2 - a_o^2)/(a_u^2/b_l^2 - a_o^2/b_o^2) \leqslant y_p^2.$$

So, $q < y_p$. Similarly, $p < x_q + 1$ □

Theorem 5.4. *If $D = D(E(a,b)) = D_o$, then both the following conditions hold [41]: (A) $a_l < a < a_u$ and (B) $b_l < b < b_u$.*

In other words, $R_{ul} = Rect((a_l, b_u), (a_u, b_l))$, the rectangle formed by (a_l, b_u) and (a_u, b_l) as diagonally opposite corners, properly contains the domain of D_o, i.e., $(a, b) \in R_{ul}$.

Proof: We first show that $b_l < b$ and $a < a_u$. We consider three cases, each of which leads to a contradiction.

Case 1: $a \leqslant a_u$ and $b \leqslant b_l$. but $(a, b) \neq (a_u, b_l)$.

This case leads to the fact $(p, y_p) \notin D$, a contradiction.

Case 2: $a \geqslant a_u$ and $b \geqslant b_l$

This case leads to the fact $(x_q, q) \notin D$, a contradiction.

Case 3: $a > a_u$ and $b < b_l$

Let $E \cap E_{ul} = (x', y')$ where $E_{ul} = E(a_u, b_l)$. If $x' > p$, we use Lemma 5.2, to show $(p, y_p) \notin D$, a contradiction. If $x' \leqslant p$, then we use Lemma 5.2 to show that $(x_q, q) \notin D$, a Contradiction.

Combining all cases $b > b_l$ and $a < a_u$, the rest follows similarly.   □

The above theorem establishes that R_{ul} is a rectangular bound on the domain of the possible a, b values such that $D(E(a, b)) = D_o$. It is shown [41] that this bound could be made tight by first providing an algorithm to find a possible a (or b) for any given b (or a) between the upper and lower limits, and then by deriving the boundaries of the domain directly.

Now we state some additional properties that we will use later.

Lemma 5.4. *[41] Let $E_{ul} = E(a_u, b_l)$ and p, q, x_q, y_p are as defined in Eq. 5.1. If $D = D_o$, then $E \cap E_{ul} = (x', y')$ such that*

$$(A)\ \ p\ \ \leqslant\ \ x' < x_q + 1\ \ and\ (B)\ q < y' \leqslant y_p\ where,$$
$$p\ \ =\ \ \max\{p' : y_{p'} = b_l\sqrt{(1 - (p'/a_u)^2)}\},\ and$$
$$q\ \ =\ \ \max\{q' : x_{q'} = a_u\sqrt{(1 - (q'/b_l)^2)} - 1)\}.$$

Proof: Consider any p and q. As $(p, y_p) \in D_Y^o = D_Y$, $y_p \leqslant b(\sqrt{(1 - p^2/a^2)}$. Now if $x' < p$, we can easily show (Lemma 5.2) that $y_p = b_l\sqrt{(1 - p^2/a_u^2)} > b\sqrt{(1 - p^2/a^2)}$. Contradiction. So, $x' \geqslant p$. Other cases follow similarly. Finally, the *maximum* is computed to take the possibility of multiple p's and/or multiple q's into consideration.   □

Lemma 5.5. *If $r = x_q + 1$, then $y_r = q - 1$ where p, q, x_q, y_p are as defined in Eq. 5.1 [41].*

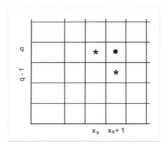

FIGURE 5.4: Proof of Lemma 5.5. "$*$" $\in E_o$ and "$.$" $\in E_{ul}$.

Reprinted from *CVGIP: Graphical Models and Image Processing*, 54(5)(1992), S. Chattopadhyay et al., Parameter Estimation and Reconstruction of Digital Conics in Normal Positions, 385–395, Copyright (1992), with permission from Elsevier.

Proof: From the definitions of D_X and D_Y, $y_r \leqslant q - 1$ (Fig. 5.4). Next, let us show that $x_{q-1} > x_q$. In general, $x_{q-1} \geqslant x_q$. If $x_{q-1} = x_q$, then from the definition of a_u, $(x_q + 1)/\sqrt{(1 - q^2/b_l^2)} < (x_q + 1)/\sqrt{(1 - (q - 1)^2/b_l^2)}$ or $q^2 < (q - 1)^2$. This is a contradiction. So, $x_{q-1} \geqslant x_q + 1 = r$. But then $y_r \geqslant q - 1$. Hence, $y_r = q - 1$. $\qquad\square$

Note that in Example 5.4, $q = 3$ and $x_q = 11$. So, $r = 12$ and $y_r = 2$.

Lemma 5.6. *$D(E_{ul})$ differs from D_o only at point(s) like $(x_q + 1, q)$ where a_u is (are) achieved [41].*

Proof: It is easy to see that $D_o \subseteq H(E_{ul})$ where $H(E)$ denotes the set of all grid points lying inside the first quadrant of an ellipse E. Consider any $(x_i, i) \in D_X^o$ such that $(x_i, i) \notin D_X(E_{ul})$. Now, if $(x_i^*, i) \in D_X(E_{ul})$, then $x_i^* \geqslant x_i + 1$. But in that case either $i = q$ and $x_i^* = x_q^* = x_q + 1$ or $i \neq q$ and $x_q^* > x_q + 1$. If $i \neq q$, then the $a_u - b_l$ iteration would have stopped with a higher a_u and lower b_l at $(x_i + 1, i)$. Hence, $i = q$. Next consider a $(i, y_i) \in D_Y^o$ and $\notin D_Y(E_{ul})$. So, there exists $(i, y_i^*) \in D_Y(E_{ul})$ such that $y_i^* \geqslant y_i$. But this contradicts the construction of b_l. Hence, the result. $\qquad\square$

Now let us introduce a definition for comparing the digitization of an arbitrary ellipse with the given digitization.

Definition 5.4. *We write $D_X = D_X(E) \leqslant D_X^o$ iff $\forall i, 0 \leqslant i \leqslant \lfloor b_o \rfloor$, $\lfloor a\sqrt{(1 - i^2/b^2)} \rfloor \leqslant x_i$ where $E = E(a, b)$ is an arbitrary ellipse for some a and b. $D_X \geqslant D_X^o, D_Y \leqslant D_Y^o$ and $D_Y \geqslant D_Y^o$ are defined analogously.* $\qquad\square$

For an arbitrary ellipse E, X- and Y-digitizations are said to be consistent iff $D_X \leqslant D_X^o$ implies $D_Y \leqslant D_Y^o$ and vice versa. The former case is represented by $D \leqslant D_o$ and the latter as $D \geqslant D_o$.

We prove, in the next theorem, the main result for reconstruction.

Theorem 5.5. *If $(a, b) \in R_{ul}$, then either $D \leqslant D_o$ or $D \geqslant D_o$ [42].*

Proof: First let us prove that if $(a, b) \in R_{ul}$, then the following hold: (A) $D_Y \leqslant D_Y^o$ or $D_Y \geqslant D_Y^o$ and (B) $D_X \leqslant D_X^o$ or $D_X^o \geqslant D_X^o$.

If both a and b are less (or greater) than a_o and b_o, respectively, then it is trivial to see that $D_X \leqslant D_X^o$ and $D_Y \leqslant D_Y^o$ (or $D_X^o \geqslant D_X^o$ and $D_Y^o \geqslant D_Y^o$).

So we have another case to settle here where one of a or b is less than a_o or b_o and the other is greater. Without any loss of generality, let us take $b_l < b < b_o$ and $a_u > a > a_o$. We prove that $D \geqslant D_o$. Let us first assume the converse of $D_Y \geqslant D_Y^o$ i.e., $\exists i \lfloor b\sqrt{(1 - i^2/a^2)} \rfloor < y_i$. Now, if $E \cap E_o = (x', y')$ and $E \cap E_{ul} = (x'', y'')$, then from Lemma 5.2, we get $x'' > x' > i$. But then $\lfloor b_l \sqrt{(1 - i^2/a_u^2)} \rfloor \leqslant \lfloor b\sqrt{(1 - i^2/a^2)} \rfloor < y_i$. So, $b_l < y_i/\sqrt{(1 - i^2/a_u^2)}$ which contradicts the definition of b_l and no such i can exist. Hence, $D_Y \geqslant D_Y^o$.

Now assume the converse of $D_X \geqslant D_X^o$, i.e., there exists some j, such that $\lfloor a\sqrt{(1 - j^2/b^2)} \rfloor < x_j = r$, say. Clearly, $(x_j, j) \notin H(E(a, b))$ and $\lfloor b\sqrt{(1 - r^2/a^2)} \rfloor < j$. Again $y_r \geqslant j$ as $r = x_j$. Therefore, there exists one r, such that $\lfloor b\sqrt{(1 - r^2/a^2)} \rfloor < y_r$, which contradicts $D_Y \geqslant D_Y^o$. Hence, $D_X \geqslant D_X^o$. Thus, $D \leqslant D_o$ if $a < a_o$ and $b > b_o$ and $D \geqslant D_o$ otherwise.   □

A reconstruction algorithm from the digitization is now available. We present it in the next subsection.

5.3.2   The Reconstruction Algorithm

A simple binary-search-like algorithm, similar to the one used for circles in [157], is presented in Algorithm 9 to find a pair of a and b such that $D(E(a, b)) = D_o$.

Algorithm 9: Reconstruction of Ellipse

Algorithm Reconstruct_Ellipse
Input: Digitization D_o
Output: A pair of a, b such that $D(E(a, b)) = D_o$
Method:

1. $b = (b_l + b_u)/2; a = (a_l + a_u)/2; a_L = a_l; a_R = a_u$

2. while $D \neq D_o$ do
   begin if $D \leqslant D_o$ then begin $a_L = a$; $a = (a + a_R)/2; end$
   else begin $a_R = a; a = (a_L + a)/2; end;$
   end;

3. write (a, b);

End Reconstruct_Ellipse

The justification of the algorithm follows from Theorem 5.5. The algorithm finds one suitable a for a fixed b. Actually the choice of this b is arbitrary so long as $b_l < b < b_u$. Finally we prove that the algorithm terminates.

Theorem 5.6. *Given D_o, algorithm Reconstruct_Ellipse terminates with a and b such that $D_o = D(E(a,b))$ [41].*

Proof: Since the value of b is held constant, two situations are possible depending on whether $b < b_o$ or $b > b_o$. We first consider the former. Without any loss of generality, we assume that at an arbitrary iteration step we have $D \leqslant D_o$. Hence, the new a will be $a' = (a + a_R)/2$. Note that $a_l < a' < a_u$. Two cases can arise here.

Case 1: $D' = D(E(a',b)) \leqslant D_o$.

Clearly, $D \leqslant D' \leqslant D_o$. Hence D' does not match D_o at most, at those components where D and D_o differ. Since in this way a' can be monotonically increased, there must exist some a' such that D' matches D_o in at least one component more than D. Eventually then, D' will converge to D_o.

Case 2: $D' \geqslant D_o$.

Clearly, $D' \geqslant D_o \geqslant D$. Consider the last iteration with a'' where $D'' \geqslant D_o$. From the algorithm $a' < a''$ and $D_o \leqslant D' \leqslant D''$. Again, similar to Case 1, the number of matches can be argued to gradually increase culminating in the convergence.

Now if $b > b_o$, from Theorem 5.5, we will always get $D \geqslant D_o$ and keep on decrementing a (actually $a \to a_l$). Since $D(E(a_l,b)) < D_o$, eventually we shall get an a where $D = D_o$.

Hence, the algorithm terminates with a correct estimation.

$\square$

As the above argument holds for any $(a,b) \in R_{ul}$, it is easy to see that there exists an a (or b) for an arbitrary $b, b_l < b < b_u$ (or $a, a_l < u < a_u$) such that $D = D_o$.

5.3.3 The Domain Theorem

In this section, we show that a variation of the iteration equations helps to compute the domain of D_o, of which R_{ul} is a tight bound. We present it in the next theorem after necessary lemmas.

Lemma 5.7. *The following hold [41]:*

(A) $a_l < a < a_u$ if and only if $b_l^*(a) < b_u^*(a)$ where $b_l^*(a) = \max_i(y_i/\sqrt{(1 - i^2/a^2)})$ and $b_u^*(a) = \min_i((y_i + 1)/\sqrt{(1 - i^2/a^2)})$.

(B) $b_l < b < b_u$ if and only if $a_l^*(b) < a_u^*(b)$ where $a_l^*(b) = \max_i(x_i/\sqrt{(1 - i^2/b^2)})$ and $a_u^*(b) = \min_i((x_i + 1)/\sqrt{(1 - i^2/b^2)})$. $\square$

Theorem 5.7. *(The Domain Theorem) The domain $Domain(D_o)$ of all possible values of (a,b) parameters to give reconstruction is given by the following* *formulas [41]:*

$$Domain(D_o) = \bigcup_{a_l < a < a_u} [b_l^*(a), b_l^*(a)); \; or$$

$$Domain(D_o) = \bigcup_{b_l < b < b_u} [a_l^*(a), a_u^*(a)).$$

$\square$

Corollary 5.1. R_{ul} *is the smallest rectangle in the (a, b)-space such that* $Domain(D_o) \subseteq R_{ul}$.

$\square$

In view of the above corollary, we immediately get an algorithm from iterative refinement to test whether a given digital set can at all be the digitization of an ellipse in canonical form or not. Mathematically it implies that we need to check whether a $Domain(D_o)$ is an empty set or not.

Corollary 5.2. $Domain(D_o) = \{\}$ *if and only if $a_l > a_u$ or $b_l > b_u$ or both. That is, if the digital set D_o is not a valid digitization of an ellipse, then there exists a k such that $a_l^k > a_u^k, b_l^k > b_u^k$.*

$\square$

Corollary 5.3. $Domain(D_o)$ *is axially convex and connected.*

$\square$

It may be noted that the above theorem provides the necessary algebraic characterization of the domain $Domain(D_o)$. But it does not help in analytically finding the boundaries of the domain. However, it helps to numerically compute the domain (up to any desired accuracy), because we find that for a given a, $(a_l < a < a_u)$ there exists a single interval of b values, i.e., $[b_l^*(a), b_u^*(a))$ corresponding to the domain. Immediately, we also conclude that $(a, b_l^*(a))$ and $(a, b_u^*(a))$ are points on the boundary of the domain. By computing these points for various values of a, we can determine the domain as we show in the following example. We can even numerically integrate some quantity like the property estimator error over this domain to derive further results. It may also be noted that Theorem 5.2 renders the reconstruction algorithm redundant. It is presented for the sake of completeness in an attempt to maintain coherence with other forms of analysis such as [157].

Example 5.5. *For the ellipse in Example 5.1 (Fig. 5.1), we estimate the domain of digitization using the above theorem. We have $a_o = 12.5$, $b_o = 10.5$ and $D_X^o = \{12, 12, 12, 11, 11, 10, 10, 9, 8, 6, 3\}$, $D_Y^o = \{10, 10, 10, 10, 9, 9, 9, 8, 8, 7, 6, 4, 2\}$. The domain is plotted in Fig. 5.5.*

5.4   Reconstruction of Hyperbola in Canonical Form

The analysis of a hyperbola in *canonical form* is similar to that of an ellipse, so we present only an outline of the analysis. First, let us define the

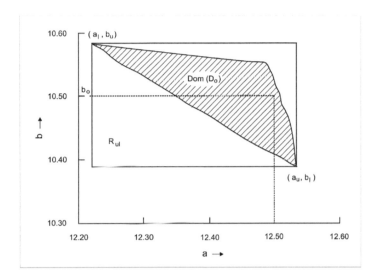

FIGURE 5.5: The domain $Dom(D_o)$ and the bounding rectangle of D_o of an ellipse for $a_o = 12.5$, $b_o = 10.5$ (Example 3.8). The domain has been computed using Theorem 3.14. The original parameter value has been shown by a ".".

digitization of the hyperbola $H_o : x^2/a^2 - y^2/b^2 = 1$. As in the case of an ellipse, we treat the OBQ image of H_o as the union of the following two sets of digital points, namely D_X and D_Y.

$$D_X^o = \{(x_i, i) : 0 \leqslant i \leqslant y_n \text{ and } x_i = \lceil a_o \sqrt{(1 + i^2/b_o^2)} \rceil\} \text{ and}$$
$$D_Y^o = \{(i, y_i) : \lceil a_o \rceil \leqslant i \leqslant n \text{ and } y_i = \lfloor b_o \sqrt{(i^2/a_o^2 - 1)} \rfloor\}.$$

The *iterative refinement* equations for computing tight bounds of a_o and b_o are given in the following theorem.

Theorem 5.8. *Let the upper and lower bounds of a_o and b_o are defined by the following iterative algorithm where $k \geqslant 0$:*

$$a_l^0 = x_0 - 1; a_l^{k+1} = \max_i((x_i - 1)/\sqrt{(1 + i^2/(b_l^k)^2)}),$$
$$b_l^0 = y_r/\sqrt{((r/(r-1)^2 - 1)} \text{ where } r = x_0,$$
$$b_l^{k+1} = \max_i(y_i/\sqrt{(i^2/(a_l^k)^2 - 1)}),$$
$$a_u^0 = x_0; a_u^{k+1} = \min_i(x_i/\sqrt{(1 + i^2/(b_l^k)^2)});$$
$$b_u^0 = (y_{r+1} + 1)/\sqrt{(((r+1)/r)^2 - 1)} \text{ where } r = x_0,$$
$$b_u^{k+1} = \min_i((y_i + 1)/\sqrt{(i^2/(a_u^k)^2 - 1)}).$$

Then there exist b_l, b_u, a_l, and a_u such that

$$\lim_{k \to \infty} b_l^k = b_l, \lim_{k \to \infty} b_u^k = b_u, \lim_{k \to \infty} a_l^k = a_l, \lim_{k \to \infty} a_u^k = a_u, \text{ and}$$

$$b_l < b_o < b_u \text{ and } a_l < a_o < a_u.$$

$\square$

The subsequent theorem establishes that the rectangle with diagonally opposite vertices (a_l, b_l) and (a_u, b_u) in the a-b space properly contains the domain of D_o.

Theorem 5.9. *If $D = D(H(a, b)) = D_o$, then both the following conditions hold: (A)$a_l < a < a_u$ and (B)$b_l < b < b_u$.* $\square$

Finally, we state the domain theorem, which enables us to compute the domain numerically.

Theorem 5.10. *(The Domain Theorem) The domain $Domain(D_o)$ of all possible values of (a, b) parameters of a hyperbola in canonical form to give reconstruction is given by the following formulas:*

$$(A) \ Domain(D_o) = \bigcup_{a_l < a < a_u} [b_l^*(a), b_l^*(a)); \text{ or}$$

$$(A) \ Domain(D_o) = \bigcup_{b_l < b < b_u} [a_l^*(a), a_u^*(a));$$

where $b_l^*(a) = \max_i(y_i/\sqrt{(i^2/a^2 - 1)})$, $b_u^*(a) = \min_i((y_i + 1)/\sqrt{(i^2/a^2 - 1)})$, $a_l^*(b) = \max_i((x_i - 1)/\sqrt{(1 + i^2/b^2)})$ and $a_u^*(b) = \min_i(x_i/\sqrt{(1 + i^2/b^2)})$. $\square$

We conclude this section after offering an example.

Example 5.6. *Consider a canonical hyperbola with $a_o = 10.5$ and $b_o = 8.5$ which is truncated at $x = 20$. $D_X^o = \{11, 11, 11, 12, 12, 13, 14, 15, 16, 17, 18, 19, 20\}$ and $D_Y^o = \{2, 4, 6, 7, 8, 9, 10, 11, 12, 13\}$. After convergence of the I_R algorithm we get $a_l = 10.2618834, a_u = 10.7322793, b_l = 7.7711441, b_u = 8.8987359$. For this hyperbola, the domain of digitization is computed using Theorem 5.10. The domain is diagrammatically presented in Fig. 5.6.*

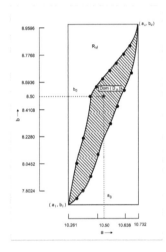

FIGURE 5.6: The domain $Dom(D_o)$ of an hyperbola for $a_o = 10.5$, $b_o = 8.5$, and truncated at x = 20.

5.5 A Restricted Class of Digitized Planar Curves

In the previous sections we addressed the reconstruction problem of conics in canonical form. We employed an iterative refinement technique to find close bounds and the domain of the parameters for these geometric objects. The success of I_R as a general methodology for solving the reconstruction problem for a diverse class of geometric objects has led to characterizing the properties of the class of digital curves where I_R may be fruitfully applied. In this section, we develop an algorithmic scheme that utilizes some general properties of the function representing the curve to solve the reconstruction problem. Moreover, we characterize the class of curves that are amenable to this generic approach of analysis [40].

5.5.1   Characterizing Properties of the Class

A planar curve with one or two unknown control parameters may be represented by the equations $f(x, y, z) = 0$ or $f(x, y, a, b) = 0$ respectively where x, y denote the special variables and a, b denote the unknown control parameters. Without any loss of generality, we can assume that the segment of the curve in the first quadrant in a rectangular mesh defined by $0 \leqslant x \leqslant n$, $0 \leqslant y \leqslant n$. Moreover, there is no grid line that does not intersect the curve. To be precise, we assume the following:

Assumption 0: The curve is continuous and differentiable in the region of interest.

Our next assumption about the class of curves is that the equation $f(x, y, a, b) = 0$ may be rearranged to write the following four equivalent equations $x = f_x(y, a, b)$, $y = f_y(x, a, b)$, $a = f_a(x, y, b)$, and $b = f_b(x, y, a)$. We call it the separability property of the curve. It is important to note that to solve the reconstruction problem using the I-R scheme, the curve has to possess this separability property.

Assumption 1. The curve possesses the separability property.

Now let us formalize the concept of monotone curves in our context. In the following definition, $\underline{x} = (x_1, .., x_n)$ denotes a vector of n variables. Also, $\underline{x}^i$ denotes the vector of $(n-1)$ variables $(x_1, .., x_{i-1}, x_{i+1}, ..., x_n)$.

Definition 5.5. *A function $g(\underline{x})$ is monotone decreasing with respect to x_i if and only if $x_i > x'_i$ implies $g(x_1, .., x_i, ..., x_n) < g(x_1, .., x'_i, ..., x_n)$ for any choice of values (in the region interest) for the other $(n-1)$ variables.*

Similarly, $g(\underline{x})$ may be defined as monotone increasing with respect to x_i. In other words, $g(\underline{x})$ is monotone decreasing if and only if $\frac{\partial g}{\partial x_i} < 0$ over our interval of attention.

Definition 5.6. *A function $g(\underline{x})$ is monotone if and only if $\forall i, 1 \leqslant i \leqslant n$, $g(\underline{x})$ is either monotone increasing or monotone decreasing with respect to x_i. Hitherto we refer to g as an increasing (a decreasing) function if g is monotone increasing (decreasing).*

Definition 5.7. *We say that a curve given by $f(\underline{x}) = 0$ is monotone if and only if the function f is monotone.* □

Assumption 2. The curve is monotone.

As we shall see in Theorem 5.11, if f is monotone and $x_i = f_i(\underline{x}^i)$, then the function f_i is also monotone for all i, $1 \leqslant i \leqslant n$ and vice versa.

Then it follows from the above discussions that we can capture the monotonicity of a particular curve $f(\underline{x}) = 0$, if we know the nature of the monotonicity of the individual functions $f_i, 1 \leqslant i \leqslant n$. This can be easily depicted by a matrix, which we shall call the **monotonicity matrix** (MM).

Definition 5.8. *We define a monotonicity matrix (MM) as an $n \times n$ matrix whose rows correspond to the functions $f_i, 1 \leqslant i \leqslant n$, and the columns correspond to variables $x_i, 1 \leqslant i \leqslant n$, respectively. The entry in a cell for such a matrix is either 'I' or 'D'. $MM(i,j) = 'I'$ (or 'D') indicates that the function f_i is increasing (decreasing) with respect to the variable x_j.* □

Although the monotonicity matrix fully describes the nature of monotonicity of a curve, the enormous number of MMs depicting different kinds of monotone curves is a severe drawback for the easy handling of MMs. Moreover, such a characterization of a monotone curve is not obtained directly from the equation of the curve. The next theorem, however, offers a solution.

Theorem 5.11. *For any separable function $f(x)$, we have $\frac{\partial f_i}{\partial x_j} > 0$, if and only if $\frac{\partial f}{\partial x_i} \cdot \frac{\partial f}{\partial x_j} < 0$ for $1 \leqslant i, j \leqslant n, i \neq j$ [40].*

Proof: Since all other variables are treated as constants, we may write $f(x) = 0$ as $f(x_i, x_j) = 0$. Restated differently, $x_i = f_i(x_j)$. We know that $\frac{\partial f_i}{\partial x_j} = -(\frac{\partial f}{\partial x_i})/(\frac{\partial f}{\partial x_j})$.

Hence, $\frac{\partial f_i}{\partial x_j} > 0$ if and only if $\frac{\partial f}{\partial x_i}$ and $\frac{\partial f}{\partial x_j}$ have opposite signs.

□

The above theorem helps us to construct the monotonicity matrix if we define a partial derivative sign vector (PDSV).

Definition 5.9. *A PDSV of a function $f(x)$ is an n-tuple where the i th element stands for the sign of the partial derivative $\frac{\partial f}{\partial x_i}$. If π is a PDSV then π_i, the i-th component of π, is '+' if $\frac{\partial f}{\partial x_i}$ is positive and '-' if it is negative.* □

If we consider a function $f(x, y, a)$, then every PDSV π of f is a three-tuple. π_1, π_2, π_3 denote the signs of $\frac{\partial f}{\partial x}, \frac{\partial f}{\partial y}$ and $\frac{\partial f}{\partial a}$, respectively. Since each of these π_i may be either a '+' or a '-' there may be at most eight PDSVs for the given function.

Every PDSV corresponds to a unique MM in the following manner. Let π be the given PDSV.

$$MM(i,j) = 'I' \text{ if } \pi_i \neq \pi_j$$
$$= 'D' \text{ if } \pi_i = \pi_j.$$

This is a direct consequence of the definition of MM(i,j) and Theorem 5.11.

Clearly then, there can be no more than 2^n (number of n-tuple PDSVs) distinct MMs defining the monotone functions. We use the following properties of PDSVs to identify equivalent MMs (with respect to our analysis). We assume that f is a d-dimensional curve with $k = n - d$ unknown parameters, i.e., $f(x_l, x_2, .., x_d, a_1, .., a_{n-d}) = 0$.

Property 1: (Complementation) The complement π_c of a π, where a '+'

TABLE 5.1: Listing of distinct PDSVs for one-parameter planar curves in the first quadrant.

$$PDSV_1 = (\text{-}, +, \text{-})$$
$$PDSV_2 = (+, +, \text{-})$$
$$PDSV_3 = (+, +, +)$$

Reprinted from *Sadhana* 18(2)(1993), S. Chattopadhyay et al., A Generalized Approach to the Reconstruction of a Restricted Class of Digitized Planar Curves, 349–364, Copyright (1993), with permission from Indian Academy of Sciences.

(or '-') in π is changed to a '-' (or '+') in π_c, denotes the same monotonicity matrix as π.

Property 2(a): (Permutation of spatial variables) Since x_i and x_j can be interchanged (with proper adjustment of the coordinate system)

$$\pi = (\pi_1, .., \pi_{i-1}, \pi_i, \pi_{i+1}, .., \pi_{j-1}, \pi_j, \pi_{j+1}, .., \pi_d, .., \pi_n), \text{ and}$$

$$\pi' = (\pi_1, .., \pi_{i-1}, \pi_j, \pi_{i+1}, .., \pi_{j-1}, \pi_i, \pi_{j+1}, .., \pi_d, .., \pi_n)$$

define equivalent monotone classes of curves.

Property 2(b): (Permutation of non-spatial variables) As a_i and a_j can be interchanged by renaming, their order also becomes immaterial in a PDSV.

In this chapter we have treated only one ($k = 1$) or two ($k = 2$) parameter planar ($d = 2$ and $n = 3$ or 4) curves. Thus, at most eight PDSVs may arise in the first case while the number of PDSVs may go up to sixteen in the latter. But there are three distinct PDSVs to consider for a planar curve with one parameter. Similarly, the number of distinct PDSVs reduce to five for curves with two parameters. The different PDSVs are listed in Tables 5.1 and 5.2. The MM for $PDSV_1$ in Table 5.1 and $PDSV_3$ in Table 5.2 are shown in Tables 5.3 and 5.4, respectively.

We have selected the OBQ scheme for our approach. Now let us revisit the formal definition of the OBQ scheme.

Definition 5.10. *$I(f)$, the OBQ image of a curve given by $f(x, y, a, b) = 0$, is the set of digital points obtained as follows: While traversing f clockwise (1) whenever f passes through a grid point P, then P belongs to $I(f)$, (2) whenever f crosses a grid line L but not a grid point, the nearest grid point to the right of the curve and on L is a point of $I(f)$, (3) no other point is included in $I(f)$.* □

Thus, for a closed contour $f = 0$, $I(f)$ is the set of grid points inside f and *nearest* to its boundary. Note also that we should collect the points on the left of f if we traverse it counter-clockwise. This definition, though precises is

TABLE 5.2: Listing of distinct PDSVs for two parameters planar curves in the first quadrant.

$$
\begin{array}{l}
PDSV_1 = (\ +, -, -, -\ ) \\
PDSV_2 = (\ +, -, -, +\ ) \\
PDSV_3 = (\ +, +, -, -\ ) \\
PDSV_4 = (\ +, +, +, -\ ) \\
PDSV_5 = (\ +, +, +, +\ )
\end{array}
$$

TABLE 5.3: Monotonicity matrix corresponding to $PDSV_1$ of Table 5.1.

	x	y	a
f_x	-	I	D
f_y	I	-	I
f_a	D	I	-

TABLE 5.4: Monotonicity matrix corresponding to $PDSV_3$ of Table 5.2.

	x	y	a	b
f_x	-	D	I	I
f_y	D	-	I	I
f_a	I	I	-	D
f_b	I	I	D	-

not algebraic in nature. Fortunately under the assumptions of monotonicity, $I(f)$ can also be expressed through simple algebraic expressions. To see this, we first define another set of digital points as D from $f = 0$ and subsequently prove its equivalence with $I(f)$.

Definition 5.11. *Given f, $D = D(f)$ is a set of grid points defined as follows: $D = D_X \cup D_Y$ where*

$$
\begin{aligned}
D_X &= \{(x_i, i)|x_i = \lfloor f_x(i, a, b)\rfloor, 0 \leqslant i \leqslant n\}, & \text{if } M(f_x, y) =' D' \\
&= \{(x_i, i)|x_i = \lceil f_x(i, a, b)\rceil, 0 \leqslant i \leqslant n\}, & \text{if } M(f_x, y) =' I' \\
D_Y &= \{(i, y_i)|y_i = \lfloor f_y(i, a, b)\rfloor, 0 \leqslant i \leqslant m\}.
\end{aligned}
$$

$\square$

The above expressions can be easily justified for the given monotonicity. We can informally say that if a curve is decreasing in the first quadrant then it bends down (or to the left) as its x-coordinate (or y-coordinate) increases. The converse holds if the curve is increasing in the first quadrant. The following theorem presents the equivalence between D and $I(f)$.

Theorem 5.12. *For a monotone curve f, $I(f) = D(f)$. Moreover, the set D_Y can be obtained from the set D_X and vice versa [40].*

Proof: The first part of the proof directly follows from the definitions of $D(f)$ and $I(f)$.

To prove the second part, we assume the MM of the curve to be the one given in Table 5.4.

The algorithm to construct D_Y from D_X is presented in Algorithm 10.

Algorithm 10: To Construct D_Y from D_X

Algorithm Construct_D_Y
Input: The set $D_X = ((x_i, i)|0 \leqslant i \leqslant n)$.
Output: The set D_Y.

$D_Y = \{\}$;

$x_{n+1} = -1$;

for i= 1 to n+1 do

for $j = (x_i + 1)$ to x_{i-1} do

$\quad D_Y = D_Y \cup \{(j, i - 1)\}$;

$D_Y = D_Y \cup \{(x_n, n)\}$;

End Construct_D_Y

To prove the correctness of the algorithm $Construct\_D_Y$, we require the following equivalence relation:

$(i, y_i) \in D_Y$ if and only if $x_{r+1} < i \leqslant x_r$, where $r = y_i$.

(If part): Let $x_{r+1} < i \leqslant x_r$. We show that $(i, y_i) \in D$. As $i \leqslant x_r$, and $x_r \leqslant f_x(r)$, we get $i \leqslant f_x(r)$. So, $f_y(i) \geqslant f_y(f_x(r)) = r$, which means that $y_i \leqslant f_y(i)$.

Again, $x_{r+1} \leqslant i - 1$ and $x_{r+1} > f_x(r+1) - 1$. So, $i - 1 > f_x(r+1) - 1$. In other words, $i > f_x(r+1)$. Hence, $f_y(i) < f_y(f_x(r+1)) = r+1$. That is, $y_i > f_y(i) - 1$.

Thus, $f_y(i) - 1 < y_i \leqslant f_y(i)$. So, $y_i = \lfloor f_y(i) \rfloor$ and $(i, y_i) \in D_Y$.

Therefore, to generate D_Y one needs to produce all (i, r) pairs satisfying $x_{r+1} + 1 \leqslant i \leqslant x_r$, $0 \leqslant r < n - 1$ where $(x_r, r) \in D_X$. This is performed in the nested for loops in the algorithm $Construct\_D_Y$. Hence, the algorithm is correct.

(Only if part), We need to show that $x_{r+1} < i \leqslant x_r$. Assume that $x_{r+1} \geqslant i$. From the definition of D_X- digitization, $x_{r+1} \leqslant f_x(r+1)$. Combining the two inequalities, we get $i \leqslant f_x(r+1)$, which implies that $f_y(i) \geqslant f_y(f_x(r+1)) = r+1$ as $M(f_y, x) =' D'$. Again, as $(i, y_i) \in D_Y$ by assumption, we get $f_y(i) < (y_i + 1) = r + 1$. This means $(r+1) \leqslant f_y(i) < r+1$, a contradiction. Hence $x_{r+1} < i$.

Now assume $x_r < i$. From the definition of D_X-digitization, $f_x(r) - 1 < x_r \leqslant i - 1$. Because of the monotonicity of the function f_y, $f_y(f_x(r)) = r > f_y(i)$. But $r = y_i$ is an integer and hence $y_i - r > \lfloor f_y(i) \rfloor = y_i$. Contradiction. Therefore, $x_r \geqslant i$.

$\square$

We can also easily separate out the sets D_X and D_Y from $I(f)$. So while $I(f)$ may be obtained experimentally from the acquired image data, it suffices to carry out the theoretical analysis using D_X and D_Y only. We note that the digitization for a one-parameter curve can be analogously defined.

5.5.2 One-Parameter Class

As already mentioned, the equation of the curve is given by $f(x, y, a) = 0$, which may be rewritten as $x = f_x(y, a), y = f_y(x, a)$, and $a = f_a(x, y)$. Henceforth we shall denote the original value of a by a_o, and D_o will denote the digitization of $f(x, y, a_o) = 0$, i.e., $D_o = D(f(x, y, a_o))$. For ease of analysis, let us consider one particular monotonicity matrix, which is given in Table 5.3. In this case, $x_i = \lceil f_x(i, a_o) \rceil$ and $y_i = \lfloor f_y(i, a_o) \rfloor$. Using the properties of $\lfloor . \rfloor$ and $\lceil . \rceil$ functions, we can write,

$$f_x(i, a_o) \leqslant x_i < f_x(i, a_o) + 1 \text{ and}$$

$$f_y(i, a_o) - 1 < y_i \leqslant f_y(i, a_o).$$

As f_a is decreasing in x,

$$f_a(x_i, i) \leqslant a_o = f_a(f_x(i, a_o), i) < f_a(x_i - 1, i).$$

Similarly, as f_a is increasing in y,

$$f_a(i, y_i) \leqslant a_o = f_a(i, f_y(i, a_o)) < f_a(i, y_i + 1).$$

Thus, the two bounds on a_o are obtained as follow:

$$a_l = \max(\max_i(f_a(x_i, i)), \max_i(f_a(i, y_i)))$$

$$a_u = \min(\min_i(f_a(x_i - 1, i)), \min_i(f_a(i, y_i + 1))).$$

For any value of a such that $a_l \leqslant a < a_u$,

$$f_a(x_i - 1, i) \geqslant a_u > a \geqslant a_l \geqslant f_a(x_i, i).$$

That is, $f_a(x_i - 1, i) > a \geqslant f_a(x_i, i)$.
Hence, $x_i = f_x(i, f_a(x_i, i)) \geqslant f_x(i, a) > f_x(i, f_a(x_i - 1, i)) = x_i - 1$.
Rearranging, $f_x(i, a) \leqslant x_i < f_x(i, a) + 1$.
Consequently, $x_i = \lceil f_x(i, a) \rceil$.
Similarly $\forall a, a_l \leqslant a < a_u, y_i = \lfloor f_y(i, a) \rfloor$.
In fact, if $a < a_l$, it is easy to see that $D(f(x, y, a))$ will miss some point (x_i, i) or (i, y_i) of D_o. Also, if $a > a_u$, then some point is included in $D(f(x, y, a))$ that is not in D_o. Thus, we have the following theorem.

Theorem 5.13. $a_l \leqslant a < a_u$ *if and only if* $D(f(x, y, a)) = D_o$. *In other words, Domain of* $D_o = [a_l, a_u)$ *[40].* ☐

Example 5.7. *Let* $f : y^2 - 4ax = 0$. *Here, the monotonicity matrix is the same as given in Table 5.3. The different functions are listed below.*

$$x = f_x(y, a) = y^2/4a$$

$$y = f_y(x, a) = \sqrt{(4ax)}$$

$$a = f_a(x, y) = y^2/4x$$

It has been shown in a previous section that

$$a_l = \max(\max_i(i^2/4x_i), \max_i(y_i^2/4i)) \text{ and}$$

$$a_u = \min(\min_i(i^2/4(x_i - 1)), \min_i((y_i + 1)^2/4i)).)$$

These formulae are the same as the ones given in this section.

5.5.3 Two-Parameter Class

As in the last section, we shall consider one particular MM to highlight the theme of our analysis. The MM we are considering is given in Table 5.4. So, $x_i = \lfloor f_x(i, a_o, b_o) \rfloor$ and $y_i = \lfloor f_y(i, a_o, b_o) \rfloor$. We present the main result in the following theorem [40].

Theorem 5.14. *Let a_l^{k+1}, b_u^{k+1}, a_u^{k+1}, and b_l^{k+1}, be defined by the following iterative algorithm for $k \geqslant 0$.*

$$
\begin{aligned}
a_l^{k+1} &= \max_i(f_a(x_i, i, b_u^k)) \\
b_u^{k+1} &= \min_i(f_b(i, y_i + 1, a_l^k)) \\
a_u^{k+1} &= \min_i(f_a(x_i + 1, i, b_l^k)) \\
b_l^{k+1} &= \max_i(f_b(i, y_i, a_u^k))
\end{aligned}
$$

With a proper choice of a_l^0, b_u^0, a_u^0, and b_l^0 that satisfies

(a) $a_l^0 < a_o < a_u^0, b_l^0 < b_o < b_u^0$ *and*

(b) $a_l^1 \geqslant a_l^0, b_u^1 \leqslant b_u^0, a_u^1 \leqslant a_u^0, b_l^1 \geqslant b_l^0,$

there exist b_l, b_u, a_l, and a_u such that

(A) $\lim\limits_{k\to\infty} b_l^k = b_l, \lim\limits_{k\to\infty} b_u^k = b_u, \lim\limits_{k\to\infty} a_l^k = a_l, \lim\limits_{k\to\infty} a_u^k = a_u,$ *and*

(B) $b_l < b_o < b_u$ *and* $a_l < a_o < a_u$

Proof: Proof of the theorem is similar to the proof of Theorem 5.3. This uses the definition of the floor function and the fact that f_b is a decreasing function in a. □

Now, if we consider the curves $f(x, y, a_l, b_u) = 0$ and $f(x, y, a_u, b_l) = 0$, we can find a few interesting properties which are summarized as the following theorems and lemmas.

The following Theorem is required to prove the tightness of the bounds of a_o and b_o.

Theorem 5.15. *The curves $f(x, y, a_l, b_u) = 0$ and $f(x, y, a_u, b_l) = 0$ intersect* [40]. □

The proof is by contradiction and is left as an exercise.

Corollary 5.4. *The open region between f_{lu} and f_{ul} (i.e., excluding the arcs of the curves) do not contain any grid point.* □

Let us define p, p', q, q' such that $a_l = f_a(x_{p'}, p', b_u)$, $b_l = f_b(p, y_p, a_u)$, $b_u = f_b(q', y_{q'} + 1, a_l)$, and $a_u = f_a(x_q + 1, q, b_l)$.

Corollary 5.5. $D(f_{lu})$ *or* $D(f_{ul})$ *differs from* D_o *only at point(s) like* $(q', y_{q'} + 1)$ *or* $(x_q + 1, q)$, *which define* b_u *and* a_u. $\square$

From the above corollary, we can say that f_{lu} or f_{ul} are very close approximations of the original curve represented by the function f. If we assume one more property about the curve, then we can prove

(i) that the Domain of D_o in the $a - b$ plane is contained in the rectangle R_{ul} whose diagonally opposite vertices are (a_l, b_u) and (a_u, b_l); and

(ii) that R_{ul} is the smallest rectangle containing the domain. This assumption is stated below.

Assumption 3. Let, $\frac{\partial f_y}{\partial x} = g(x, a, b)$. Then for $a_1 < a_2$ and $b_1 > b_2$ either

$$\forall x, \ g(x, a_1, b_1) < g(x, a_2, b_2), \quad \text{or}$$

$$\forall x, \ g(x, a_1, b_1) > g(x, a_2, b_2).$$

In terms of the limiting and original curves, the above assumption helps us in proving the following properties.

Lemma 5.8. *If* f *satisfies Assumption 3 then* $f_1 : f(x, y, a_1, b_1) = 0$ *and* $f_2 : f(x, y, a_2, b_2) = 0$ *cannot intersect more than once for* $a_1 < a_2$ *and* $b_1 > b_2$ *[40].*

Proof: If f_1 and f_2 do not intersect at all, then the lemma trivially holds. So we assume that f_1 and f_2 intersect at least twice. But Assumption 3 is violated. $\square$

In the sequel, we shall use only the first part of Assumption 3. Symmetric results may be obtained using the other part.

Similar to the case of the ellipse, the following lemma relates the grid points that are hit by a_u and b_l at the end of iterative refinement.

Lemma 5.9. *If* p *and* q *are defined such that* $a_u = f_a(x_q + 1, p, b_l)$ *and* $b_l = f_b(p, y_p, a_u)$, *then* $q < y_p$ *and* $p < x_q + 1$ *[40].* $\square$

So we can say that the domain of D_o is properly contained in the rectangle R_{ul}, which is defined by the diagonally opposite points (a_l, b_u) and (a_u, b_l) in the parametric space. This is formally stated in the next theorem.

Theorem 5.16. $D(f(x, y, a, b)) = D_o$ *implies that* $a_l < a < a_u$ *and* $b_l < b < b_u$.

Proof: This proof is similar to Theorem 5.4. $\square$

The next theorem claims that R_{ul} is the smallest rectangle enclosing the domain of D_o.

Theorem 5.17. *If we select some* a *so that* $a_l < a < a_u$, *then there exists some* b *for which* $D(f(x, y, a, b)) = D_o$ *[40].* $\square$

The domain of D_o can be computed using the following theorem.

Lemma 5.10. *The following hold [40]:*

(A) $a_l < a < a_u$ *if and only if* $b_l^*(a) < b_u^*(a)$
     where $b_l^*(a)$ $=$ $\max\limits_i (f_b(i, y_i, a))$ *and* $b_u^*(a) = \min\limits_i (f_b(i, y_i + 1, a))$

(B) $b_l < b < b_u$ *if and only if* $a_l^*(b) < a_u^*(b)$
     where $a_l^*(b)$ $=$ $\max\limits_i (f_a(x_i, i, b))$ *and* $a_u^*(b) = \min\limits_i (f_a(x_i + 1, i, b))$.

$\square$

Using the above lemma, we can compute the domain that is stated in the following theorem [40].

Theorem 5.18.

$$
\begin{aligned}
Domain(D_o) &= \cup_{a_l < a < a_u} [b_l^*(a), b_u^*(a)) \\
&= \cup_{b_l < b < b_u} [a_l^*(b), a_u^*(b)).
\end{aligned}
$$

$\square$

We complete this section by considering the following two examples.

Example 5.8. *Let f be an ellipse with center at origin and axes parallel to the coordinate axes. So, $f : x^2/a^2 + y^2/b^2 - 1 = 0$ and the parameters to estimate are a and b. Clearly, f is continuous and its MM is the same as given in Table 5.2. f is also separable where the separated functions are:*

$$
x = f_x = a\sqrt{(1 - y^2/b^2)}, y = f_y = b\sqrt{(1 - x^2/a^2)}
$$

$$
a = f_a = x/\sqrt{(1 - y^2/b^2)}, b = f_b = y/\sqrt{(1 - x^2/a^2)}.
$$

Further, $g(x, a, b) = \frac{\partial f_y}{\partial x} = -bx/(a^2\sqrt{(1 - x^2/a^2)})$. *So* $g(x, a_1, b_1) < g(x, a_2, b_2)$ *for* $a_1 < a_2$, *and* $b_1 > b_2$ *and f satisfies Assumption 3 as well.*

We have dealt thoroughly with this curve in the previous section. It is seen that the initial choice of the lower and upper bounds of a and b for Theorem 5.14 can be given by the following set of equations:

$$
a_l^0 = x_0, a_u^0 = x_0 + 1, b_u^0 = y_0, b_l^0 = y_0 + 1.
$$

We have presented an analysis of the class of curves whose nature of monotonicity is depicted by the particular MM as given in Table 5.2. However, along a similar line the other four MMs may be treated to achieve similar results. In the next example we consider a straight line because it has a different MM than the earlier one.

Example 5.9. Let $f : y - mx - c$. Its $PDSV = (-,+,-,-)$, i.e., a variant of $PDSV_1$ in Table 5.2. Also we consider the segment from $x = 0$ to $x = n$. We can develop a similar iterative refinement scheme using

$$m_l^0 = -1/n, m_u^0 = (n+1)/n, c_l^0 = y_0, c_u^0 = y_0 + 1$$

as initial choices of the bounds of the parameters.

It is easy to verify that f satisfies all assumptions. Consequently, the analysis presented to introduce I_R follows as a special case now.

5.6 Summary

In this chapter we addressed the reconstruction and the domain construction problems for digital conics in canonical form and a class of digitized planar curves having one or two parameters. In Chapter 3, a new framework of analysis has been introduced to determine the domain of a given DSLS. Through this analysis, a general methodology for reconstruction, namely the iterative refinement (I_R)technique, has also been developed.

The I_R technique is first applied to digital conics in canonical form. A detailed analysis of such digital ellipses has been carried out to obtain the smallest rectangle in the parametric space that encloses the domain. For the sake of brevity, relevant results are only presented for digital hyperbola. Since the method is iterative, heuristics have also been suggested to enhance the convergence of the I_R algorithm.

Next, the I_R technique is developed as a unified methodology to solve the reconstruction problem for a class of digital curves. It is shown that the domain of the given digitization can be exactly formulated for separable, monotone curves with one unknown parameter. In case of two parameters, the domain can be numerically computed if the curve satisfies another additional property.

Exercises

1. Prove Lemma 5.1

2. Prove Theorem 5.8.

3. Prove Theorem 5.10.

4. Prove Theorem 5.15.

5. Prove Lemma 5.9.

6. Prove Theorem 5.17.

7. Prove Theorem 5.18.

8. The speed of convergence of the iterative refinement algorithm to compute the bounds of a and b of an Ellipse may be improved if more recent estimates are used in the equations to compute the $(k+1)$-th estimate of the lower or upper bounds. Modify the iterative refinement algorithm in this line and check by experiments that the speed is really improved.

9. The domain of a quarter circle with a known radius is a region in $(\alpha - \beta)$-plane where $(\alpha_1 - \beta_1)$ is the center of the circle. Suppose that we some how know $\lfloor \alpha_1 \rfloor$ and $\lfloor \beta_1 \rfloor$. Can you modify the iterative refinement algorithm to obtain a scheme to compute the domain of a quarter circle with a known radius?

10. Consider an isothetic ellipse E defined by the following equation $(\frac{x-a}{A})^2 + (\frac{y-b}{B})^2 = 1$. The set of digital points resulting from its digitization may be defined as

$$H(E) = \{i, j | (\frac{i-a}{A})^2 + (\frac{j-b}{B})^2 \leqslant 1, i, j, \text{ integers}\}.$$

The discrete moment of a $H(E)$ may be defined as

$$\mu_{k,l}(H(E)) = \sum_{(i,j \in H(E))} i^k j^l.$$

Show that $H(E_1, H(E_2))$ are equivalent iff

$$(\mu_{0,0}(H(E_1)), \mu_{1,0}(H(E_1)), \mu_{0,1}(H(E_1)), \mu_{2,0}(H(E_1))) =$$
$$(\mu_{0,0}(H(E_2)), \mu_{1,0}(H(E_2)), \mu_{0,1}(H(E_2)), \mu_{2,0}(H(E_2)))$$

11. Refer to the previous problem. Present an algorithm to convert $H(E)$ to the Object Boundary Quantization and to Freeman's chain code.

12. Let A and B be two sets of points in 2-D. A and B are linearly separable if there exists a straight line such that A lies entirely to one side of that line while B lies entirely on the other side. Formulate iterative refinement technique to solve this problem of linear separability.

Chapter 6

Medial Axis Transform

The Medial Axis Transform (MAT) [21] is an attractive representation scheme for spatial occupancy of objects in 2-D and 3-D. In its lowest form, an object is represented as a set of points in an integral coordinate space that it occupies. We know that a point in this form of representation is called *pixel* in 2-D and *voxel* in 3-D. The MAT of these objects provides a relatively higher level of structural description, as it represents the object as a set of disks (circles in 2-D and spheres in 3-D). To reduce the number of such circles (or spheres), in this representation, only those are considered that are not totally contained in any one of them. These are called *medial disks* (or centers of maximal disks (CMD)) of the pattern or object. However, even with these medial disks, there is a scope of redundancy in the set. The disks may be overlapping. Moreover a medial disk may be contained by more than one member from the remaining set. The representation of a binary object using medial disks is called its MAT. In this representation, individual disks are denoted by their centers and radii. In this chapter we discuss the MAT of binary objects in 2-D and 3-D, and its application to various geometric computation on images.

In chapter 2 (refer to Section 2.5), we illustrated the properties of digital disks of some of the interesting classes of distance functions, in particular the hyperspheres of *octagonal distances* [59]. In our discussion, we consider these distance functions for variation in the shape of hyperspheres of different octagonal distances, and the ease of computation of the distance transform (refer to Section 6.1) using them.

6.1 Distance Transform

The distance transform [181][23][57] of a bilevel image provides the distance to the nearest background point at every pixel of the image. The computation of distances of the object points from the nearest background is performed by a simple iterative approach. In this case for an image of N points it requires $O(N^2)$ computation. However, this transform may efficiently be computed in $O(N)$ by performing a fixed number of scans over the images, a process known as *chamfering*, as discussed below. First we discuss the iterative algorithm for computing the distance transform given an octagonal distance $d(B)$, where B is the sequence of m-neighbor types, say, $\{b(1), b(2), \ldots, b(p)\}$ (refer to Section 2.4.1 of Chapter 2). We denote distance of a point q from the background in the distance transformed image as $D(q; B)$. Let us represent foreground pixels (object points) and background pixels of a digital image as Σ and Ψ, respectively.

6.1.1 Distance Transform through Iterative Scan

In this technique, the computation of distance proceeds from the boundary points of the object (Σ) and moves toward its inner layers iteratively. First, the distance of an object point ($\in \Sigma$) is set to 1, which has a neighboring background pixel ($\in \Psi$) and the point is also flagged for acting as the wavefront at a distance 1 for the next iteration. In the subsequent kth iteration, distance of any unflagged object point is computed, if it has a neighboring point in the approaching wavefront flagged in the previous iteration (with a distance value of $k - 1$) and its distance is set to k. The process continues till all the object points are assigned a distance value. For an octagonal distance with a neighborhood sequence B, neighborhood definitions change following the periodic ordering of the neighborhood types. For example, in our notations, the order of neighborhood types to be used are $b(1)$, $b(2)$, ..., $b(p)$, $b(1)$, $b(2)$, etc. The computation is explained in Fig. 6.1. The boundary points of the polygonal object in the figure are computed using the neighborhood definition of type $b(1)$ and they are set at a distance of 1. In the figure, these are shown by the color red. Its immediate neighbors with the neighborhood type $b(2)$ are set at a distance of 2 and they are shown by the color blue. The process continues with changing neighborhood definitions following B till the distance values of all the object points are computed. The algorithm runs in $O(D_{max}N)$ times where D_{max} is the maximum distance of an object point from the background and N is the number of points in an image. As D_{max} could be of $O(N)$, it has a quadratic time complexity.

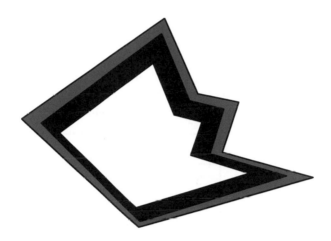

FIGURE 6.1: Computation of DT using iterative scan from boundary points. (See color insert.)

6.1.2 Chamfering Algorithm

Borgefors [22] discussed a linear time algorithm for computing the distance transform of an object. In this technique, an extended neighborhood mask enumerating the distance values from its center using the distance function under consideration, is scanned by placing its center at every pixel and the distance values of the pixel (from the background) is updated by observing distances of its neighboring pixels, visited before. This computation is performed by two scans following the forward and reverse ordering of scans. The nature of the scan depends upon the dimension of image. For example, in 2-D, the forward scan involves an ordering from left to right and top to bottom of the image array, while the reverse scan goes in an ordering from right to left and bottom to top. Similarly, in 3-D we have another added dimension imposing an ordering of object planes from front to back. So in this case, the forward scan involves an ordering from left to right, top to bottom, and front to back. The reverse order is also similarly defined as it is done for the case 2-D. Borgefors

[22] used *weighted distance functions* (refer to Section 2.3.3 of Chapter 2) in defining the masks, which are very convenient in expressing the distances of neighboring pixels around the central pixel. However, in our discussion, we restrict ourselves to octagonal distances, as their digital disks are easily computable using Theorems 2.26 and 2.27. We also restrict our discussion to 2-D, as it could be trivially extended to 3-D.

6.1.2.1 Designing Masks

For a neighborhood sequence of length p, a mask should be taken of size $(2p+1) \times (2p+1)$. This is to ensure that all possible paths of length p around the neighborhood of a point are considered in the computation. Let us denote a mask as M and the distance value of a pixel $x \in M$ as $d(x; B)$. A typical mask formed for the octagonal distance function {112} is shown in Fig. 6.2. In the figure, distance values from the center mask using the distance function {112} are shown. For pixels that are at a distance greater than p, values are marked as $*$, as they are irrelevant in the computation.

*	*	3	3	3	*	*
*	3	3	2	3	3	*
3	3	2	1	2	3	3
3	2	1	0	1	2	3
3	3	2	1	2	3	3
*	3	3	2	3	3	*
*	*	3	3	3	*	*

FIGURE 6.2: Chamfering mask of {112}.

6.1.3 Forward and Reverse Scans

Initially, all the object points are set to a very high distance value (at least more than D_{max}) and all the background points are assigned to 0. In the forward scan, the mask is placed at every pixel (say, q) while scanning from left to right and top to bottom. Hence the visited neighboring pixels around q belong to the unshaded zone of Fig. 6.3(a). Let the set of visited pixels be

F. Hence, the distance at q from the background can be computed as follows:

$$D(q; B) = \min_{\forall x \in F} (D(x; B) + d(x; B).) \qquad (6.1)$$

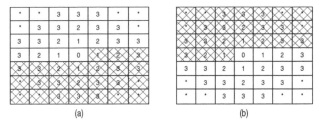

FIGURE 6.3: Pixels of chamfering mask of $\{112\}$ used in (a) forward scan, and (b) reverse scan.

Similarly, in the reverse scan, while visiting the pixels in an order from bottom to top and right to left, the unshaded zones of Fig. 6.3(b) are considered for updating the distance values. Let the corresponding visited neighboring pixels in the mask form the set R. In that case, at a pixel q, the distance value from the background is updated as follows:

$$D(q; B) = \min_{\forall x \in R} (D(x; B) + d(x; B)). \qquad (6.2)$$

Once these two scans are completed, we obtain the distance transform of the image.

An example of the generalized octagonal distance transforms is shown in Figure 6.4 (b) using the distance function $\{112\}$. Each distance value in the distance-transformed image is taken as the gray-scale value. The brighter color represents the larger distance and the darker color represents the smaller distances.

6.1.4 Euclidean Distance Transform

There are algorithms [82] for computing distance transforms of digital images using Euclidean metrics. However, computation of Euclidean distance transform (EDT) is a nontrivial problem. In this case, the Voronoi regions [169] of background points may not be connected as explained in [51]. For this property, local propagation of distance values, as followed for digital distance transforms discussed in previous sections, is not recommended for computing the EDT. There are algorithms for computing exact EDTs based on efficient computation of discrete Voronoi regions [185, 138] of background points. However, all these algorithms are computationally costlier than digital distance transforms. We discuss here an approximate computation of EDT

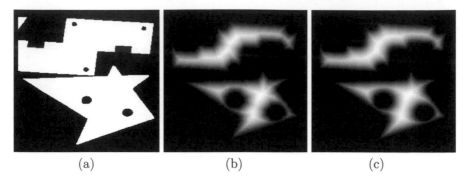

(a)   (b)   (c)

FIGURE 6.4: (a) An object (set-square) (b) distance transform using {112}, (c) Euclidean distance transform using 8SED algorithm [53].

[53], which is based on local propagation of distance values and uses a scheme similar to chamfering. Danielsson reported this algorithm in the early eighties. Subsequently, other researchers [133, 170] reported many variations and improvements of this scheme.

In the computation of EDT, initially all the object points are assigned to a large distance value and all the background points are set to zeroes. However, in the representation of Euclidean distance at an object point, a two-dimensional element such as (a, b) is used, where the distance is expressed as $\sqrt{a^2 + b^2}$. The final distance-transformed array for the object contains these elements. The technique follows a vectorial propagation scheme from the neighborhood of a point. In [53], two different propagation schemes based on 4-neighbors and 8-neighbors were proposed. The first technique was referred to as the $4SED$ algorithm and the latter was named the $8SED$ algorithm. As the $8SED$ algorithm provides better approximation of the EDT, it is discussed here. Similar to the chamfering scheme, in this algorithm, the computation involves two scans, one in the forward directions (from left to right and top to bottom), and the other in the reverse directions (from right to left and top to bottom). Each scan uses three masks for computing the minimum distance value at a pixel from the distances of its neighbors by adding the propagated distance toward it. For example, for each column of the forward scan, the mask of Fig. 6.5(a) is used, followed by the other two masks in Figs. 6.5(b) and (c), respectively. Similarly, in the reverse scan, each column first uses the mask of Fig. 6.5(d). Then the values are further updated by scanning of the masks of Figs. (e) and (f), respectively. In the figure, the coordinate convention is shown by x and y axes. The example of EDT obtained using this algorithm is shown in Fig. 6.4. We observe that the result appears to be similar to what was obtained from the octagonal distance {112} (refer Fig. 6.4(b)).

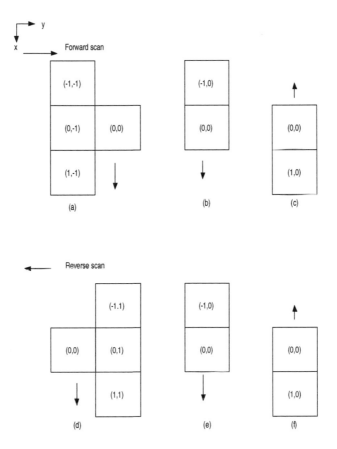

FIGURE 6.5: Masks for Euclidean distance transform used in the 8SED algorithm [53].

6.2   Medial Axis Transform (MAT)

Medial Axis Transform (MAT)[21] of an image consists of a set of maximal disks (or hyperspheres) that could be contained within the pattern (or foreground). An interesting property of a medial point (the center of a maximal disk (CMD)) is that it touches the boundary of the object at two or more points (refer Fig. 6.6). This property leads to the development of a simple

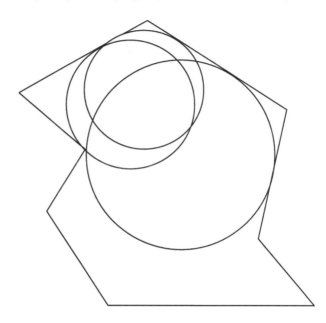

FIGURE 6.6: A few examples of maximal disks of an object in 2-D Euclidean space.

but straightforward algorithm for the computation of MAT. Let the boundary points of the object form a set Δ. For every foreground point q ($\in \Sigma$), its closest boundary point(s) in Δ are computed. If p has more than one such point, it is declared as a medial point and the distance of the boundary point is taken as the radius of the medial (or maximal) disk at that point. In the formation of this set of medial points, the distance function plays an important role. The number of medial points and the shape of axes depend upon the choice of the distance function. The MAT could be computed more efficiently using distance transforms. This we discuss in the following subsections.

6.2.1 MAT from the Distance Transform

The MAT of an image can be obtained by the following theorem.

Theorem 6.1. *Given a distance transform D of an image, its local maxima form the set of medial points and one less than the distance value at that point provides the radius of the medial disk.* ☐

For proving the above theorem, we prove that a digital disk with a radius of $k - 1$, where k is the distance value at a local maxima, is totally contained in the foreground of the image, and it is not completely covered by any other digital disk containing only foreground points. Let $r(q)$ be the minimum distance of the point $q \in \Sigma$ from the boundary. We consider the digital disk $H(q, r(q); B)$ and its boundary points $S(q, r(q); B)$ at q in the metric space of $d(B)$, where $B = \{t(1), t(2), \ldots, t(p)\}$ is the neighborhood sequence. Clearly, $H(q, r(q); B) \subset \Sigma$. Let $N_{t(i)}(q)$ be the set of neighbors of q of type $t(i)$. Hence, for any path from a point $u \in S(q, r(q); B)$ to q, the neighborhood type at q is $t(j)$ where $j = (r(q) - 1) \bmod p$. Hence, clearly, $\forall x \in N_{t(j)}(q), r(x) \leqslant r(q) + 1$. It implies that if $\exists x \in N_{t(j)}(q)$ such that $r(x) > r(q)$, then $r(x) = r(q) + 1$. From this property, we can state that the local maximum in the distance transform provides the center of the maximal block contained in the pattern as given in the following theorem.

Theorem 6.2. *If $\exists x \in N_{t(j)}(q), r(x) \geqslant r(q)$, then $H(q, r(q); B) \subset H(x, r(x); B)$.*

Proof: $\forall z, z \in S(q, r(q); B), d(z, q; B) = r(q)$ (by definition).
If $\exists x \in N_{t((r(q)-1) \bmod p)}(q)$, there exists a path from $z \in S(q, r(q); B)$ to the point x, whose length is $r(q) + 1$.
Therefore, $\forall z \in S(q, r(q); B), d(z, x) \leqslant r(q) + 1$.
Now, $H(x, r(x); B) = \{y \mid d(y, x) \leqslant r(q) + 1\}$.
Therefore $S(q, r(q); B) \subset H(x, r(x); B))$.
Since the $H(q, r(q); B)$ is convex (due to metricity) and the boundary of $H(q, r(q); B)$ is contained in $H(x, r(x); B)$, all the internal points of $H(q, r(q); B)$ are contained in $H(x, r(x); B)$.
Hence, $H(q, r(q); B) \subset H(x, r(x); B)$. ☐

The algorithm [127] for the computation of the MAT of images is presented below:

The above algorithm is presented for any arbitrary dimension. Typical results of adaptation of these algorithms in 2-D and 3-D are shown in Figs. 6.7 (a)-(c), and Figs. 6.8 (b)-(d), respectively. The variations of the set of medial points (also called *centers of maximal disks* (CMD)) are also observed with the changes of metric space in 2-D and 3-D in those figures. In subsequent sections, we discuss a few applications of MAT in image processing. We should also note that the MAT from the EDT is a nontrivial problem. In the Euclidean space, the extent of the neighborhood for searching the local maxima is not precise. In [53], a local search method is proposed. In this case, a mapping

---

Algorithm 11: Medial Axis Transform (MAT) from Distance Transform (DT) of an Object

---

Algorithm: *MAT_through_DT* (MATDT)
Input: Bilevel image.
Output: A set of medial points with centers and radii of the maximal disks.

1. Compute the distance transform of the image.

2. Scan the distance mapped array as follows:

   2a. For each point q in the distance mapped array, check whether its distance value $(D(q))$ is a maximum among its neighbors. The neighborhood type is determined as $(D(q) - 1)$ **mod** p, where p is the length of the neighborhood sequence B.

   2b. If the point is a local maximum, declare it as a medial point and store the center of the disk and the *radius* at that point as $D(q) - 1$.

End *MAT_through_DT* (MATDT)

---

from the values of the radius of a Euclidean circle (specified in the form of (a, b) as discussed in Section 6.1.4), the maximum radii of its 4-neighbors and 8-neighbors, which could be contained in the circle, are computed and used. It requires additional storage space for storing the tables or arrays of these mappings.

6.2.2 Reduced Centers of Maximal Disks (RCMD)

As shown in Fig. 6.6, medial disks in a MAT, may overlap among themselves. One of the properties of these disks is that none of them are contained in any other from the set. However, it does not ensure its coverage by more than one disk from the remaining members. This implies that a disk that is covered by a few other medial disks, becomes redundant in the reconstruction of the object from the MAT. A set of medial disks that can fully reconstruct the object and has no redundant disk, is called a *reduced set of centers of maximal disks* (RCMD) [204]. An algorithm based on a greedy approach for obtaining an RCMD from the MAT is discussed in [26]. We outline the approach in the context of 2-D images here.

In this technique a temporary array is used for recording the number of medial disks covering a pixel. We refer to this array as the *count array*. This is computed while performing the reconstruction of object points from the medial disks of the MAT. Next, each medial circle is checked to see whether it has all the pixels inside with counts of more than one. In that case, the circle is

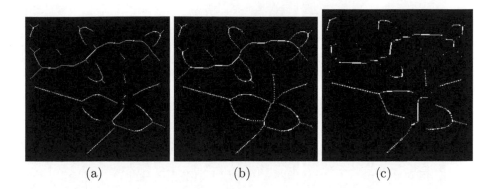

(a) (b) (c)

FIGURE 6.7: Medial Axis Transform of set square from distance functions: (a) {1}, (b) {112}, and (c) {2}.

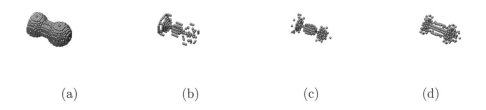

(a) (b) (c) (d)

FIGURE 6.8: Medial Axis Transform: (a) original image, centers of maximal disks (CMD) of (a) using distances : (b) {113}, (c) {123}, and (d) {3}.

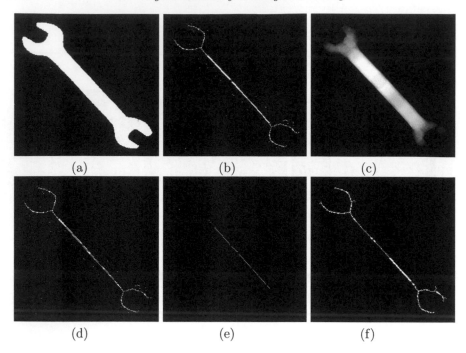

(a)   (b)   (c)

(d)   (e)   (f)

FIGURE 6.9: Computation of RCMD and thinned pattern: (a) spaner, (b) the MAT using {112}, (c) count array showing the number of medial disks covering a pixel, the number being proportional to intensity values, (d) the RCMD, (e) the set of redundant centers of maximal disks, and (f) skeleton of the object.

declared redundant and removed from the list of medial circles. Subsequently, the count array is updated by decreasing the counts of pixels lying within the redundant circle. The process continues till all the medial circles are tested. For retaining medial circles of larger radii in preference to the smaller ones, the testing of redundancy is carried out with the circles in the increasing order of their radii. A typical result on the computation of the RCMD is shown in Fig. 6.9.

6.3 Skeletonization Using MAT

In Chapter 1 (refer to Section 1.4.1) we discussed skeletonization of the binary pattern, which is the process of obtaining skeletal representation of an object consisting of a set of points forming the spine of the object. The structure of the skeleton has a thickness of a unit pixel and it preserves the topological properties of the objects, such as connectivity of patterns and the cavities or holes inside. As discussed there, skeletonization is usually carried out by iterative deletion of points lying at the present outermost layer of the skinned object. If the removal of a point does not introduce any decomposition, opening in a hole, or erosion of the pattern, the point is *safe* for deletion. Such a point is said to be *simple*. The checking for condition of removal of a point is a sequential operation of considering the effect of progressive deletion of points. One such condition for testing a point to be declared as simple in an 8-connected pattern is given in Section 1.4.1, in the discussion of the *safe point thinning algorithm* (SPTA). We consider here another example of the safe point testing condition using the *Hilditch number* [98]. It is the number of components among the 8-neighbors of a point. If this number is more than one, the point is not safe for deletion, otherwise it is declared as a simple point. The method for obtaining the Hilditch number is explained with the help of Fig. 6.10. Let m_i, $1 \leq i \leq 8$ be variables with the values from the neighboring pixels as either 1 (belonging to foreground) or 0 (belonging to background). Then the Hilditch number $X_H(q)$ is given by the following [204]:

$$X_H(q) = \sum_{j=1}^{4} b_j \qquad (6.3)$$

where,

$$b_j = \begin{cases} 1 & \text{if } (m_{2j-1} = 0) \text{ and } ((m_{2j} = 1) \text{ or } (m_{2j+1} = 1)). \\ 0 & otherwise. \end{cases} \qquad (6.4)$$

In the above computation m_9 is equated to m_1.

In the conventional thinning algorithms [130] (refer to the safe point thinning algorithm (SPTA) discussed in Section 1.4.1 of Chapter 1) the boundary points are classified into several groups depending upon the position of the background pixel in its neighborhoods. For example, if any of the 4 neighbors belongs to the background, the pixel is called an *edge boundary point*. Moreover, if its left edge neighbor is 0, it is also called a *left edge boundary point*. Similarly, other types are *right*, *top*, and *bottom* edge boundary points. A boundary pixel could belong to more than one categories as a combination of edge points may be empty. Usually at each iteration the deletion of boundary points are carried out sequentially for the same types of boundary points. Thus an iteration consists of scanning all groups of boundary points, one after another. The strategy is followed to avoid any bias or preferential treatment in the deletion or retainment of pixels along a specific direction in the image. This aims at keeping the thinned pattern in equidistant regions of the objects from the boundary points. However, layer-wise removal of simple points do not require any check on the types of edge points, as candidates for deletion. The kth layer of the DT provides a set of pixels, which could be checked for their safe removal without bothering about the directionality of removal. In this case, unless all the pixels in the kth layer are checked, the deletion of pixels in a deeper layer (i.e., $(k + 1)$th layer) cannot start. Further, to ensure the quality of the thinned pattern, medial points act as anchor points that are never deleted. They are connected at the end of the computation. In [204], these anchor points are taken from the RCMD of the object. The computation is described in Algorithm 12.

m_4	m_3	m_2
m_5	q	m_1
m_6	m_7	m_8

FIGURE 6.10: Neighborhood variables of a point q.

Algorithm 12: Skeletonization from MAT of a 2-D Object

Algorithm: *Skeletonization_from_MAT* (SMAT)
Input: The Object Σ, The distance Transform DT, The MAT M. Let D_{max} be the maximum distance value in DT.
Output: Skeleton S.

1. Obtain the RCMD of the object. Let us denote the set as Γ.

2. For $k = 1$ to D_{max} {

   2a. For each pixel p,

      i. If $(DT(p) = k)$ and p has a 4-neighbor in the background, compute $X_H(p)$.

      ii. If $(X_H(p) \neq 1)||(p \in \Gamma)$ retain the point and include it in S, else delete it.

   2b. Perform step 2a till all the points at level k are checked.

   }

End *Skeletonization_from_MAT* (SMAT)

It was observed in [204], that there could be at most two scans required for each level k in Step 2 of the above algorithm. The other advantage of having a distance transform in this case is that it is convenient to perform morphological operations for preprocessing of the object. For example, instead of starting from level 1 in Step 2 of the algorithm SMAT, we may initiate processing from a deeper level and delete all the object points in the outer layer till that level in the beginning of Step 1. An example of a thinned pattern is shown in Fig. 6.9(f). We have deleted all the object pixels that are at a distance of 1 before applying the algorithm. As a result, we find that isolated points are removed in the thinned pattern. However, we should note that the thickness of the pattern may be more than one pixel, due to the retainment of the RCMD.

6.4   Geometric Transformation

Geometric transformation of a binary object could be trivially carried out by applying the transformation on the coordinates of each object point (foreground point). However, it involves a significant amount of computation, which is proportional to the number of object points. Moreover, due to discretization of transformed coordinates, there could be gaps or empty pockets created in the transformed objects as shown in Fig. 6.11(b).

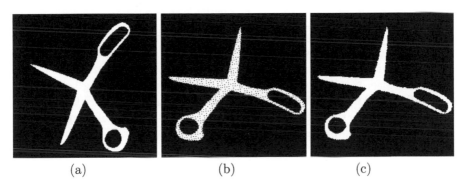

(a)   (b)   (c)

FIGURE 6.11: (a) Scissors, (b) rotated by 75° with pixel representation, and (c) rotated by 75° with MAT.

The MAT is a useful representation for performing geometric transformation [129] efficiently and rendering the object using efficient display routines of polygon filling in a graphics environment. A typical example is shown in Fig. 6.11 (c). In this case, the same transformation is applied to the vertices of medial circles, which are convex polygons in the digital metric space. If an object contains n pixels, geometric transformation takes place in $O(n)$ time complexity, whereas for an object with k medial disks, it runs in $O(k)$ time complexity. As $k \ll n$, the latter algorithm runs much faster. However, for the purpose of reconstruction, it is required to fill the polygons. This would effectively make the time complexity of $O(n)$. As polygon filling [84] is done efficiently with the help of special hardware and architecture by graphics processors, we may consider it as a unit task. Moreover, since the polygons of the disks are convex, the computation could be further optimized. In our discussion, we consider the cost of polygon filling is a unit cost, ignoring its coverage.

Since the vertices of the rotated convex polygons are discretized, and the vertices themselves are computed using approximations as discussed in Section 2.5.5.2 of Chapter 2, it is of interest to observe how the reconstructed objects differ from the true rotations of each and every pixel of our object. As the latter acts as the reference, the rotated objects are smoothed by filling the erroneously created pockets or gaps using simple checking of its 4-neighbors. Experiments [129] were carried out to observe the quality of reconstruction by rotating an object about the center of an image by a degree θ. Suppose the set of object points after rotation is $\Sigma_{pix}(\theta)$. The object is similarly rotated by rotating its medial disks and reconstructed through polygon filling. Suppose this set is denoted by $\Sigma_{MAT}(\theta)$. Then the error of rotation $E_{rot}(\theta)$ is expressed as the percentage of pixels that differ with respect to the pixels of $\Sigma_{pix}(\theta)$. This is expressed in the following:

$$E_{rot}(\theta) = \frac{|\Sigma_{pix}(\theta) - \Sigma_{MAT}(\theta)| + |\Sigma_{MAT}(\theta) - \Sigma_{pix}(\theta)|}{|\Sigma_{MAT}(\theta)|} \times 100\%. \quad (6.5)$$

It has been reported in [129] that the average $E_{rot}(\theta)$ of objects (in an image of size 256×256) lies within 3%.

Similar experimentation was also carried out for transforming 3-D objects. In this case, rotation around the z-axis was performed using both voxel and MAT (in 3-D) representations. For reconstructing objects with the transformation over MAT representation, digital disks in the form of convex polyhedra are filled by computing intersections with planes parallel to the XY-plane. The image sizes were taken as $128 \times 128 \times 128$. The reported error margins ($E_{rot}(\theta)$) lie within 8%.

6.4.1 Approximate Transformation by Euclidean Disks

If the digital disks are replaced by Euclidean counterparts with the same set of centers and radii, the computation becomes much simpler and faster. In this case, the transformation is applied to the centers of disks only. In 2-D, the polygon filling is replaced by computing the set of pixels inside a circle and similarly, 3-D computation involves to get the point set in the interior of a sphere. Naturally, the closer a digital disk is to a Euclidean one (of the same radius), the better the approximation of transformation. In [129], the same experiments were carried out by replacing digital disks with a Euclidean one using MAT of different octagonal distance functions, in both 2-D and 3-D. Average errors of the transformations for different metrics are shown in Tables 6.1 and 6.2 for 2-D and 3-D, respectively.

TABLE 6.1: Percentage average error between the rotated image using pixel representation and the rotated reconstructed image using approximated Euclidean circles for medial disks in MATs.

B	Avg. $E_{rot}(\theta)$	B	Avg. $E_{rot}(\theta)$
{1}	9.98	{12}	7.01
{1112}	4.94	{122}	10.23
{112}	4.58	{1222}	11.84

TABLE 6.2: Percentage average error between the rotated image using voxel representation and the rotated reconstructed image using approximated Euclidean spheres for medial disks in MATs.

B	Avg. $F_{rot}(\theta)$	B	Avg. $E_{rot}(\theta)$
{1}	12.8	{13}	12.6
{112}	7.7	{133}	17.5
{113}	8.6	{2}	23.2
{12}	9.7	{223}	23.5
{122}	13.3	{23}	24.7
{123}	14.6	{233}	26.3

The experimental results confirm that distance metrics with octagonal neighborhood sequences {112} and {1112} are good approximations for the representation of 2-D data. Similarly in 3-D, good octagonal distances are found to be {112} and {113}.

6.5 Computation of Normals at Boundary Points of 2-D Objects

In this section, we discuss how MAT is useful in computing normals at boundary points of a 2-D binary object. In Fig. 6.12, the principle behind this computation is illustrated. Let p be a point on the contour of a 2-D object and let a medial disk touch the contour at p. Let its center be o. Then $\vec{po}$ forms the inward normal at point p. This concept is extended for computing normals using digital disks. It can easily be seen that the technique requires correspondences between boundary points and their touching medial disks. Additional computation is required to derive this necessary information from the MAT representation. However, the technique as opposed to analytical

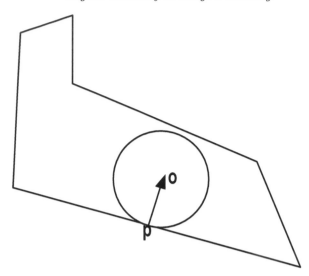

FIGURE 6.12: Schematic diagram for normal computation.

regression-based techniques [206] of fitting boundary curves, is distinguished by two characteristics. First, it does not require contour tracing, including the computation for determination of inward or outward direction of normal. Next, the whole computation can be performed using integer arithmetics only. We discuss the algorithm below.

6.5.1 Algorithm for Normal Computation

Let the set of contour points of a 2-D object be denoted as C, and let S be the set of medial disks for the object. The algorithm works in two stages. First, the correspondence of a contour point and a medial disk is established by checking its distance from the center. In an ideal case, the distance should be equal to the radius of the disk. However, to provide tolerance in the computation, a margin of error is allowed. Hence, a boundary point may have a number of corresponding medial disks. The normal is computed as the resultant of all these vectors. The algorithm [153] is briefly presented below:

6.5.2 Use of Octagonal Distances

As it is observed that digital circles of octagonal distances such as {112}, and {1112} in 2-D, closely resemble Euclidean circles, the *NCUM* algorithm should use one of these octagonal metrics. To observe the accuracy of normal computation, in [153], experiments were carried out with objects of known geometry, such as circles, squares, and rectangles. A typical example is shown

---

Algorithm 13: Computation of Normals at Boundary Points of 2-D Object

---

Algorithm: *Normal-Computation-Using-MAT (NCUM)*
Input: A set of contour points (C), the MAT of an object (S) derived by the metric $d(.)$.
Output: Normals at the boundary points.

1. For each boundary point $p \in C$, find the set of corresponding medial disks Q_p as follows:

   1a. Compute the distances from the center o_M of a medial disk M. Let the distance be denoted as $d(p, o_M)$.

   1b. Let the radius of the disk be r_M. Then the disk corresponds to p (i.e., $M \in Q_p$) if
   $$\mid d(p, o_M) - r_M \mid < Nthresh$$
   where $NThresh$ is the threshold for declaring M as approximately touching the contour at p.

   1c. Perform Steps 1a and 1b for all the medial disks in S.

2. Normal at p is expressed by the vector along $\vec{N_p}$, where
   $$N_p = \sum_{\forall M \in Q_p} (\vec{po}_M)$$

End *Normal-Computation-using-MAT (NCUM)*

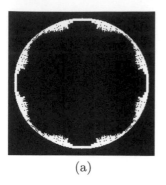
(a)

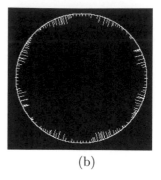
(b)

FIGURE 6.13: Normal computation at the contour points for the image circle for B={112} (a) dense normal map, (b) sparse normal map.

Reprinted from *Pattern Recognition Letters*, 23(2002), J. Mukherjee et al., Use of Medial Axis Transforms for Computing Normals at Boundary Points, 1649–1656, Copyright (2002), with permission from Elsevier.

in Fig. 6.13 by reproducing it from [153]. In Fig. 6.13(a), the normals computed at the contour points of the circle using digital distances for $B = \{112\}$ are shown. In this figure the lengths of the normals are proportional to the magnitudes of $\vec{N_p}$'s as computed by the algorithm *NCUM*. Also, for the clarity of the presentation, only sparse normal maps are shown in Fig. 6.13(b). The value of *Nthresh* was kept as 1 in this case.

Typical examples of computed normals for some of the objects of known geometry using different octagonal distances are also shown in Figs. 6.14, 6.15, and 6.16. In these figures, only results with good distances such as $\{112\}$, $\{1112\}$, and $\{12\}$ are shown.

6.5.3 Quality of Computation

In [153], a quantitative measure is used for judging the quality of the normal computation. In this measure, analytical maps of normals at the boundary points of objects of known geometry are used. For example, in Fig. 6.17 (a), (b), and (c), analytical normal maps for circle, square, and rectangle, respectively, are shown. Let $\vec{n_p}$ be the unit normal vector at a point $p \in C$ obtained from the algorithm *NCUM*. Let the analytical value of unit normal vector at a point $p \in C$ be denoted as $\vec{m_p}$ (Fig. 6.18). Then an error measure E_n is defined as

$$E_n = 1 - \sum_{\forall p \in C} \mid \vec{n_p} \cdot \vec{m_p} \mid / \mid C \mid$$

where $\mid C \mid$ is the number of points in C.

The value of E_n becomes zero when $\vec{n_p}$ and $\vec{m_p}$ lie along the same direction. The larger the deviation, the greater is the contribution in the aggregated error measure (E_n). However, if the algorithm *NCUM* fails to compute $\vec{n_p}$ at any

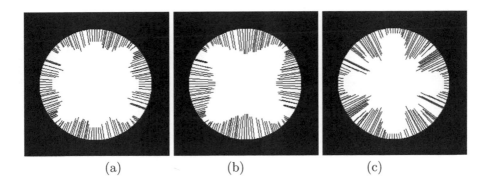

FIGURE 6.14: Normal computation for an image CIRCLE using digital circles for different distance functions (a) {112}, (b) {1112}, and (c) {12}.

Reprinted from *Pattern Recognition Letters*, 23(2002), J. Mukherjee et al., Use of Medial Axis Transforms for Computing Normals at Boundary Points, 1649–1656, Copyright (2002), with permission from Elsevier.

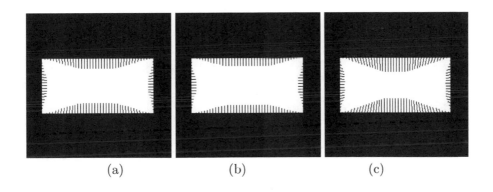

FIGURE 6.15: Normal computation for an image RECTANGLE using digital circles for different distance functions (a) {112}, (b) {1112}, and (c) {12}.

Reprinted from *Pattern Recognition Letters*, 23(2002), J. Mukherjee et al., Use of Medial Axis Transforms for Computing Normals at Boundary Points, 1649–1656, Copyright (2002), with permission from Elsevier.

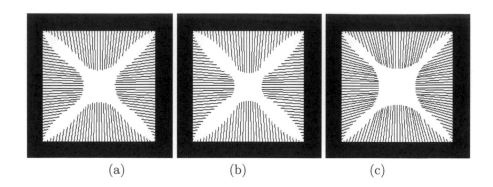

(a)   (b)   (c)

FIGURE 6.16: Normal computation for an image SQUARE using digital circles for different distance functions (a) {112}, (b) {1112}, and (c) {12}.

Reprinted from *Pattern Recognition Letters*, 23(2002), J. Mukherjee et al., Use of Medial axis Transforms for Computing Normals at Boundary Points, 1649–1656, Copyright (2002), with permission from Elsevier.

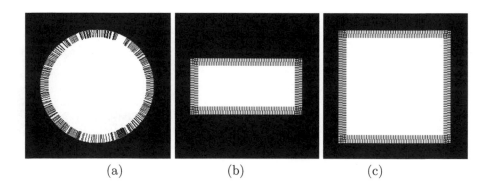

(a)   (b)   (c)

FIGURE 6.17: Analytical normal maps: (a) for circle, (b) for rectangle, (c) for square.

Reprinted from *Pattern Recognition Letters*, 23(2002), J. Mukherjee et al., Use of Medial Axis Transforms for Computing Normals at Boundary Points, 1649–1656, Copyright (2002), with permission from Elsevier.

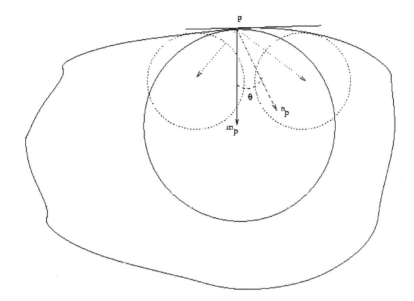

FIGURE 6.18: Computation of errors between analytical and computed normals.

Reprinted from *Pattern Recognition Letters*, 23(2002), J. Mukherjee et al., Use of Medial Axis Transforms for Computing Normals at Boundary Points, 1649–1656, Copyright (2002), with permission from Elsevier.

TABLE 6.3: Average percentage normal computational errors using MAT with digital metric (for $Nthresh=1$).

Octagonal Distances	Average E_n	Octagonal Distances	Average E_n
{1}	1.83	{12}	3.19
{1112}	1.32	{122}	5.15
{112}	1.67	{1222}	6.08

point $P \in C$, its contribution at that point toward the error measure (E_n) is not considered, and subsequently the point p is excluded from the set C.

In Table 6.3, the average error measures for the geometric objects of Fig. 6.17 as reported in [153] are shown for normal computation using MAT with different digital distances. As expected, computation with {112} and {1112} distances yield low errors.

To demonstrate the effectiveness of the technique, in Fig. 6.19, examples of normal computation for arbitrary 2-D objects are shown. In this case, the distance function used is {112}.

6.6 Computation of Cross-Sections of 3-D Objects

One of the important tasks of rendering a 3-D object is to show its cross-section on an intersecting plane. This reveals internal structures of 3-D objects along that plane. In a voxel-based representation, the computation is performed by selecting the set of voxels lying on the plane. In this case, each voxel's distance from the plane is computed, and if it lies within a threshold, the point is considered to be part of the plane. However, due to the discretization of the point set, there would be gaps in the computed set, which needs to be smoothed out in a post-processing stage. Typical examples are shown in Figs. 6.20 and 6.21 to demonstrate this effect. In Figs. 6.20 and 6.21, a set of 3-D objects and their corresponding cross-sections along a plane with the normal in a direction $(1, -1, 1)$ passing through the center point of the voxel array are displayed, respectively. The size of the synthetically generated 3-D objects of Fig. 6.20 is $128 \times 128 \times 128$. The cross-section images computed from the voxel representation are marred by the presence of empty pockets, which should have been otherwise occupied by an object point.

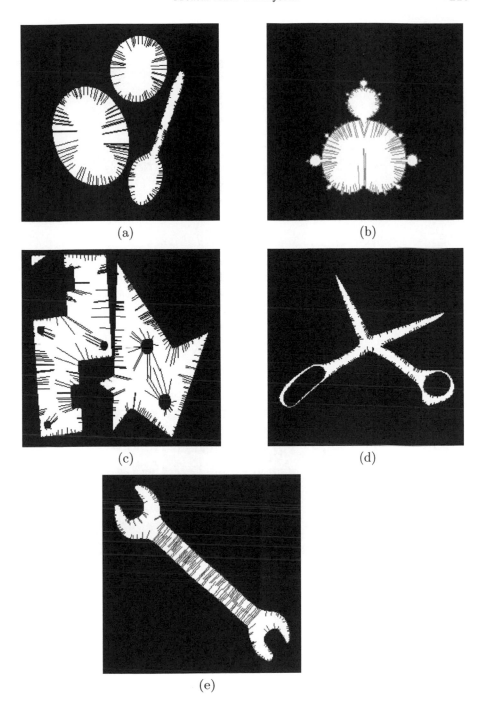

FIGURE 6.19: Normal maps of different images using digital octagonal distance {112}.

Reprinted from *Pattern Recognition Letters*, 23(2002), J. Mukherjee et al., Use of Medial Axis Transforms for Computing Normals at Boundary Points, 1649–1656, Copyright (2002), with permission from Elsevier.

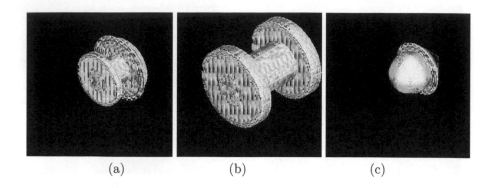

FIGURE 6.20: Objects in voxel representation displayed with shading: (a) DUMBCONE, (b) DUMBLE, and (c) SPHCONE.

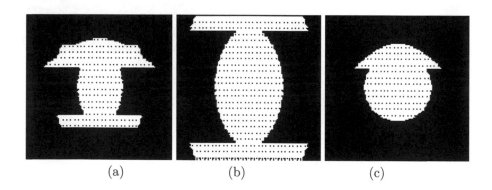

FIGURE 6.21: Cross-sections obtained by intersecting a plane passing through the center and with a normal along the direction $(-1, 1, -1)$ with the objects in voxel representation: (a) DUMBCONE, (b) DUMBLE, and (c) SPHCONE.

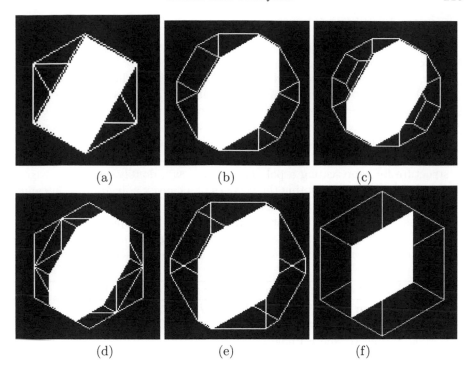

(a) (b) (c)

(d) (e) (f)

FIGURE 6.22: Intersections of digital spheres with a plane for octagonal metrics: (a) {1}, (b) {12}, (c) {112}, (d) {113}, (e) {122}, and (f) {3}.

Reprinted from *Pattern Recognition Letters*, 21(2000), J. Mukherjee et al., Fast Computation of Cross-Sections of 3-D Objects from Their Medial Axis Transforms, 605–613, Copyright (2000), with permission from Elsevier.

6.6.1 Computation with MAT

Like geometric transformation, computation with a MAT of a 3-D object could be carried out by computing the intersection of the given plane with individual medial spheres. The union of all these intersections would result in forming the cross-section along that plane. As the digital spheres are convex polytopes, the intersections with a plane form a convex polygon. The vertices of these polygons are the intersecting points of edges of the polyhedra with the plane. In Fig. 6.22, typical examples of these intersections of digital spheres of different octagonal metrics with a plane are shown. Once the intersecting points with edges of the polyhedra are computed, the convex polygon could be obtained by carrying out Graham's scan [169] through them. In this scan, the vertices are angularly sorted to provide a *star polygon* [169], which is convex in this context.

6.6.2 The Algorithm

For computing the cross-sections of 3-D objects with a given plane P, the objects are rotated in such a way that the normal of the plane is aligned with the z-axis of the coordinate space. For applying this transformation to a 3-D object represented by MAT, computation is carried out only with the vertices of its medial spheres. Next, the intersection points between edges of a medial sphere and the cross-sectional plane (i.e., the plane whose normal is now along the z-axis) are obtained. Finally, the intersecting convex polygon is constructed by angularly sorting these points. We may use another data structure for representing a polyhedra (such as a doubly connected edge list (DCEL) [169]) for computing the vertices of a convex polygon by a linear scan (of the number of faces and edges of the polyhedra). However, as the number of edges are small and constant for an individual MAT representation, the cost due to sorting could be taken as constant. The algorithm for the computation of an intersection of a digital sphere with a plane is described below.

Algorithm 14: Computation of Cross-Sections of 3-D Objects

Algorithm: *Cross_Sectioning_Using_MAT* (CSUM)
Input: The MAT M of an object, the digital distance function used in computing M, and a Plane P.
Output: A Convex polygon

1. For each medial sphere $s \in M$ with its center as o_s and radius as r_s, compute the following:

   1a. Compute the edges and the vertices of s in the 3-D coordinate system.

   1b. Apply the transformation to the vertices of the polyhedron so that the cross-sectional plane P lies along the xy plane.

   1c. For each edge, compute the intersection point with xy ($z = 0$) plane.

   1d. Perform angular sorting of the intersection points about their centroid. The sequence provides the vertex sequence of the convex polygon

End *Cross_Sectioning_Using_MAT* (CSUM)

Since the intersection of a plane and a convex polytope is a convex polygon, in Step 1d of the above algorithm, it is sufficient to sort the intersecting points by their angles formed at their centroid with any arbitrary reference axis. This computational step is a special case of Graham's convex hull computation [169]. As a medial sphere is represented in the continuous domain (as a convex polyhedra), the intersecting polygon also represents an area in the continuous

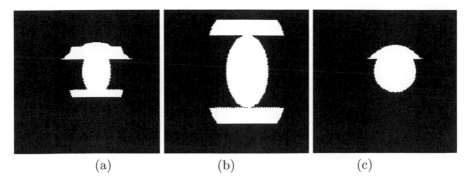

(a)   (b)   (c)

FIGURE 6.23: Cross-sections obtained by intersecting a plane passing through the center and with a normal along the direction $(-1, 1, -1)$ with the objects using MAT representation from {3}: (a) DUMBCONE, (b) DUM-BLE, and (c) SPHCONE.

domain. Hence, the points, which are missed in the cross-sections computed from voxel representation, are covered by the algorithm CSUM. This feature improves the quality of the results (see Figs. 6.23).

The time complexity of the voxel-based algorithm for computing the cross-sections of the 3-D voxel data of size $n \times n \times n$ is $O(n^3)$. But the algorithm $CSUM$ takes linear time in the number of medial spheres. As the number of medial spheres in the MAT is much less than the number of voxel data, the computation time is sufficiently reduced. In this case also, it is advisable to use the distance function, which usually provides less number of medial spheres. In [152], it was shown experimentally that the octagonal distance {3} yields the fewest medial spheres in most cases in the family of octagonal distances. It has this feature, because the volume of a medial sphere (in the form of a cube) is the largest in the family. It has also the least number of edges (12 in number), thus saving time in the computation of intersecting points with edges of medial spheres. Considering the above two properties, it is recommended to use {3} as the distance function for deriving MATs of objects to compute cross-sections. To demonstrate the varying numbers of medial spheres for different distance functions, a few typical examples (from [152]) are shown in Table 6.4. We observe that {3} does not necessarily provide the least number of medial spheres in all objects. The number also depends on the shape of the object. The other interesting fact, we should note here, is that the quality of the result does not depend on the choice of an octagonal distance. In this case, the results are almost the same with those obtained from voxel representation after the inclusion of missing or unmapped points [152].

TABLE 6.4: Number of medial spheres for different objects.

Distance Function	Dumble	Dumb-Cone	Sph-Cone
{1}	31947	16356	4695
{112}	26995	15818	4652
{113}	25371	15696	4963
{12}	25253	15378	4509
{122}	23509	15278	4678
{123}	21813	14964	4625
{13}	22877	15046	4933
{133}	19941	14816	5037
{2}	21305	14210	5179
{223}	19825	13956	4961
{23}	19289	13646	4929
{233}	18025	13620	5107
{3}	15977	13020	5331

6.6.3 Using Euclidean Approximation

We know that computation of the intersection with a Euclidean sphere and the xy-plane is trivial. Suppose (t_x, t_y, t_z) be the center of an Euclidean sphere of radius r. Its intersection with a plane parallel to the xy-plane at $z = k$, is given by a Euclidean circle, whose center is at (t_x, t_y) and radius is $\sqrt{r^2 - (k - t_z)^2}$. Hence, for computing the intersection with Euclidean MAT, in the step $1a$ of the algorithm, it is required only to transform its center instead of vertices of a polyhedron. Moreover, there is no need to compute intersecting points and form a convex polygon from them in Steps 1b to 1d of the algorithm. It is sufficient to note the center and radius of the intersecting circle as discussed above. The union of all such circles from the medial spheres results in the formation of a cross-section of the objects. The same concept can be extended to MATs of digital distances, assuming their spheres as Euclidean. Naturally, the distances whose disks are closer to Euclidean disks, should provide better results in this computation. A measure of the quality of approximation is used in [152], which is discussed below. Let $S_d(i, j)$ and $S_e(i, j)$ be the cross-section images computed from digital medial spheres using the algorithm CSUM and their Euclidean approximations as discussed in this section, respectively. We define an approximation error E_{csect} of the computation as follows:

$$E_{csect} = \frac{\Sigma_i \Sigma_j e(i, j)}{\Sigma_i \Sigma_j S_d(i, j)} \times 100\% \tag{6.6}$$

where $e(i, j) = \begin{cases} 1 & \text{if } S_d(i, j) \neq S_e(i, j) \\ 0, & \text{otherwise.} \end{cases}$

In Table 6.5, typical values of E_{csect} for computation of cross-sections using Euclidean approximation are reported from [152]. We note that for some distance functions like {113}, {112}, and {12}, the values are small. In partic-

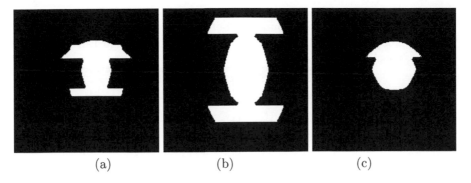

(a)   (b)   (c)

FIGURE 6.24: Cross-sections obtained by intersecting a plane passing through the center and with a normal along the direction $(-1, 1, -1)$ with the objects using Euclidean approximation of medial spheres of MAT representation from {113}: (a) DUMBCONE, (b) DUMBLE, and (c) SPHCONE.

Reprinted from *Pattern Recognition Letters*, 21(2000), J. Mukherjee et al., Fast Computation of Cross-Sections of 3-D Objects from Their Medial Axis Transforms, 605–613, Copyright (2000), with permission from Elsevier."

ular, for the distance {113} these values are reported to be within 5% in most cases. In Fig. 6.24, typical results obtained from Euclidean approximation are shown using the MAT obtained from the distance function {113}. The results are quite similar to those shown in Fig. 6.23.

6.7 Shading of 3-D Objects

The MAT is also useful for rendering 3-D objects, and in [151] it is reported to be the better representation for volume rendering compared to the schemes using an octree [186] or voxels. From the MAT of an object, each medial sphere is rendered independently. The process may be aided by the z-buffering features [84] of the graphics hardware environment, so that the hidden surface elimination is managed. For rendering a medial sphere, the shaded color values of each face of the polyhedron are computed following different shading interpolation techniques, such as flat rendering, and Goraud's [93] and Phong's [164] interpolation techniques. For applying the shading interpolation, it is required to compute the normals at the faces and vertices of the polyhedron. Computation of normals at faces is trivial, as it is the vector formed by the centers of the face and the sphere, respectively. Similarly, the direction of the normal at a vertex is simply given by the vector from the center of the sphere and the vertex itself. The technique may be further refined by approximating each digital sphere as Euclidean and the shading is performed as it is done for

TABLE 6.5: % Error computation between the slices obtained from MSR using a digital sphere and the MSR using an approximated Euclidean sphere for cross-section plane (-1,1,-1).

Distance Functions	Dumbel	Dumb-Cone	Sph-Cone
{1}	11.24	21.15	32.01
{112}	4.14	6.48	6.99
{113}	**3.88**	**4.42**	4.46
{12}	5.18	5.34	**3.80**
{122}	8.02	6.40	5.89
{123}	9.74	7.56	7.98
{13}	5.98	5.06	8.40
{133}	9.79	8.42	13.70
{2}	16.14	10.46	17.99
{223}	16.07	11.10	16.66
{23}	16.02	11.47	15.42
{233}	16.31	11.95	16.87
{3}	17.36	12.19	19.82

Reprinted from *Pattern Recognition Letters*, 21(2000), J. Mukherjee et al., Fast Computation of Cross-Sections of 3-D Objects from Their Medial Axis Transforms, 605–613, Copyright (2000), with permission from Elsevier.

a sphere using conventional techniques. Once again in [151], it was reported that MATs obtained from distances such {113}, {123}, and {13} are better in approximating such rendering results. Typical results on volume rendering of objects with known geometry, such as CYLINDER, CONE, and SPHERE, using diffrent representations are shown in Fig. 6.25. The representations under consideration are voxel, octree, and MAT of the distance function {123}. For rendering voxels, surfaces of each unit cube of the 3-D array are rendered independently. Similarly, the leaf nodes of an octree represent a cubic box. Hence, they are shaded following the same principle. The shading was carried out assuming an illumination from parallel incident rays along the direction $(-1, -1, 0)$. In all the cases, all other shading parameters, such as reflection coefficients of surfaces, shading interpolation techniques, etc., are kept the same. For a better perception of the quality of shading, benchmark shaded images are also rendered from an analytical boundary surface representation of these objects. In this representations, the algebraic expressions for representing a cylinder, a cone, and a sphere are used. It could be easily observed that the quality of shaded images using MAT is much better than the other two representation schemes.

6.8   Summary

Medial Axis Transform (MAT) is a useful representation for 2-D and 3-D objects. It exploits the advantage of shape representation in terms of overlapping circles and spheres in 2-D and 3-D, respectively. In the discrete space,

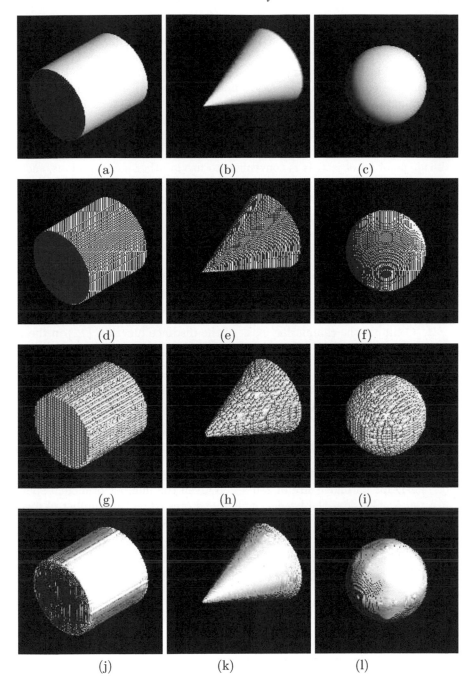

FIGURE 6.25: Rendering of CYLINDER, CONE, and SPHERE: (a)-(c) analytical boundary surface representation, (d)-(f) voxel representation, (g)-(i) octree representation, and (j)-(l) MAT using {123}.

Reprinted from *Pattern Recognition Letters*, 20(1999), J. Mukherjee et al., Discrete Shading of Three-Dimensional Objects from Medial Axis Transform, 1533–1544, Copyright (1999), with permission from Elsevier."

shapes of digital disks are different from their Euclidean counterparts. But there are distance functions that provide good approximations of Euclidean disks. For example, digital circles defined using distance functions like {112} and {1112} in 2-D, are closer to Euclidean circles. Similarly, in 3-D, the {113} distance function provides a good approximation of a Euclidean sphere. As the computation of MAT from digital distances is simpler than techniques using the Euclidean metric, digital distance transforms are widely used in this case. There are several applications of MAT in the processing and shape analysis of images. It could be used in obtaining a skeletal representation of the shape of an object. It is convenient to use in other computations, such as geometric transformation, computation of boundary normals of 2-D objects, computation of cross-sections of 3-D objects, and volume rendering of 3-D objects.

Exercises

1. Compute the chamfering mask of the distance function {12} in the 2-D space. Can you design a chamfering mask for the inverse square root weighted t-cost function (refer to Section 2.4.3 of Chapter 2)? Justify.

2. Find the maximum relative error in percentage for the Euclidean distance transform computed using the 8-SED algorithm.

3. Explain why MAT has redundancy in representing an object. Suggest a method to reduce this redundancy. Find the complexity of the algorithm.

4. How many bytes are required for storing the MAT of a binary object in 2-D for an image size of 256×256 given the number of medial spheres is N. Consider the header of the file stores to be the number of medial spheres in two bytes. Design an efficient scheme for representing the MAT that would require fewer bytes than its simple storage scheme. Implement the scheme using any programming language and find the average percentage of reduction of storage size by experimenting with a number of binary images of size 256×256. Out of the set of octagonal distances of neighborhood length less than 4, which distance function would be efficient in this respect. Justify and verify experimentally.

5. Suppose the volume of the sphere of a distance function $d_1(\cdot)$ in 3-D with the radius r is greater than the volume of a sphere of the same radius with another distance function $d_2(\cdot)$. Will the number of medial spheres of an object using $d_1(\cdot)$ be smaller than that using $d_2(\cdot)$? Justify.

6. Discuss the merits and demerits of using MAT for computing-skeleton of binary objects.

7. What is the overhead of computing normals at boundary points of 2-D objects using MAT? What is its advantage compared to regression-based analytical techniques? Modify the algorithm of the computation of normal using MAT so that all the operations are performed using integer arithmetics only.

8. Design an algorithm for contour-following and boundary-curve segmentation using the similar concept of normal computation at boundary points using the MAT in 2-D.

9. Write a program for computing the intersection of a plane and a digital sphere of radius R for the octagonal distance {113}. Compute the area of intersection.

10. Out of the set of octagonal distances in 3-D with the length of period of the neighborhood sequence being less than 4, which distance function should be used to compute the cross-section of a 3-D object using MAT? Justify.

11. Design a parallel volume-rendering algorithm in a suitable computing environment using the MAT representation of an object.

12. The volume-rendering algorithm using MAT presented in this text does not use explicit computation of surface normals at boundary points. Design an algorithm that computes surface normals at points of the boundary surface of a 3-D object from its MAT. *Hints: Extend the 2-D algorithm of computation of normals at boundary points using MAT.*

Chapter 7

Modeling of a Voxelated Surface

Modeling of 3-D objects and surfaces is the prerequisite of many applications in computer graphics and computer vision. It should follow a scientific or mathematical representation that conforms to optimal storage and efficient computation. Modeling is primarily done in two broad categories: one for regular geometric objects (polyhedra, platonic solids, etc.) or well-defined mathematical surfaces, and another for real-world objects (sculptures, day-to-day objects, living organisms, etc.). See Fig. 7.1 for some typical examples of 3-D models with triangulated surfaces.

In order to define the shape of a polyhedral object, a polygon or wireframe mesh is used. Such a mesh is a collection of vertices, edges, and faces. The faces usually consist of triangles, but may also consist of quadrilaterals or other simple convex polygons so as to simplify the rendering in 3-D computer graphics, but depending on need and application, concave polygons, orthogonal polygons, or polygons with holes are also sometimes used.

In the domain of digital geometry, a surface is expressed as a set of voxels. To define the voxelated surface, in this chapter we expand some preliminary concepts introduced in Chapter 1. A *3-D digital object* A is defined as a finite subset of $\mathbb{Z}^3$, with all its constituent points (i.e., voxels) having integer coordinates and connected in 26-neighborhood. Each voxel is equivalent to a *3-cell* [115] centered at the concerned integer point (Fig. 7.2(a)). Note that a

FIGURE 7.1: Some typical examples of 3-D models (triangular faces shown randomly colored). From top-left to bottom-right: `icosahedron` (20 identical equilateral triangular faces), `teapot`, `turbine`, `cow`, `pickup-van`. (See color insert.)

3-cell is a unit cube consisting of six unit squares called *2-cells*, twelve unit edges called *1-cells*, and eight vertices called *0-cells* [115].

In this chapter, we define the *isothetic distance* between two points $p(x_1, y_1, z_1)$ and $q(x_2, y_2, z_2)$ as the Minkowski norm L_∞ given by $d_\top(p, q) = \max\{|x_1 - x_2|, |y_1 - y_2|, |z_1 - z_2|\}$. This metric, along with other distance metrics, are discussed in detail in Chapter 2. The (isothetic) distance of a point p from an object A is therefore defined as $d_\top(p, A) = \min\{d_\top(p, q) : q \in A\}$, and the distance between two connected components A_1 and A_2 is $d_\top(A_1, A_2) = \min\{d_\top(p, q) : p \in A_1, q \in A_2\}$.

In Chapter 1 (refer to Section 1.2), we discussed the digital grid and presented it as a point set. In this chapter, we provide a fuller exposition of this grid in 3-D by considering it as a cellular complex of cubic cells or 3-cells. Some of the basic definitions related to this complex are introduced here. In this representation, since the isothetic cover of an object is obtained with respect to an underlying grid in 3-D digital space, we define it as follows. A *digital grid* in 3-D consists of three orthogonal sets of equispaced grid lines,

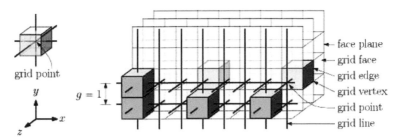

FIGURE 7.2: 3-D digital space and 26N [115]. Left: A 3-cell and its corresponding grid point. Right: Three pairs of α-adjacent 3-cells for $\alpha \in \{0, 1, 2\}$, $\alpha \in \{0, 1\}$, and $\alpha = 0$ (from left to right). The 3-cells in each of these three pairs are connected in 26N.

$\mathbb{G}_{yz}$, $\mathbb{G}_{zx}$, and $\mathbb{G}_{xy}$, such that

$$
\mathbb{G}_{yz} = \{
$$
$$
\vdots
$$
$$
\ldots, l_x(j - g, k - g), l_x(j, k - g), l_x(j + g, k - g), \ldots
$$
$$
\ldots, l_x(j - g, k), l_x(j, k), l_x(j + g, k), \ldots
$$
$$
\ldots, l_x(j - g, k + g), l_x(j, k + g), l_x(j + g, k + g), \ldots
$$
$$
\vdots
$$
$$
\} \subset \mathbb{Z}^3.
$$

Similarly, $\mathbb{G}_{zx}$ and $\mathbb{G}_{xy}$ can be represented in terms of l_y and l_z for a grid size $g \in \mathbb{Z}^+$. Here, $l_x(j, k) = \{(x, j, k) : x \in \mathbb{R}\}$, $l_y(i, k) = \{(i, y, k) : y \in \mathbb{R}\}$, and $l_z(i, j) = \{(i, j, z) : z \in \mathbb{R}\}$ denote the *grid lines* (Fig. 7.2) along x-, y-, and z-axes, respectively, where i, j, and k are integral multiples of g. The three orthogonal lines $l_x(j, k)$, $l_y(i, k)$, and $l_z(i, j)$ intersect at the point $(i, j, k) \in \mathbb{Z}^3$, which is called a *grid point*; a shift of $(\pm 0.5g, \pm 0.5g, \pm 0.5g)$ with respect to a grid point designates a *grid vertex*, and a pair of adjacent grid vertices defines a *grid edge* [115] (Fig. 7.2).

A grid, as defined above, is characterized by several elements. These are shown in Fig. 7.2 and explained next. A *unit grid cube* (UGC) is a (closed) cube of length g whose vertices are *grid vertices*, edges constituted by *grid edges*, and faces constituted by *grid faces*. Each face of a UGC lies on a *face plane* (henceforth referred as a *UGC-face*), which is parallel to one of three coordinate planes. Clearly, each *face plane*, containing coplanar UGC-faces, is at a distance of an integer multiple of g from its parallel coordinate plane. A UGC-face, f_k, has two adjacent UGCs, U_1 and U_2, such that $f_k = U_1 \cap U_2$. The *interior* of a UGC is the open cubical region lying strictly inside the UGC. A smaller (larger) value of g implies a denser (sparser) grid. For $g = 1$, the grid $\mathbb{G}$ essentially corresponds to $\mathbb{Z}^3$. As each grid point p is equivalent to a

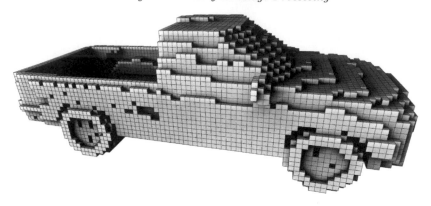

FIGURE 7.3: Voxelation of `pickup-van` shown in Fig. 7.1. (See color insert.)

3-cell c_p centered at p for $g = 1$, each face of c_p is a *grid face* lying on a *face plane*, which is parallel to one of the three coordinate planes. It may be noted that a UGC consists of $g \times g \times g$ voxels and each UGC-face consists of $g \times g$ voxels.

Two 3-cells c_1 and c_2 having centers at (x_1, y_1, z_1) and (x_2, y_2, z_2), respectively, are α-*adjacent* if and only if $c_1 \neq c_2$ and $c_1 \cap c_2$ contains an α-cell ($\alpha \in \{0,1,2\}$) [115]. The grid points (x_1, y_1, z_1) and (x_2, y_2, z_2) are in a k-*neighborhood* where $k \in \{6, 18, 26\}$; $k = 6$ denotes *2-adjacency*, $k = 18$ denotes *1-adjacency* and $k = 26$ denotes *0-adjacency* with respect to the cell model. As introduced in Chapter 1 and Chapter 2, the set of 6-adjacent points of a point $p(x_1, y_1, z_1)$ is called the 6-neighborhood of p, which is given by $N_6(p)$. The set of 18-adjacent points is called 18-neighborhood given by $N_{18}(p)$. The set of 26-adjacent points is called 26-neighborhood given by $N_{26}(p)$. Each point in $N_k(p)$ is said to be a k-neighbor of p where $k \in \{6, 18, 26\}$. Two points p and q are k-connected in a digital set $A \subset \mathbb{Z}^3$ if and only if there exists a sequence $\langle p := p_0, p_1, ..., p_n := q \rangle \subseteq A$ such that $p_i \in N_k(p_{i-1})$ for $1 \leqslant i \leqslant n$. For any point $p \in A$, the set of points that are k-connected to $p \in A$ is called a k-connected component of A. In other words, a k-connected component of a nonempty set $A \subseteq \mathbb{Z}^3$ is a maximal k-connected set of A. If A has only one connected component, it is called a k-connected set.

7.1   Voxelation and Approximation of 3-D Surface

Voxelation (also called *voxelization* or *3-D scan conversion*) has been addressed in the literature since the 1980s [47, 48, 106, 107, 108, 145]. Recently, several interesting works have been done on *discrete volume polyhedrization*, which deals with construction of a polyhedron P enclosing a set of voxels S

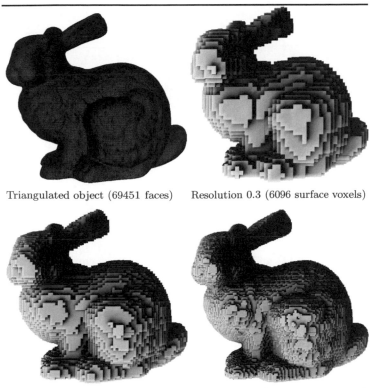

Triangulated object (69451 faces)   Resolution 0.3 (6096 surface voxels)

Resolution 0.5 (15488 surface voxels) Resolution 1.0 (58560 surface voxels)

FIGURE 7.4: Voxelation of bunny at different resolutions. (See color insert.)

[31, 45]. In general, these algorithms follow the principle of *marching cube* that triangulates a 3-D iso-surface based on cubic cell decomposition, their local configurations, and displacement [45, 136, 158]. Approaches to reduce computational time have been suggested by several researchers [91, 135, 212]. Nevertheless, the algorithm can be enhanced to handle multi-resolution rectilinear data [211] and data sets in higher dimensional space [12, 11]. More recently, the problem of approximating a 3-D digital object by a polyhedral surface has also been investigated by several researchers [189, 190]. It has been proved that obtaining an optimal polyhedral reconstruction with trapezoidal or triangular facets is strongly NP-hard [33]. Several other methods have been used for digitization of 3-D objects including *majority interpolation*, a modification of the *marching cubes algorithm, ball union, tri-linear interpolation, bcc and fcc grids*, etc. [199, 198, 200].

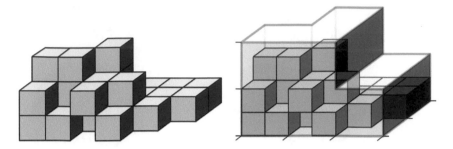

FIGURE 7.5: An object A in $\mathbb{Z}^3$ (left) and its outer isothetic cover for $g = 2$ (right). (See color insert.)

7.1.1   3-D Isothetic Covers

Given a grid $\mathbb{G}$ and a 3-D digital object A, we can approximate A by its *3-D isothetic cover*, which is defined as the minimum-volume isothetic polytope $\overline{P}_{\mathbb{G}}(A)$ that contains A (Fig. 7.5). Mathematically, the following conditions are satisfied:

- $A \subseteq \overline{\mathbf{P}}_{\mathbb{G}}(A)$

- for each $p \in \overline{P}_{\mathbb{G}}(A)$, $0 \leqslant d_\top(p, A) < g$

Here, g denotes the grid size, and $\overline{\mathbf{P}}_{\mathbb{G}}(A)$ denotes the entire cover including its surface $\overline{P}_{\mathbb{G}}(A)$ and interior region. The first condition implies that each point of A lies on or inside $\overline{\mathbf{P}}_{\mathbb{G}}(A)$, and the second condition addresses the minimum volume of $\overline{\mathbf{P}}_{\mathbb{G}}(A)$. Also note that an *isothetic polytope* is a polytope with all its vertices as grid vertices, all its edges made of grid edges, and all its faces lying on face planes. Each face of an isothetic polytope is an isothetic polygon whose alternate edges are isothetic and constituted by the grid edges of $\mathbb{G}$.

The algorithm to construct the outer isothetic cover $\overline{P}_{\mathbb{G}}(A)$ of A is given in detail in [105]. From a well-defined *start vertex*, eligible UGC-faces are determined by BFS (breadth first search) [49], which are stored in a *doubly connected edge list* (DCEL) [9]. The eligible UGC-faces, which are coplanar and contiguous, are later merged into isothetic polygons of maximal size, parallel to the yz-, zx-, and xy-planes, which together constitute the isothetic cover. The DCEL stores topological information in the form of a vertex list, edge list, and face list. Each vertex of the polytope is stored only once in the vertex list. The edge list is maintained as sets of four half-edges per face of UGC. Each half-edge is represented by its source vertex, and four consecutive half-edges are assigned the face number that denotes the face to which the edges belong. The edge list also records the pairing of all half-edges. The *id* (a unique number) of a half-edge, considered as the first half-edge corresponding to a face, and the plane (yz, zx, or xy) to which the face is parallel, is stored in the face list for each face of the polytope. From this half-edge, the face can be traversed by referring to the previous and next pointers in the edge list [9, 169].

The start vertex $v_s(i_s, j_s, k_s)$ of $\overline{P}_{\mathbb{G}}(A)$ is obtained from $p_0(i_0, j_0, k_0)$, the top-left-front point of A. The point p_0 is the top-left-front point of A if and only if for each other point $(i, j, k) \in A$, $k \leqslant k_0$ and $j \leqslant j_0$ and $i > i_0$. Then it can be shown that

$$i_s = (\lceil i_0/g \rceil - 1) \times g, \; j_s = (\lfloor j_0/g \rfloor + 1) \times g, \; k_s = (\lfloor k_0/g \rfloor + 1) \times g. \quad (7.1)$$

A UGC-face f_k is a part of the outer cover if and only if exactly one of its two adjacent UGCs has object containment. Hence, it can be shown that if p is a point lying on $\overline{P}_{\mathbb{G}}(A)$, then $0 \leqslant d_{\top}(p, A) < g$. Using this property, we can construct the set of UGC-faces comprising the surface of $\overline{P}_{\mathbb{G}}(A)$. In order to organize these UGC-faces as a set of isothetic polygons, the DCEL is constructed. For each face in face list F, we maintain its id, the id of its incident edge, and the corresponding coordinate plane to which it is parallel. The edge information in edge list E includes the edge id, the face on which it is incident, its source vertex, ids of its next and previous edges, and its paired half-edge. Vertex information in vertex list V consists of the vertex id corresponding to each vertex.

The start vertex v_s is shared by three UGC-faces, parallel to the yz-, zx-, and xy-planes. Out of these, the one parallel to the xy-plane, i.e., the front face is enqueued in a queue Q assigning it a unique id. The UGC-faces are traversed by BFS [49] whereby a face f_i is dequeued from Q and each face f_k incident on the edges $e_{ij} \in f_i$ is checked for eligibility. The vertices of f_i are enlisted in V, avoiding repetition. Consequently, the source vertex of $e_{ij} \in f_i$, and its next and previous edges are recorded in E, and f_i is inserted in F along with e_{ij} as its starting edge and the plane to which f_i is parallel. If f_k is an eligible face, it is assigned a unique id and enqueued in Q if it has not already been enqueued. Otherwise, if f_k is coplanar with f_i and has been previously enqueued and dequeued, then f_k can be merged with f_i by deleting their common edge e_{ij}. Hence, $pair[e_{ij}]$ in E is set to the id of the edge $\bar{e}_{ij} \in f_k$ whose start and end vertices are the respective end and start vertices of e_{ij}. Once dequeued, f_i is not suitable to be enqueued again, and another face is dequeued from Q. The procedure continues until Q is empty.

While merging the UGC-faces, their co-planarity and adjacency are tested through half-edge information in the edge list E. The basic steps include deletion (from E) of (half-)edges shared by two coplanar-adjacent faces, updating face information in F, and readjusting the pointers of the edges associated with the merged faces. A brief outline of the algorithm to find the 3-D isothetic cover of an object A embedded in grid $\mathbb{G}$ is given in Iso-Cover-3D (Algorithm 15).

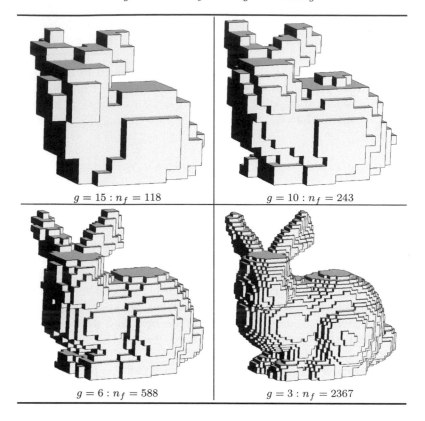

$g = 15 : n_f = 118$ $g = 10 : n_f = 243$

$g = 6 : n_f = 588$ $g = 3 : n_f = 2367$

FIGURE 7.6: Isothetic covers of `Stanford Bunny` for different grid sizes; n_f = number of cover faces defined as isothetic polygons. (See color insert.)

7.1.2 Test Results

A set of results by the algorithm on the object `Stanford Bunny`[1] is presented in Fig. 7.6. It is quite apparent that as the grid size g is decreased, a tighter approximation of the object is obtained; consequently, the number of vertices, edges, and faces increase in a quadratic manner. In 3-D, at most eight UGCs can be incident at a grid vertex. Depending on the object occupancy of these UGCs, it is decided whether the grid vertex is a vertex of the isothetic cover. For instance, if all eight UGCs have object occupancy, then the grid vertex lies within the object, and hence it is not a vertex of the cover. Based on the object occupancy and the nature of arrangement of the occupied UGCs, the vertices can be classified into five categories: Type 1, Type 3, Type 4, Type 5, and Type 7 [1], the denotation corresponding to the number

[1]Source: http://graphics.stanford.edu/data/3Dscanrep.

Algorithm 15: Major Steps of the Algorithm to Find the 3-D Isothetic Cover of an Object A Embedded in Grid $\mathbb{G}$.

Algorithm ISO-COVER-3D($A, \mathbb{G}$)
01. Find the start vertex, v_s (Eq. 7.1).
02. Initialize an empty queue, Q.
03. Determine the front face f_s of the UGC having v_s as a vertex.
04. Enqueue f_s in Q.
05. **while** Q is not empty
06.   Dequeue face f_i from Q.
07.   Insert the vertex, edge, and face information of f_i into DCEL.
08.   **for** each 'eligible face' f_k incident on some edge of f_i
09.   **if** f_k is not already enqueued in Q
10.   **then** enqueue f_k in Q.
11.   **else if** f_k is dequeued **and** f_i and f_k are coplanar
12.   **then** set half-edges $e_{ij} \in f_i$ and $\bar{e}_{ij} \in f_k$ as pairs.
13. Merge the contiguous coplanar faces that are parallel to each coordinate plane.

---

of neighboring UGCs occupied by object voxels. The complexity of the structure of the vertices increases with the category as mentioned, i.e., Type 7 is more complex in structure than Type 1. For a particular grid size, the total number of vertices of a higher category increases with the complexity of the object topology.

7.2 Voxelation of Surface of Revolution

The traditional millennium-old artistry of wheel throwing for creation of potteries has gained an enduring popularity with today's state-of-the-art ceramic technology [35, 213].[2] With the proliferation of digitization techniques in our computerized society, the digital creation and artistic visualization of potteries is a call of the day. The existing graphic tools are mostly based on complex trigonometric procedures involving computation in real space, which have to be tuned properly to suit the discrete nature of the 3-D digital space [90, 99]. The method of *circularly sweeping* a shape/polygon/polyline/generating curve is rotated about the axis of revolution in discrete steps, which requires a discrete approximation, and choosing k equi-spaced angles about the y-axis re-

---

[2]From a historic point of interest, a recreation of wheel-thrown potteries may be seen at Conner Prairie (http://www.connerprairie.org). Conner Prairie is an interactive history park, or living history museum, located in Fishers, Indiana, USA. With a noble objective to preserve the William Conner home, which is listed on the National Register of Historic Places, it recreates part of life in Indiana in the 19th century on the White River.

quires the following transformation matrices for a real-geometric realization:

$$
M_i = \begin{pmatrix} \cos\theta_i & 0 & \sin\theta_i & 0 \\ 0 & 1 & 0 & 0 \\ -\sin\theta_i & 0 & \cos\theta_i & 0 \\ 0 & 0 & 0 & 1 \end{pmatrix}
$$

where, $\theta_i = 2\pi i/k, i = 0, 1, \ldots, k-1$ [99]. Clearly, in order to generate circles of varying radii describing the surface of revolution, the number of steps, k, becomes a concerning issue.

In the domain of digital geometry, the idea of *digital wheel-throwing* has recently been proposed in [126, 125]. The algorithm works purely in the digital domain and banks on a few primitive integer computations only. Its input is a *digital generatrix*, taken in the form of an *irreducible digital curve segment* (Sec. 4.1). The digital surface produced from the digital generatrix by digital wheel-throwing is both connected and irreducible in a digital-geometric sense. This, in turn, ensures its successful rendition with a realistic finish that involves conventional processing of quad decomposition, texture mapping, illumination, etc. The method is robust and efficient, and guarantees easy implementation in Java3D^{TM} and OpenGL^{TM} with all relevant features incorporated. The rendered visualization is found to be absolutely free of any bugs or degeneracies, regardless of the zoom factor. Further, producing a monotone or a non-monotone digital surface of revolution is also feasible without destroying its digital connectivity and irreducibility by respective input of a monotone or a non-monotone digital generatrix. To create the exquisite digital products having the desired resemblance with on-the-shelf potteries in toto, a double-layered generatrix can be, therefore, supplied as the generatrix, which is its ultimate benefit. Some typical products from the software developed by Kumar et al. [125] and their corresponding statistical figures including CPU times to generate them, have been furnished in Section 7.2.6 of this book to exhibit the efficiency and elegance of the proposed technique.

7.2.1 Various Techniques

Almost all methods for modeling potteries start with an initial cylindrical piece of clay and use deformation techniques, as described in [104, 120]. The deformation is based either on shifting the individual points horizontally or on using devices that accept feedback by means of touch sensation at multiple points [207]. As a result, the user has direct control over the geometry of the shape to be constructed, which, poses more and more difficulties to represent the surface mathematically. In addition, it requires large memory for display and storage. Some of the major approaches are as follows.

Polyhedra Representation: The whole surface of a pottery is represented by a finite set of surface polygons. Its major drawback is that many polygons are often required to create a realistic feel. For example, in the interactive

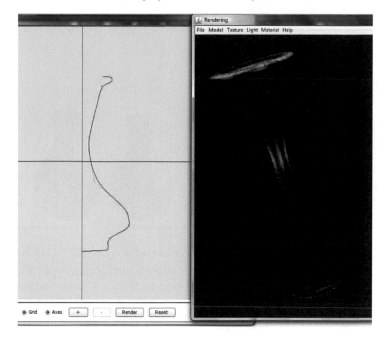

FIGURE 7.7: A snapshot of a part of the algorithm in action: The digital generatrix (shown in the left pane) and the corresponding digital surface resembling a flowerpot generated in the right pane. (See color insert.)

Reprinted from *International Journal of Arts and Technology*, **4**: 196–215, G. Kumar et al., Copyright 2011, with permission from Inderscience Publishers.

modeling technique proposed in [90], a simple teapot consists of 9244 polygons. The voxel data is converted to a polygonal surface using a marching-cubes algorithm. Although the technique is appropriate for modeling a boulder or a tooth, it cannot not create an exact surface of revolution such as a crankshaft or a pottery, since the circular property of the pottery is not utilized. Texture can be applied readily, but storing the object requires more memory.

Finite Element Method: This is the naive approach for representing any 3-D object [96]. The entire volume of the object is composed from cubes of small dimension. The shape is modified by eroding/depositing the finite elements using a suitable interface. Too many points are required for good resolution, which have to stored explicitly, thereby requiring large memory. The circular property of potteries is not utilized at all.

Cylindrical Element Method: The entire pottery volume of a deformable object is discretized with a set of thin cylinders for its boundary representation as described in [96]. Shape control is performed using collision detection between the Haptic Interface Point (HIP) and the virtual pottery. The drawback

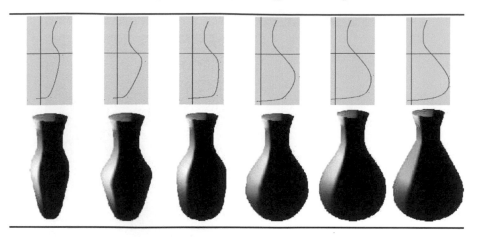

FIGURE 7.8: Simulating the local effect of a potter's hand by inserting the control points (from left to right). Other operations (e.g., deletion and repositioning of control points) are also incorporated in the algorithm. (See color insert.)

Reprinted from *International Journal of Arts and Technology*, **4**: 196–215, G. Kumar et al., Copyright 2011, with permission from Inderscience Publishers.

is that to achieve a good resolution for large objects, the number of cylinders will be large, thereby requiring a large storage space.

Circular Sector Element Method: The pottery is composed from a number of circular sector elements with small internal angle and thickness. A pottery can be deformed from outside to inside (contraction mode) or from inside to outside (expansion mode) [132]. This is done by collision detection between the individual sector elements and HIP. This method generates realistic and efficient models, but the simplicity of collisions limits the types of decorations that can be made on the surface. Moreover, for rendering several objects with high resolution, this method takes more time because of increase in elements and subsequent collisions.

We discuss an efficient technique here, in which the whole surface can be represented by a finite number of control points and the axis of rotation.

7.2.2 Digital Curves and Surfaces

A 2-D digital curve segment $\mathcal{G} := \{p_i : i = 1, 2, \ldots, n\}$ is a finite sequence of digital points in 2-D (i.e., 2-D points with integer coordinates) [115]. To ensure that $\mathcal{G}$ is *simple, irreducible,* and *open-ended,* each point $p_i(i = 2, 3, \ldots, n-1)$ should have exactly two neighbors, and p_1 and p_n should each have one, from $\mathcal{G}$. Thus, the chain code of a point $p_i \in \mathcal{G}$ with respect to its previous point $p_{i-1} \in \mathcal{G}$ is given by $c_i \in \{0, 1, 2, \ldots, 7\}$ (proposed in [86], see also Section

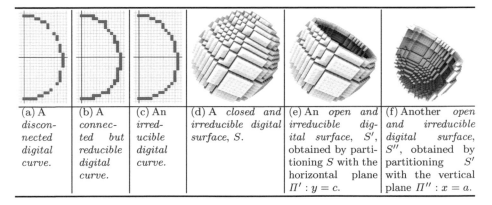

(a) A discon-nected digital curve.	(b) A connec-ted but reducible digital curve.	(c) An irred-ucible digital curve.	(d) A *closed and irreducible digital surface, S.*	(e) An *open and irreducible dig-ital surface, S', obtained by parti-tioning S with the horizontal plane $\Pi' : y = c$.*	(f) Another *open and irreducible digital surface, S'', obtained by partitioning S' with the vertical plane $\Pi'' : x = a$.*

FIGURE 7.9: An illustration of connectivity and irreducibility of digital curves and surfaces. (See color insert.)

Reprinted from *International Journal of Arts and Technology*, **4**: 196–215, G. Kumar et al., Copyright 2011, with permission from Inderscience Publishers.

1.3.3.2 of Chapter 1). In order to generate a digital surface of revolution, S, we consider a digital curve segment G as the *digital generatrix*, as shown in Fig. 7.7. The digital generatrix may be taken as input, either as a sequence of chain codes or as a sequence of control points. Given $m(\geqslant 4)$ control points as input, the digital generatrix is considered as the digital irreducible curve segment that approximates the sequence of $m - 3$ uniform non-rational cubic B-spline segments interpolating the sequence of these control points [84, 99]. The reasons for B-spline interpolation of the generatrix is its unique character-istic of having both parametric and geometric continuities, which ensure the smoothness and the optimal exactness of the fitted curve against the given set of control points. In addition, a local change (insertion/deletion/repositioning) of control point(s) has only a local effect on the shape of the digital generatrix (and a local effect on the generated surface, thereof). This, in fact, simulates the local effect of a potter's hand rolling and maneuvering the clay on his rotating wheel (Fig. 7.8).

Using the same notation discussed before, each point in $N_k(p)$ is said to be a *k-neighbor* of p ($k \in \{4, 8\}$ if $p \in \mathbb{Z}^2$ and $k \in \{6, 26\}$ if $p \in \mathbb{Z}^3$). We consider $k = 8$ and $\bar{k} = 4$ in $\mathbb{Z}^2$, and $k = 26$ and $\bar{k} = 6$ in $\mathbb{Z}^3$. If a connected digital set S is such that $\bar{S} := \mathbb{Z}^2 \smallsetminus S(\mathbb{Z}^3 \smallsetminus S)$ contains at least two points, $\bar{p} \in \bar{S}$ and $\bar{q} \in \bar{S}$, which are not $\bar{k}$-connected, then S is said possess a *hole*. A hole H is a finite and maximal subset of $\bar{S}$ such that every pair of its points are $\bar{k}$-connected and none of its points is k-connected with any point from $\bar{S} \smallsetminus H$. Clearly, S contains $h - 1(h \geqslant 1)$ holes if and only if $\bar{S}$ contains at least h points, namely, $\bar{p}_1, \bar{p}_2, \ldots, \bar{p}_h$, no two of which are $\bar{k}$-connected.

If $S \subset \mathbb{Z}^2(\mathbb{Z}^3)$ is a connected digital set containing exactly one hole, namely H, then S is said to be a *closed digital curve (surface)*, of possibly arbitrary thickness. A closed digital curve (surface) S is *irreducible* (i.e., of unit thickness) if and only if exclusion of *any* point p from S (and its inclusion in $\bar{S}$) gives rise to a $\bar{k}$-connected path from each point of H to each point of $\bar{S} \smallsetminus H$.

Let L be a real line that partitions an irreducible and closed digital curve S into two or more (k-connected) components such that all points in each component are on the same side of L or lying on L, and no two components share a common point. Then each such component of S is an *open digital curve* (irreducible segment). Partitioning of an open digital curve segment, in turn, gives rise to several open segments. Clearly, if the closed digital curve S is such that any horizontal (vertical) line L_y (L_x) always partitions S into at most two components, then S is a y-monotone (x-monotone) digital curve. Similarly, partitioning an irreducible and closed digital surface S by a plane Π produces two or more components such that all points in each component are on the same side of Π or lying on Π, and no two components share a common point. Each component obtained by partitioning S with Π is an *open digital surface* (irreducible). If the closed digital surface S is such that any plane Π_y (Π_x, Π_z) orthogonal to the y-axis (x-, z-axis) always partitions S into at most two components, then S is a y-monotone (x-, z-monotone) digital surface. It may be mentioned here that, if the generatrix is an open, irreducible, and y-monotone digital curve segment, C, then the digital surface of revolution, S, produced by the algorithm is also a y-monotone digital surface.

Figure 7.9 illustrates the notions of connectivity and irreducibility of a digital curve [115], which, in turn, produces an irreducible digital surface. Since there exists no $k(=8)$-connected path from the topmost point to the bottommost point of the digital curve in Fig. 7.9a, the concerned curve is *disconnected*. Again, in Fig. 7.9b, there exist multiple paths from the topmost point to the bottommost point, which implies that there are some redundant points in the corresponding digital curve, wherefore it is *reducible*. The set of reducible points, once eliminated, produces an *irreducible* digital curve (Fig. 7.9c). In Fig. 7.9d, a *closed and irreducible digital surface* S is shown, which contains a hole H inside it. The surface S is connected in the $k(=26)$-neighborhood as mentioned earlier. *Irreducibility* of S implies that no voxel $\bar{p} \in H$ is $\bar{k}(=6)$-connected to any voxel p lying outside S (i.e., $p \in \mathbb{Z}^3 \smallsetminus (S \cup H)$); further, if we remove *any* voxel $q \in S$ from S, then each voxel $\bar{p} \in H$ has a $\bar{k}$-connected path to each voxel $p \in \mathbb{Z}^3 \smallsetminus (S \cup H)$. Figure 7.9e shows an *open and irreducible digital surface*, namely S', which is obtained by partitioning S with a horizontal (integer) plane $\Pi' : z = c'(\in \mathbb{Z})$. Since S itself is irreducible, S' is also irreducible, as explained earlier. If we again partition S by some other orthogonal integer plane, then we again get a smaller irreducible digital surface, S'', as shown in Fig. 7.9f.

7.2.3 Algorithm to Wheel-Throw a Single Piece

We generate a wheel-thrown piece by revolving the generatrix $\mathcal{G} := \{p_i : i = 1, 2, \ldots, n\}$ about an axis of revolution given by $\alpha : \langle z = -c, x = a \rangle$, where

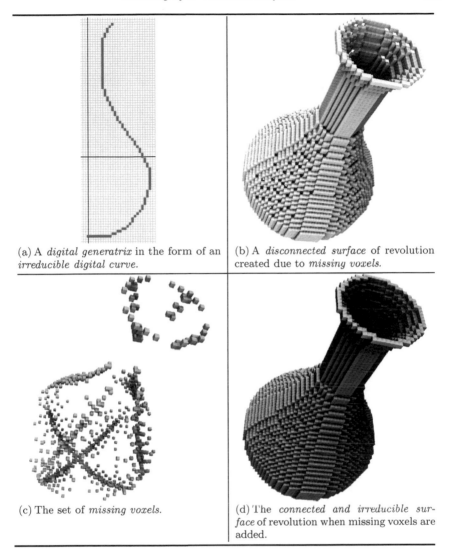

(a) A *digital generatrix* in the form of an *irreducible digital curve.*

(b) A *disconnected surface* of revolution created due to *missing voxels.*

(c) The set of *missing voxels.*

(d) The *connected and irreducible surface* of revolution when missing voxels are added.

FIGURE 7.10: How a *connected and irreducible surface* of revolution is created by fixing the missing voxels, failing which a disconnected surface would be produced. (See color insert.)

Reprinted from *International Journal of Arts and Technology*, **4**: 196–215, G. Kumar et al., Copyright 2011, with permission from Inderscience Publishers.

a and c are two positive integers. In order to achieve this, for each digital point $p_i \in \mathcal{G}$, we construct the digital circle $\mathcal{C}_1^{\mathbb{Z}}$ by revolving p_i around the specified axis of revolution, α. However, as $\mathcal{G}$ is irreducible in nature, two consecutive points p_i and p_{i+1} have their distances, namely r_i and r_{i+1}, measured from α,

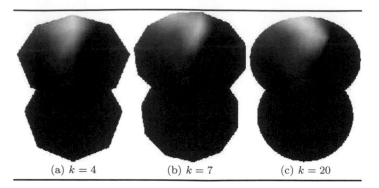

Reprinted from *International Journal of Arts and Technology*, **4**: 196–215, G. Kumar et al., Copyright 2011, with permission from Inderscience Publishers.

FIGURE 7.11: Results on quad-decomposition for a digital jug: Approximating generating digital circles by regular $2k$-gons. (See color insert.)

differing by at most unity. If their distances from α are the same, then it is easy to observe that the surface $\mathcal{C}^{\mathbb{Z}}_i \cup \mathcal{C}^{\mathbb{Z}}_{i+1}$ is digitally connected and irreducible. The problem arises when p_i and p_{i+1} have their respective distances from α differing by unity. Then there may arise some missing voxels trapped between $\mathcal{C}^{\mathbb{Z}}_i$ and $\mathcal{C}^{\mathbb{Z}}_{i+1}$, which results in digital disconnectedness in the surface $\mathcal{C}^{\mathbb{Z}}_i \cup \mathcal{C}^{\mathbb{Z}}_{i+1}$, as shown in Fig. 7.10. Detection of these missing voxels is performed to achieve a digitally connected surface, namely $S_{\mathcal{G}} := \mathcal{C}^{\mathbb{Z}}_1 \cup \mathcal{C}^{\mathbb{Z}}_2 \cup \ldots \cup \mathcal{C}^{\mathbb{Z}}_n$, as follows.

Case 1 ($r_{i+1} > r_i$): While generating the digital circle $\mathcal{C}^{\mathbb{Z}}_{i+1}$ parallel to the zx-plane corresponding to the point $p_{i+1} \in \mathcal{G}$, there is either an east (E) transition or southeast (SE) transition from the current point $q(x, y, z)$ in Octant 1 ($z \leqslant x \leqslant r_{i+1}$) [84]. If we take the respective projections, $\mathcal{C}^{\mathbb{Z}'}_i$ and $\mathcal{C}^{\mathbb{Z}'}_{i+1}$, of $\mathcal{C}^{\mathbb{Z}}_i$ and $\mathcal{C}^{\mathbb{Z}}_{i+1}$ on the zx-plane, then $\mathcal{C}^{\mathbb{Z}'}_i$ and $\mathcal{C}^{\mathbb{Z}'}_{i+1}$ become concentric with their radii differing by unity. As $r_{i+1} > r_i$, each run-length $\lambda^{(k)}_{i+1}(k \geqslant 0)$ of $\mathcal{C}^{\mathbb{Z}'}_{i+1}$ in Octant 1 is either the same as the corresponding run-length $\lambda^{(k)}_i$ of $\mathcal{C}^{\mathbb{Z}'}_i$ or greater by unity [14]. Hence, a *missing voxel* between $\mathcal{C}^{\mathbb{Z}}_i$ and $\mathcal{C}^{\mathbb{Z}}_{i+1}$ is formed, only if there is a transition toward SE (a change in run, thereof) from a point/pixel in $\mathcal{C}^{\mathbb{Z}'}_{i+1}$ giving rise to a missing pixel between $\mathcal{C}^{\mathbb{Z}'}_i$ and $\mathcal{C}^{\mathbb{Z}'}_{i+1}$. We detect such missing pixels by determining whether or not there is a "miss" during each SE transition for $\mathcal{C}^{\mathbb{Z}'}_{i+1}$. More precisely, if the point next to the current point $q(x, z) \in \mathcal{C}^{\mathbb{Z}'}_{i+1}$ in Octant 1 is $(x - 1, z + 1)$ and the point $(x - 1, z)$ does not belong to $\mathcal{C}^{\mathbb{Z}'}_i$, then we include the point $(x - 1, y, z)$ in $\mathcal{C}^{\mathbb{Z}}_{i+1}$ between $(x, y, z) \in \mathcal{C}^{\mathbb{Z}}_{i+1}$ $(x - 1, y, z + 1) \in \mathcal{C}^{\mathbb{Z}}_{i+1}$.

Case 2 ($r_{i+1} < r_i$): If the point next to the current point $q(x, z) \in \mathcal{C}^{\mathbb{Z}'}_{i+1}$ in Octant 1 is $(x - 1, z + 1)$ and the point $(x, z + 1)$ does not belong to $\mathcal{C}^{\mathbb{Z}'}_i$,

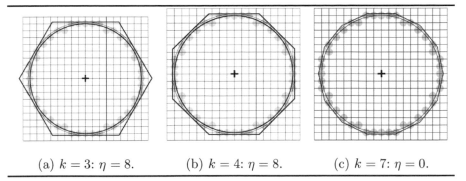

(a) $k = 3$: $\eta = 8$.　　　　(b) $k = 4$: $\eta = 8$.　　　　(c) $k = 7$: $\eta = 0$.

FIGURE 7.12: Approximate regular $2k$-gons for $k = 3$, 4, and 7 corresponding to the digital circle of radius 8. An error point is a point that lies inside the polygon but outside the digital circle [15]. Error points for $k = 4$ are given by $\{(i, j) : \{|i|, |j|\} = \{3, 8\}\}$. For other values of k, the corresponding errors are $\eta(k = 2) = 68$, $\eta(k = 3) = 18$, $\eta(k = 5) = 8$, $\eta(k = 6) = 4$, and $\eta(k \geqslant 7) = 0$.

then we include the point $(x, y, z + 1)$ in $\mathcal{C}^{\mathbb{Z}}_{i+1}$ between $(x, y, z) \in \mathcal{C}^{\mathbb{Z}}_{i+1}$ $(x - 1, y, z + 1) \in \mathcal{C}^{\mathbb{Z}}_{i+1}$.

Thus, depending on Case 1 and Case 2, we detect and include the missing voxels for $S_{\mathcal{G}}$. Finally, the 8-axis symmetry of digital circles [84] is used to complete the digitally connected and irreducible surface of revolution from Octant 1 (Fig. 7.10).

7.2.4　Number-Theoretic Approach

In this approach, certain number-theoretic concepts have been used in designing an efficient algorithm for the construction of the digital surface of revolution. These number-theoretic concepts are discussed in detail in [14]. The number-theoretic algorithm is particularly useful (apropos its speed and efficiency) when we have to construct a large number of digital circles that are mostly of high radii. Essentially, the algorithm is different in principle from the algorithms that were developed in the early period of scan-conversion [29, 52, 75, 100, 124, 165, 193], and it has also some distinguishing features that are notably different from the ones proposed in later periods [20, 30, 101, 140, 203, 214, 218]. The work in [222] has also shown that the notion of number theory is indeed useful for generating a large number of digital circles. The previous algorithms essentially resort to appropriate digitization (i.e., computation of numeric differences) of 1st-order and 2nd-order derivatives, which are mostly useful to analyze and solve problems involving curves and curve segments in the Euclidean/real plane.

As shown in [14], for a radius exceeding 100 or so, the number-theoretic technique has appreciable margin over Bresenham's algorithm [29]. Further,

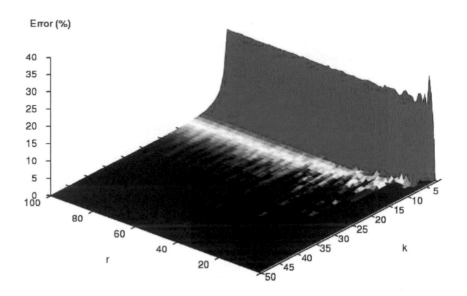

FIGURE 7.13: 3-D plot of error versus k and radius r of a digital circle. Note that with higher values of k, the resultant $2k$-gon covering the digital circle has fewer errors. (See color insert.)

the gain for the run length finding approach using the number-theoretic technique continues to increase as the radius increases. For a very large radius, exceeding 1000 or so, the number-theoretic technique contributes substantial improvements to a circle generation procedure. The technique is based on the distribution of *perfect squares* (square numbers) in integer intervals. Owing to such a unique characterization, the problem of constructing a digital circle or a circular arc maps to the domain of number theory. Given an integer radius and an integer center, the corresponding digital circle can be constructed efficiently by using number-theoretic properties only. Some of these relevant number-theoretic properties discussed here.

While generating the digital circle $\mathcal{C}_1^{\mathbb{Z}}(o, r)$ with center at $o = (0,0)$ and radius r as a positive integer, a decision is made to select between east pixel $p_{\mathrm{E}}(i+1, j)$ or southeast pixel $p_{\mathrm{SE}}(i+1, j-1)$, standing at the current pixel $p(i, j)$, depending on which one between p_{E} and p_{SE} is closer to the point of intersection of the next ordinate line (i.e., $x = i + 1$) with the real circle

$\mathcal{C}^{\mathbb{R}}(o, r)$. It can be shown that a tie between p_E and p_{SE} is possible only if there is any computation error; hence we have the following lemma.

Lemma 7.1. *A tie for selecting one of the two candidate pixels (p_E and p_{SE}) can never occur.* □

If $p(i, j)$ is a grid point that lies in $\mathcal{C}_1^{\mathbb{Z}}(o, r)$, then the point of intersection p' of $\mathcal{C}^{\mathbb{R}}(o, r)$ with the vertical grid line $x = i$ should have a distance of less than $\frac{1}{2}$ from p. Hence, if $(i, j - \delta)$ are the coordinates of p', then we have $-\frac{1}{2} < \delta < \frac{1}{2}$, the strict inequality being evident from Lemma 7.1.

Now, since $p'(i, j - \delta)$ lies on the real circle $\mathcal{C}^{\mathbb{R}}(o, r)$, we can show that

$$r^2 - j^2 - j \leqslant i^2 < r^2 - j^2 + j. \tag{7.2}$$

Equation 7.2 reveals the pattern of the grid points that would represent the digital circle $\mathcal{C}^{\mathbb{Z}}(o, r)$ in the first octant. Since the first grid point in the first octant is always $(0, r)$ (considering clockwise enumeration), Eq. 7.2 for the topmost grid points (i.e., $j = r$) becomes $0 \leqslant i^2 < r^2 - r^2 + r = r$, or, $0 \leqslant i^2 \leqslant r - 1$. This implies that in the first octant, the grid points, having their ordinates as r, will have the squares of their abscissae in the (closed) interval $[0, r - 1]$. Let us denote the interval $[0, r - 1]$ by I_0 (zero-th interval).

Let I_1 be called the first interval, which is obtained by substituting $j = r - 1$ in Eq. 7.2. So, $I_1 = \left[r^2 - (r - 1)^2 - (r - 1), r^2 - (r - 1)^2 + (r - 1) - 1 \right] = [r, 3r - 3]$, which contains the squares of abscissae of all the grid points (in the first octant) whose ordinates are $r - 1$. Thus, in general, if I_k denotes the k-th ($k \geqslant 1$) interval, given by

$$
\begin{aligned}
I_k &= \left[r^2 - (r - k)^2 - (r - k), r^2 - (r - k)^2 + (r - k) - 1 \right] \\
&= \left[(2k - 1)r - k(k - 1), (2k + 1)r - k(k + 1) - 1 \right], \tag{7.3}
\end{aligned}
$$

which is obtained by substituting $j = r - k$ in Eq. 7.2, then proceeding in this way, for a digital circle with radius $r \geqslant 1$ and center at $(0, 0)$, we get the following lemma.

Lemma 7.2. *The interval $I_k = [(2k - 1)r - k(k - 1), (2k + 1)r - k(k + 1) - 1]$ contains the squares of abscissae of the grid points of $\mathcal{C}_1^{\mathbb{Z}}(o, r)$ whose ordinates are $r - k$, for $k \geqslant 1$.* □

The length l_k of the interval I_k ($k \geqslant 1$) is, therefore, given by

$$
\begin{aligned}
l_k &= ((2k + 1)r - k(k + 1) - 1)) - ((2k - 1)r - k(k - 1)) + 1 \\
&= 2r - 2k. \tag{7.4}
\end{aligned}
$$

Using Eq. 7.4, the length l_{k+1} for interval I_{k+1} is given by $l_{k+1} = 2r - 2(k+1) = l_k - 2$, and hence the following lemma.

Lemma 7.3. *The lengths of the intervals containing the squares of equi-ordinate abscissae of the grid points in $\mathcal{C}_1^{\mathbb{Z}}(o, r)$ decrease constantly by 2, starting from I_1.* □

7.2.4.1 Algorithm DCS (Digital Circle Using Squares)

An algorithm for construction of digital circles can be designed, therefore, based on (the number of) square numbers in the intervals

$I_0 = [0, r-1]$,
$I_1 = [r, 3r-3]$,
$I_2 = [3r-2, 5r-7], \ldots$,
$I_k = [(2k-1)r - k(k-1), (2k+1)r - k(k+1) - 1], \ldots$.

The length of the first interval I_1 is greater than that of I_0 by $r-2$, as given by Eq. 7.4. More interestingly, for $k \geqslant 1$, the length of each interval I_{k+1} is less than that of I_k by 2, which is a constant. Hence, in the algorithm DCS, using square numbers, we search for the number of perfect squares in each interval $I_k, k \geqslant 0$, which, in turn, gives the number of grid points with ordinate $r - k$. The following theorem contains the above facts and findings in a concise way.

Theorem 7.1. *The squares of abscissae of grid points lying on $\mathcal{C}_1^{\mathbb{Z}}(o, r)$ and having ordinate $r - k$, lie in the interval $[u_k, v_k := u_k + l_k - 1]$, where u_k and l_k are given as follows.*

$$
u_k = \begin{cases} u_{k-1} + l_{k-1} & \text{if } k \geqslant 1 \\ 0 & \text{if } k = 0 \end{cases} \tag{7.5}
$$

$$
l_k = \begin{cases} l_{k-1} - 2 & \text{if } k \geqslant 2 \\ 2r - 2 & \text{if } k = 1 \\ r & \text{if } k = 0 \end{cases} \tag{7.6}
$$

$\square$

The proof follows from Lemma 7.2 and Lemma 7.3.

Using Theorem 7.1, therefore, the algorithm DCS (Digital Circle using Squares) is designed as shown in Algorithm 16. It may be noted that the $(i+1)$th square number $S_{i+1} = (i+1)^2$ can be obtained easily (without using any multiplication) from the previous square number $S_i = i^2$, since $S_{i+1} = (i+1)^2 = S_i + 2i + 1$, which is equivalent to adding a "gnomon" [192]. This is incorporated in Algorithm DCS (Steps 5–7), where the gnomon addition of $2i + 1$ is realized by adding i with the previous square s, followed by adding an incremented value $(i + +)$ of i, in order to optimize the primitive arithmetic operations. It is evident that, in Step 4, the procedure *include_8_sym_points* (i, j) includes the set of eight symmetric grid points, namely, $\{(\pm x, \pm y) : \{x\} \cup \{y\} = \{i, j\}\}$, in $\mathcal{C}^{\mathbb{Z}}(o, r)$.

7.2.4.2 Run Length Properties of Digital Circles

Using elementary tricks of number theory and mathematical induction, we have the following lemma.

Algorithm 16: Algorithm DCS.

Algorithm DCS (int r)

1. int $i \leftarrow 0, j \leftarrow r, s \leftarrow 0, w \leftarrow r - 1$
2. int $l \leftarrow 2w$
3. **while** $(j \geqslant i)$ **do**
4.   *include_8_sym_points* (i, j)
5. $s \leftarrow s + i$
6. $i \leftarrow i + 1$
7. $s \leftarrow s + i$ **while** $(s \leqslant w)$
8. $w \leftarrow w + l$
9. $l \leftarrow l - 2$
10. $j \leftarrow j - 1$

Lemma 7.4. *The number of perfect squares in a closed interval $[v, w]$ is at most one more than the number of perfect squares in the preceding closed interval $[u, v - 1]$ of equal length, where the intervals are taken from the non-negative integer axis.* $\square$

Now, from Lemma 7.3 it is obvious that the length of the interval containing the squares of abscissae (of grid points of $\mathcal{C}_1^{\mathbb{Z}}(o, r)$) with ordinate $j - 1$, is 2 less than that corresponding to ordinate j. Even if the interval corresponding to ordinate $j - 1$ had been equal in length to the preceding interval corresponding to ordinate j, then from Lemma 7.4, the maximum number of grid points with ordinate $j - 1$ would not have exceeded one more than the number of grid points with ordinate j. Hence we have the following theorem.

Theorem 7.2. *The run length of grid points of $\mathcal{C}_1^{\mathbb{Z}}(o, r)$ with ordinate $j - 1$ never exceeds one more than the run length of its grid points with ordinate j.* $\square$

Theorem 7.2 provides a good and useful upper bound on the number of grid points with ordinate j with respect to that corresponding to ordinate $j - 1$. The derivation of the lower bound that we have obtained is, however, not as so straightforward. See [14] for a detailed discussion on the lower bound. The first useful lemma for this is as follows.

Lemma 7.5. *If $[u, v - 1]$ be the interval $I_k, k \geqslant 1$, and $[v, w]$ be the interval of same length as $[u, v - 1]$, then the number of perfect squares in $[v, w]$ is at least (floor of) half the number of perfect squares less one in $[u, v - 1]$.* $\square$

Putting together the above findings, therefore, we get the following theorem, which captures the run length properties of a digital circle in a precise form.

Theorem 7.3. *If $\lambda(j)$ is the run length of grid points of $\mathcal{C}_1^{\mathbb{Z}}(o, r)$ with ordinate j, then the run length of grid points with ordinate $j - 1$ for $j \leqslant r - 1$ and*

Algorithm 17: Algorithm DCR Using Run Length Properties in Part (See Text for Explanation).

Algorithm DCR (**int** r)
1.   **int** $i \leftarrow 0, j \leftarrow r, s \leftarrow 0, w \leftarrow r - 1, t \leftarrow r, m$
2.   **int** $l \leftarrow 2w$
3.   **while** $(j \geqslant i)$
4.   **while** $(s < t)$
5. $m \leftarrow s + t$
6. $m \leftarrow m/2$
7.   **if** $(w \leqslant square[m])$
8. $t \leftarrow m$
9.   **else**
10. $s \leftarrow m + 1$
11.   **if** $(w < square[s])$
12. $s \leftarrow s - 1$
13. $s \leftarrow s + 1$
14. $include\_run\ (i, s - i, j)$
15. $t \leftarrow 2s - i + 1$
16. $i \leftarrow s$
17. $w \leftarrow w + l$
18. $l \leftarrow l - 2$
19. $j \leftarrow j - 1$

$r \geqslant 2$, *is given by*

$$\lambda(j - 1) \geqslant \left\lfloor \frac{\lambda(j) - 1}{2} \right\rfloor - 1.$$

$\square$

Combining Theorem 7.2 and Theorem 7.3, therefore, we obtain Eq. 7.7, which can be used to derive the horizontal run of grid points with ordinate $j - 1$, from the previous run with ordinate j, for $j \leqslant r$.

$$\left\lfloor \frac{\lambda(j) - 1}{2} \right\rfloor - 1 \leqslant \lambda(j - 1) \leqslant \lambda(j) + 1 \tag{7.7}$$

The algorithm DCR (Digital Circle using Run-lengths), which incorporates the relation between the consecutive runs, as captured in Eq. 7.7, is given in Algorithm 17. In this algorithm, only the upper limit $(\lambda(j) + 1)$ of the term $\lambda(j-1)$ has been considered for simplicity. The lower limit $(\lfloor \frac{1}{2}(\lambda(j) - 1)\rfloor - 1)$ of $\lambda(j - 1)$ may be considered in a similar fashion. Since we search for $\lambda(j - 1)$ in an integer interval whose upper and lower limits are computed every time using its preceding run length, $\lambda(j)$, resorting to both the limits in the run-finding procedure is computationally effective for a sufficiently large value of $\lambda(j - 1)$. Computation of the lower (or upper) limit involves a fixed number of operations, whereas the operations in a binary search is logarithmic on the length of the corresponding interval. As r increases sufficiently, the top runs also increase in length, whose interval searching would, therefore, be improved by constraining the concerned intervals using both the upper and the lower limits.

In order to find the exact value of $\lambda(j-1)$, a binary search has been used. The binary search assumes the minimum value (lower limit) as 1 and the maximum value (upper limit) as $\lambda(j) + 1$ to start with, for finding $\lambda(j-1)$. The binary search is performed on the Look-Up-Table, implemented in the form of a 1-dimensional array, namely *square*[ ], which contains the square S_n of each integer $n = 0, 1, 2, \ldots, N$, where N^2 is the largest square not exceeding the maximum value R of radius r. For example, for $R = 1000$, N (and the size of the Look-Up-Table, thereof) equals 31.

It may be noted that, the binary search procedure incorporated in Algorithm DCR is a modified one, based on these requirements, from the conventional one found in the literature [131]. In accordance with the conventional procedure, one has to check at first whether the middle element (m in this case) equals the search key, failing which, one between two other checks (smaller or greater) is performed. This modification is introduced (Steps 7–10) to reduce the number of comparisons in each iteration of the inner **while** loop.

A demonstration of the algorithm DCR for $r = 106$ is graphically shown in Fig. 7.14 until a run of unit length is found. For each row, the binary search is illustrated by circular dots, where each dot corresponds to the middle element (m) of the respective sub-array (*square*[$s..t$]). It may be noticed that, m in Algorithm DCR (Step 6) denotes the abscissa (vertical) line $x = m - 1$ in Fig. 7.14. As the binary search associated with a particular row proceeds and converges to produce the final run of the corresponding row, the respective dots have gradually been darkened to depict the effect of the run length finding procedure. The end of a run at each row is emphasized by highlighted abscissa lines passing through the endpoint of that run for visual clarity. For instance, for the topmost row, m starts with 53, followed by 26, 13, and so on, until $s = t = 11$; hence m finally becomes 11, thereby yielding the run length equal to 11. Similarly, for the next row, since the start values of s and t are 11 and 23, respectively, the subsequent values of m are 17, 20, 19, and (finally) 18, which makes the corresponding run length to $18 - 11 = 7$. Thus, the run length of a particular row ($y = j$) is given by the cumulative run length up to that row minus the preceding cumulative run length, which is realized by the function *include_run* $(i, s - i, j)$ in Step 6 of the algorithm. The *square numeric code* of a circle with radius 106 in the first octant, therefore, turns out to be $\langle 11, 7, 5, 5, 3, 3, 3, 3, 2, 2, 2, 3, 1, 2, 2, 2, 1, 2, 1, 2, 1, 1, 2, 1, 1, 1, 2, 1, 1, 1, 1, 1 \rangle$, which can be compressed to $\langle 11, 7, 5^2, 3^4, 2^3, 3, 1, 2^3, 1, 2, 1, 2, 1^2, 2, 1^3, 2, 1^5 \rangle$, which, in turn, can be used to easily derive the chain code representation of the circular arc.

Implementation of the Look-Up-Table, as mentioned above, avoids floating-point operations at any stage, such as the square root operation to find the run for $j = r$ in other existing algorithms, e.g., [222]. Also, the above

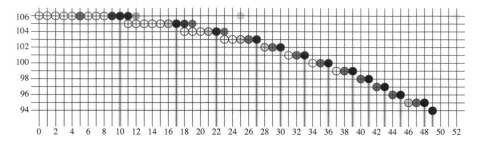

FIGURE 7.14: Demonstration of algorithm DCR for radius 106 (see text for explanation).

algorithm is not as efficient for circles (1st octant) of small radii, or to find a run length of small value, such as when $j/i \leqslant 4$ or so, since the operations (comparisons, etc.) needed in the binary search for small run lengths would raise the overall runtime of the algorithm. Hence, the algorithm DCR may be used in some models of hybridization, whose one possible realization can be seen in [14].

7.2.5 Creating Realistic Potteries

As explained in Sec. 7.2.3, the surface $S_{\mathcal{G}}$ is generated from the digital generatrix $\mathcal{G}$ as an ordered set of voxels. For each digital circle $\mathcal{C^Z}_i$ corresponding to each voxel $p_i \in \mathcal{G}$, the voxels inclusive of the missing voxels are generated in a definite order, starting from Octant 1 and ending at Octant 8. All the circles, namely $\mathcal{C^Z}_1, \mathcal{C^Z}_2, \ldots, \mathcal{C^Z}_n$, in turn, are also generated in order. The above ordering to represent a wheel-thrown piece helps to map textures in a straightforward way. We map adjacent and equal-area rectangular parts from the texture image to each of the quads. The circle is approximated as a regular polygon. Clearly, there is always a trade-off between the number of vertices of the approximate polygon and the rendering speed, apart from the fact that a coarse approximation gives a polyhedral effect (Fig. 7.11). The nature of approximation error has been studied in a recent work [15] (a part of which is illustrated in Fig. 7.2.4 and also through the 3-D plot in Fig. 7.13), which shows that if we have polygons with $k \geqslant 20$ or so, then for $r \leqslant 100$, we have 2% or less error. For higher radii, of course, we should increase k accordingly to have the desired error control. Note that an error point is that which lies inside the polygon but outside the digital circle [15].

The number of sides, *numSides*, in the polygon approximation is kept the same for each of the digital points of $\mathcal{G}$. We find the maximum among the distances of the points on $\mathcal{G}$ from the axis of revolution, say, *maxDis*. We use *maxDis* to determine the value of *numSides*. We want to map a square portion of texture-image to each of the quads. The distance between two adjacent points in $\mathcal{G}$ is either 1 or $\sqrt{2}$). We determine the value of *numSides*

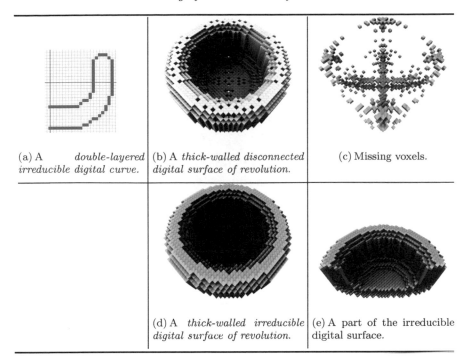

(a) A double-layered irreducible digital curve.	(b) A thick-walled disconnected digital surface of revolution.	(c) Missing voxels.
	(d) A thick-walled irreducible digital surface of revolution.	(e) A part of the irreducible digital surface.

FIGURE 7.15: A thick- and hollow-walled irreducible digital surface of revolution. (See color insert.)

Reprinted from *International Journal of Arts and Technology*, **4**: 196–215, G. Kumar et al., Copyright 2011, with permission from Inderscience Publishers.

by approximating the circle of radius *maxDis* with a regular polygon having side length, $d = 1$. Thus,

$$numSides = round\left(\frac{\pi}{sin^{-1}\left(\frac{d}{2maxDis}\right)}\right). \qquad (7.8)$$

Now, for each point on $\mathcal{G}$, we approximate the circle of revolution with a regular polygon with *numSides* sides. For every pair of adjacent points on $\mathcal{G}$, we join the corresponding vertices of the approximate polygon corresponding to the circle of revolution. In this way, we obtain a simple quad-decomposition of the surface of revolution. For a faster rendering, we can approximate the digital generatrix/B-spline by selecting alternate points (or 1 in every $h > 1$). Then the value of d has to be changed to h. In our experiments, we have taken $h = 4$.

In this way, the digital continuity of the original image is easily maintained on the texture map, with satisfactory results. Note that we resorted to floating-point computations because of the artistic finish on the digital pot-

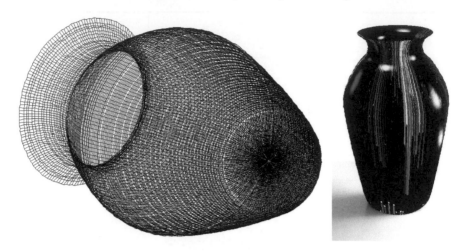

FIGURE 7.16: Quad decomposition of a wheel-thrown digital vase for texturing. (See color insert.)

Reprinted from *International Journal of Arts and Technology*, **4**: 196–215, G. Kumar et al., Copyright 2011, with permission from Inderscience Publishers.

tery. Fig. 7.16 demonstrates the quad decomposition of a wheel-thrown vase and the resultant texture mapping.

To create realistic potteries (shown in Sec. 7.2.6) with thick walls, we need to supply (connected and irreducible) digital generatrices having two parallel segments—one for the outer wall and another for the inner wall including the rim top. The procedure of generating digital surfaces of revolution is then the same as that explained in Sec. 7.2.3. As the digital generatrix, in its entirety, essentially consists of two parallel segments, which do not touch or intersect themselves, we get a single irreducible digital curve segment. Such a generatrix is an open digital curve segment (Chapter 4) with both its endpoints lying on the axis of revolution, α, and having a "turn" corresponding to the upper rim of the pottery to be generated. The inner segment defines the inner digital surface (of revolution) and the outer, the outer surface. The inner surface is mostly not visible. However, for further applications, the inner digital surface, which is an irreducible surface by definition (Sec. 7.2.2), may be required. Since we have a double-sided (i.e., thick- and hollow-walled) digital pottery, irreducible in nature, it becomes mathematically well-defined, and hence would be compliant as an input to an algorithm in any relevant application. An instance of a small yet representative thick-walled digital surface of revolution, which is irreducible in nature, is shown in Fig. 7.15.

FIGURE 7.17: A digital wheel-thrown uni-voxel thick bowl created by an irreducible digital curve segment as the digital generatrix. The surface is then decomposed into quads for texture mapping with suitable illumination and shadow formation. (See color insert.)

Reprinted from *International Journal of Arts and Technology*, **4**: 196–215, G. Kumar et al., Copyright 2011, with permission from Inderscience Publishers.

7.2.6   Some Examples

Snapshots of some example potteries, taken from [125], are given in Figs. 7.16, 7.17, and 7.18. These are generated by a software developed by the authors of [125] in Java3D^{TM} API, version 1.5.2. Figure 7.16 shows how a *digitally connected and irreducible surface of revolution* (Sec. 7.2.2) is decomposed into appropriately small quads in accordance with Eq. 7.8 for subsequent texture mapping (Sec. 7.2.5). The irreducible digital surface in $\mathbb{Z}^3$, which is an ordered set of voxels, is mapped to the corresponding real surface in $\mathbb{R}^3$ for quad-decomposition, which is quite fast, efficient, and bug-free by dint of the very property of connectivity and irreducibility of the digital surface of revolution.

Fig. 7.17 shows another result for a "bowl" with a relatively larger zoom factor. When a texture is mapped onto the surface in $\mathbb{R}^3$, we get a flawless surface resembling a real-world pottery, although uni-voxel thick. A thick-walled set of potteries, created out of double-layered digital generatrices (Sec. 7.2.5), is shown in Fig. 7.18. In this model, potteries of varying colors and textures can thus be developed as per requirement.

Some statistical figures and CPU times to generate a few potteries are presented in Table 7.1. The total CPU time required to create a digital pottery consists of two parts: T_1 and T_2. The time T_1 is for constructing the digital surface $S_{\mathcal{G}}$ from a given digital generatrix, $\mathcal{G}$, and the time T_2 is for quad-decomposition, followed by texture mapping coupled with necessary rendering. The number of voxels constituting $S_{\mathcal{G}}$ not only increases with the length of $\mathcal{G}$

FIGURE 7.18: Potteries with thick walls. (See color insert.)

Reprinted from *International Journal of Arts and Technology*, **4**: 196–215, G. Kumar et al., Copyright 2011, with permission from Inderscience Publishers.

and the distance of $\mathcal{G}$ from the axis of revolution, α, but it also depends on the shape of $\mathcal{G}$. If each pair of consecutive points comprising $\mathcal{G}$ has the radius difference unity, then the number of missing voxels increases, which, in turn, increases the total number of voxels defining $S_{\mathcal{G}}$.

7.3   Summary

The algorithm for construction of an outer cover executes in $O(n)$ time for a given grid size and requires computations in the integer domain only, thereby providing an effective approach for 3-D shape analysis through voxel-level approximation. Multi-grid analysis of the cover reveals a gradual increase in the topological complexity as the grid size decreases. This is a specialty of the algorithm as it reveals the characteristics of the object surface based on the concavities of the cover. Extension of the algorithm to an adaptive grid system may also prove to be useful for applications related to rough sets in the 3-D domain.

The algorithm on wheel-throwing in digital space shows how certain efficient techniques of digital geometry can be used for graphics and computer-vision problems in 3-D space. Some test results given in this chapter demonstrate its efficiency and robustness in creating a digital pottery defined as a dig-

♯Control Points	♯Points in original B-Spline	♯Points in approxi-mate B-Spline	♯Sides in poly-gon	♯Quads	T_1	T_2	Total time $(T_1 + T_2)$
						in millisecs.	
6	147	37	16	576	37	364	401
4	103	26	163	4075	26	449	475
4	100	25	181	4344	25	580	605
5	131	33	185	5920	32	642	674
7	227	57	185	10360	55	750	805
10	508	127	87	10962	127	665	792
20	563	141	114	15960	133	763	896
15	599	150	113	16837	137	872	1009
12	314	79	216	16848	74	891	965
28	615	154	127	19431	146	846	992
10	403	101	207	20700	98	1001	1099
16	418	105	216	22464	101	961	1062
20	592	148	186	27342	136	1488	1624
24	607	152	246	37146	145	1815	1960

TABLE 7.1:   Statistical information and CPU times for some digital potteries.

Reprinted from *International Journal of Arts and Technology*, **4**: 196–215, G. Kumar et al., Copyright 2011, with permission from Inderscience Publishers.

itally connected and irreducible surface of revolution, along with its readiness to generate realistic thick-walled potteries that satisfy both mathematical and aesthetic criteria. Future possibility lies in generating irregular/bumpy surfaces and subsurfaces of revolution, and in defining various morphological/set-theoretic operations on digital surfaces of revolution so as to generate more interesting potteries out of the "digital wheel."

Exercises

1. Prove the correctness of Equation 7.1 for all possible combinatorial cases of object occupancy in the eight UGCs incident at the start vertex.

2. Prove that a UGC-face f_k is a part of the outer cover if and only if exactly one of two adjacent UGCs of f_k has object containment. Hence, show that if p is a point lying on $\overline{P}_\mathbb{G}(A)$, the maximum error of approximation, $d_T(p, A)$, is bounded by $d_T(p, A) < g$.

3. Prove that the outer cover $\overline{P}_\mathbb{G}(A)$ has minimum volume. (Hint: Since each enqueued face has been verified for *eligibility* while being enqueued, any point $p \in \overline{P}_\mathbb{G}(A)$ satisfies $0 \leqslant d_T(p, A) < g$.)

4. Prove that while generating the digital circle $\mathcal{C}_1^{\mathbb{Z}}(o, r)$, there cannot arise any tie for selecting one of the two candidate pixels (p_{E} and p_{SE}).

5. Prove that the interval $[\max\{0, (2k-1)r - k(k-1)\}, (2k+1)r - k(k+1) - 1]$ contains the squares of abscissae of (all and only) the grid points constituting the $k(\geqslant 0)$th run in Octant 1 of $\mathcal{C}^{\mathbb{Z}}(o, r)$.

6. Prove that λ_0 is the length of the top run of a digital circle $\mathcal{C}^{\mathbb{Z}}(o, r)$ if and only if r lies in the interval $[(\lambda_0 - 1)^2 + 1, \lambda_0^2]$.

7. Let $|\mathrm{E}|$ and $|\mathrm{SE}|$ denote the respective number of east (E) and southeast (SE) transitions in $\mathcal{C}_1^{\mathbb{Z}}(o, r)$. Then show that, over all iterations of DCS, the total number of comparisons (n_c), additions (n_a), and increments (n_i) are given by

$$
\begin{aligned}
n_c &= |\mathrm{E}| + 2|\mathrm{SE}| + 1, \\
n_a &= 2|\mathrm{E}| + 4|\mathrm{SE}|, \\
n_i &= |\mathrm{E}| + 2|\mathrm{SE}|.
\end{aligned}
$$

References

[1] A. Aguilera. Isothetic polyhedra: Study and application. PhD thesis, Universitat Politécnica de Catalunya, 1998.

[2] J. R. Van Aken and M. Novak. Curve-drawing algorithms for raster display. *ACM Trans. Graphics*, 4(2):147–169, 1985.

[3] I. M. Anderson and J. C. Bezdek. Curvature and tangential deflection of discrete arcs: A theory based on the commutator of scatter matrix pairs and its application to vertex detection in planar shape data. *IEEE Trans. PAMI*, 6:27–40, 1984.

[4] T. A. Anderson and C. E. Kim. Represention of digital line segments and their preimages. *Computer Vision Graphics and Image Processing*, 30:279–288, 1985.

[5] C. Arcelli and A. Massarotti. Regular arcs in digital contours. *Computer Vision Graphics and Image Processing*, 4:339–360, 1975.

[6] C. Arcelli and A. Massarotti. On the parallel generation of straight digital lines. *Computer Vision Graphics and Image Processing*, 7:67–83, 1978.

[7] T. Asano and N. Katoh. Number theory helps line detection in digital images. In *Proceedings of the 4th International Symposium on Algorithms and Computation* (ISAAC'93), pages 312–322, 1993.

[8] F. Attneave. Some informational aspects of visual perception. *Psychological Review*, 61(3):183–193, 1954.

[9] M. D. Berg, M. V. Kreveld, M. Overmars, and O. Schwarzkopf. *Computational Geometry Algorithms and Applications*. Springer-Verlag, Berlin, 2000.

[10] J. C. Bezdek and I. M. Anderson. An application of the c-varieties clustering algorithms to polygonal curve fitting. *IEEE Trans. Sys., Man & Cybern.*, 15:637–641, 1985.

[11] P. Bhaniramka, R. Wenger, and R. Crawfis. Isosurfacing in higher dimensions. In *Proc. Visualization, Salt Lake City, Utah, USA*, pages 267–273, 2000.

[12] P. Bhaniramka, R. Wenger, and R. Crawfis. Isosurface construction in any dimension using convex hulls. *IEEE Trans. Visualization and Computer Graphics*, 10:130–141, 2004.

[13] P. Bhowmick and B. B. Bhattacharya. Fast polygonal approximation of digital curves using relaxed straightness properties. *IEEE Trans. PAMI*, 29(9):1590–1602, 2007.

[14] P. Bhowmick and B. B. Bhattacharya. Number-theoretic interpretation and construction of a digital circle. *Discrete Applied Mathematics*, 156(12):2381–2399, 2008.

[15] P. Bhowmick and B. B. Bhattacharya. Real polygonal covers of digital discs: Some theories and experiments. *Fundamenta Informaticae*, 91(3-4):487–505, 2009.

[16] P. Bhowmick, A. Biswas, and B. B. Bhattacharya. Isothetic polygons of a 2D object on generalized grid. In *Proc. 1st Intl. Conf. Pattern Recognition and Machine Intelligence (PReMI)*, volume 3776 of *LNCS*, pages 407–412. Springer, Berlin, 2005.

[17] A. Biswas, P. Bhowmick, and B. B. Bhattacharya. TIPS: On finding a Tight Isothetic Polygonal Shape covering a 2D object. In *Proc. 14th Scandinavian Conf. Image Analysis (SCIA)*, volume 3540 of *LNCS*, pages 930–939. Springer, Berlin, 2005.

[18] P. K. Biswas, S. S. Biswas, and B. N. Chatterji. An SIMD algorithm for range image segmentation. *Pattern Recognition Letters*, 28(2):255–267, 1995.

[19] P. K. Biswas, J. Mukherjee, P. P. Chakrabarti, and B. N. Chatterji. Qualitative description of 3-D surfaces. *International Journal of Artificial Intelligence and Pattern Recognition*, 6(4):651–672, 1992.

[20] J. F. Blinn. How many ways can you draw a circle? *IEEE Computer Graphics and Applications*, 7(8):39–44, 1987.

[21] H. Blum. A transformation for extracting new descriptors of shape. In Weiant Wathen-Dunn, editor, *Models for the Perception of Speech and Visual Form*, pages 362–380. MIT Press, Cambridge, 1967.

[22] G. Borgefors. Distance transformations in arbitrary dimensions. *Computer Vision, Graphics and Image Processing*, 27:321–345, 1984.

[23] G. Borgefors. Distance transformations in digital images. *Computer Vision, Graphics and Image Processing*, 34:344–371, 1986.

[24] G. Borgefors. Applications of distance transforms. In C. Arcelli et al., eds., *Aspects of Visual Form Processing*, pages 83–108, 1994.

[25] G. Borgefors. On digital transforms in three dimensions. *Computer Vision and Image Understanding*, 64:368–376, 1996.

[26] G. Borgefors and I. Nystrom. Efficient shape representation by minimizing the set of centres of maximal discs spheres. *Pattern Recognition Letters*, 18:465–472, 1997.

[27] Carl B. Boyer. *A History of Mathematics (2nd ed.)*. John Wiley & Sons, Inc., New York, 1991.

[28] J. E. Bresenham. An incremental algorithm for digital plotting. In *Proc. ACM Natl. Conf.*, 1963.

[29] J. E. Bresenham. A linear algorithm for incremental digital display of circular arcs. *Communications of the ACM*, 20(2):100–106, 1977.

[30] J. E. Bresenham. Run length slice algorithm for incremental lines. In R. A. Earnshaw, ed., *Fundamental Algorithms for Computer Graphics*, volume F17 of *NATO ASI Series*, pages 59–104. Springer-Verlag, New York, 1985.

[31] V. Brimkov. Discrete volume polyhedrization: Complexity and bounds on performance. In *Proc. International Symposium CompIMAGE '06*, Computational Methodology of Objects Represented in Images: Fundamentals, Methods and Applications, pages 117–122. Taylor and Francis, Coimbra, Portugal, 2006.

[32] V. E. Brimkov, D. Coeurjolly, and R. Klette. Digital planarity: A review. *Discrete Applied Mathematics*, 155(4):468–495, 2007.

[33] V.E. Brimkov. Digitization scheme that assures faithful reconstruction of plane figures. *Pattern Recognition*, 42:1637–1649, 2009.

[34] R. Brons. Linguistic methods for description of a straight line on a grid. *Comput. Graphics Image Process.*, 2:48–62, 1974.

[35] Victor Bryant. Web tutorials for potteries. http://www.victor.bryant.hemscott.net/, 2004.

[36] O. P. Buneman. A grammar for the topological analysis of plane figures. In B. Meltzer and D. Michie, eds., *Machine Intelligence*, volume 5, pages 383–393. Edinburgh Univ. Press, Edinburgh, 1969.

[37] R. Stranda, C. Fouarda, and G. Borgefors. Weighted distance transforms generalized to modules and their computation on point lattices. *Pattern Recognition*, 40:2453–2474, 2007.

[38] S. Chattopadhyay and P. P. Das. A new method of analysis for digital straight lines. *Pattern Recognition Letters*, 12(12):747–755, 1991.

[39] S. Chattopadhyay and P. P. Das. Estimation of the original length of a straight line segment from its digitization in three dimensions. *Pattern Recognition*, 25(8):787–798, 1992.

[40] S. Chattopadhyay and P. P. Das. A generalized approach to reconstruction of a restricted class of digitized planar curves. *Sadhana (Journal of Indian Science Academy)*, 1992.

[41] S. Chattopadhyay and P. P. Das. Parameter estimation and reconstruction of digital conics in normal positions. *Computer Vision Graphics and Image Processing: GMIP*, 54(5):385–395, 1992.

[42] S. Chattopadhyay and P. P. Das. Digital plane segments. In *Proceedings of SPIE International Conference on Vision Geometry II*, pages 289–300, 1993.

[43] T. C. Chen and K. L. Chung. A new randomized algorithm for detecting lines. *Real Time Imaging*, 7:473–481, 2001.

[44] S. Climer and S. K. Bhatia. Local lines: A linear time line detector. *Pattern Recognition Letters*, 24:2291–2300, 2003.

[45] D. Coeurjolly and I. Sivignon. Reversible discrete volume polyhedrization using Marching Cubes simplification. *SPIE Vision Geometry XII*, 5300:1–11, 2004.

[46] D. Coeurjolly, I. Sivignon, F. Dupont, and F. Feschet. On digital plane preimage structure. *Discrete Applied Mathematics*, 151(1-3):78–92, 2005.

[47] D. Cohen and A. Kaufman. Scan conversion algorithms for linear and quadratic objects. In *Proc. Volume Visualization*, pages 280–301. IEEE CS Press, 1991.

[48] Daniel Cohen-Or and Arie Kaufman. Fundamentals of surface voxelization. *Graph. Models Image Process.*, 57(6):453–461, 1995.

[49] T. H. Cormen, C. E. Leiserson, and R. L. Rivest. *Introduction to Algorithms*. Prentice Hall of India, New-Delhi, 2000.

[50] E. Creutzburg, A. Hübler, and V. Wedler. On-line recognition of digital straight line segments. In *Proc. 2nd Intl. Conf. AI and Inf. Control Systems of Robots*, pages 42–46, 1982.

[51] O. Cuisenaire and B. Macq. Fast Euclidean distance transformation by propagation using multiple neighborhoods. *CVGIP: Image Under.*, 76(2):163–172, 1999.

[52] P. E. Danielsson. Comments on circle generator for display devices. *Computer Graphics and Image Processing*, 7(2):300–301, 1978.

[53] P. E. Danielsson. Euclidean distance mapping. *Computer Graphics and Image Processing*, 14:227–248, 1980.

[54] P. P. Das. Paths and distances in digital geometry. PhD dissertation, Indian Institute of Technology, 1988.

[55] P. P. Das. Best simple octagonal distances in digital geometry. *Journal of Approximation Theory*, 68(2):155–174, February 1992.

[56] P. P. Das and B. N. Chatterji. Octagonal distances for digital pictures. *Information Sciences*, 50:123–150, 1990.

[57] P. P. Das and B. N. Chatterji. A note on distance transformations in arbitrary dimensions. *Computer Vision, Graphics and Image Processing*, 43:368–385, 1988.

[58] P. P. Das, J. Mukherjee, and B. N. Chatterji. The t-cost distance in digital geometry. *Information Sciences*, 59:1–20, 1992.

[59] P. P. Das, P. P. Chakrabarti, and B.N Chatterji. Distance functions in digital geometry. *Information Sciences*, 42:113–136, 1987.

[60] P. P. Das, P. P. Chakrabarti, and B.N Chatterji. Generalized distance in digital geometry. *Information Sciences*, 42:51–67, 1987.

[61] P. P. Das. Metricity preserving transforms. *Pattern Recognition Letters*, 10:73–76, 1989.

[62] P. P. Das. More on path generated digital metrics. *Pattern Recognition Letters*, 10:25 31, 1989.

[63] P. P. Das. Lattices of octagonal distances in digital geometry. *Pattern Recognition Letters*, 11:663–667, 1990.

[64] P. P. Das. Hypersphere of n-sequence distances. *Proc. of SPIE Conf. on Vision Geometry*, Boston, USA, Nov. 15–20, 1832:61–67, 1992.

[65] P. P. Das. A note on "distance function in digital geometry." *Information Sciences*, 64:181–190, 1992.

[66] P. P. Das. The real m-neighbor distance. *Proc. of SPIE Conf. on Vision Geometry*, Boston, USA, Nov. 15–20, 1832:68–78, 1992.

[67] P. P. Das and B. N. Chatterji. Knight's distance in digital geometry. *Pattern Recognition Letters*, 7:215–226, 1988.

[68] P. P. Das and B. N. Chatterji. A note on "distance transformations in arbitrary dimensions." *Computer Vision, Graphics and Image Processing*, 43:368–385, 1988.

[69] P. P. Das and B. N. Chatterji. Estimation of errors between Euclidean and m-neighbour distance. *Information Sciences*, 48:1–26, 1989.

[70] P. P. Das and B. N. Chatterji. Hyperspheres in digital geometry. *Information Sciences*, 50:73–93, 1990.

[71] P. P. Das and B. N. Chatterji. Digital distance geometry: A survey. *Sadhana*, 18:159–187, 1993.

[72] P. P. Das and J. Mukherejee. Metricity of super-knight's distances in digital geometry. *Pattern Recognition Letters*, 11:601–604, 1990.

[73] L. S. Davis, A. Rosenfeld, and A. K. Agrawala. On models for line detection. *IEEE Trans. Sys., Man & Cybern.*, 6:127–133, 1976.

[74] I. Debled-Rennesson and J. P. Reveilles. A linear algorithm for segmentation of digital curves. *Intl. J. Patt. Rec. Artif. Intell.*, 9:635–662, 1995.

[75] M. Doros. Algorithms for generation of discrete circles, rings, and disks. *Computer Graphics and Image Processing*, 10:366–371, 1979.

[76] L. Dorst. *Discrete Straight Line Segments: Parameters, Primitives and Properties*. PhD thesis, University of Technology, 1986.

[77] L. Dorst and R. P. W. Duin. Spirograph theory: A framework for digitized straight lines. *IEEE Transactions on Pattern Analysis and Machine Intelligence*, 6:634–639, 1984.

[78] L. Dorst and A. W. M. Smeulders. Discrete representation of straight lines. *IEEE Transactions on Pattern Analysis and Machine Intelligence*, 6:450–463, 1984.

[79] L. Dorst and A. W. M. Smeulders. Best linear unbiased estimators for properties of digitized straight lines. *IEEE Transactions on Pattern Analysis and Machine Intelligence*, 8:276–282, 1986.

[80] J. G. Dunham. Optimum uniform piecewise linear approximation of planar curves. *IEEE Trans. PAMI*, 8:67–75, 1986.

[81] D. Eberly, J. Lancaster, and A. Alyassin. On gray scale image measurements: Surface area and volume. *Computer Vision Graphics and Image Processing: GMIP*, 51(6):550–562, 1991.

[82] R. Fabri, L. D. F. Costa, J. C. Torelli, and O. M. Bruno. 2D Euclidean distance transform algorithms: A comparative survey. *ACM Computing Surveys*, 40(1):2:1–2:44, February 2008.

[83] Martin A. Fischler and Helen C. Wolf. Locating perceptually salient points on planar curves. *IEEE Trans. PAMI*, 16(2):113–129, 1994.

[84] J. D. Foley, A. Van Dam, S. K. Feiner, and J. F. Hughes. *Computer Graphics: Principles and Practice*. Addison Wesley, second edition in C edition, 1997.

[85] S. Forchhammer. Digital plane and grid point segments. *Computer Vision Graphics and Image Processing*, 47(6):373–584, 1989.

[86] H. Freeman. On the encoding of arbitrary geometric configuration. *IRE Trans. Electronics Computers*, 10:260–268, June 1961.

[87] H. Freeman. Techniques for the digital computer analysis of chain-encoded arbitrary plane curves. In *Proc. National Electronics Conf.*, volume 17, pages 421–432, 1961.

[88] H. Freeman. Boundary encoding and processing. *Picture Processing and Psychooptics*, pages 241–266, 1970.

[89] H. Freeman and L. S. Davis. A corner finding algorithm for chain-coded curves. *IEEE Trans. Computers*, 26:297–303, 1977.

[90] Tinsley A. Galyean and John F. Hughes. Sculpting: An interactive volumetric modeling technique. *ACM SIGGRAPH Computer Graphics*, 25(4):267–274, 1991.

[91] M. Giles and R. Haimes. Advanced interactive visualization for CFD. *Computing Systems in Engineering*, 1:51–62, 1990.

[92] R. C. Gonzalez and R. E. Woods. *Digital Image Processing*. Addison-Wesley, California, 1993.

[93] H. Gouraud. Continuous shading of curved surfaces. *IEEE Trans. Computers*, C-20(6):623–629, June 1971.

[94] S. B. Gray. Local properties of binary images in two dimensions. *IEEE Trans. Comput.*, 20:551–561, 1971.

[95] D. S. Guru, B. H. Shekar, and P. Nagabhushan. A simple and robust line detection algorithm based on small eigenvalue analysis. *Pattern Recognition Letters*, 25:1–13, 2004.

[96] G. Han, J. Kim, and S. Choi. Virtual pottery modeling with force feedback using cylindrical element method. In *Intl. Conf. Next-Generation Computing (ICON-C)*, pages 125–129, 2007.

[97] A. Held, K. Abe, and C. Arcelli. Towards a hierarchical contour description via dominant point detection. *IEEE Trans. Sys., Man & Cybern.*, 24:942–949, 1994.

[98] C. J. Hilditch. Linear skeletons from square cupboards. *Mach. Intelligence*, 4:403–420, 1969.

[99] Francis S. Hill, Jr. and Stephen M. Kelley. *Computer Graphics Using OpenGL*. Prentice Hall, 2007.

[100] B. K. P. Horn. Circle generators for display devices. *Computer Graphics and Image Processing*, 5(2):280–288, 1976.

[101] S. Y. Hsu, L. R. Chow, and C. H. Liu. A new approach for the generation of circles. *Computer Graphics Forum 12*, 2:105–109, 1993.

[102] S. H. Y. Hung. On the straightness of digital arcs. *IEEE Transactions on Pattern Analysis and Machine Intelligence*, 6:203–215, 1985.

[103] H. Imai and M. Iri. Computational geometric methods for polygonal approximations of a curve. *Computer Vision, Graphics, and Image Processing*, 36:31–41, 1986.

[104] Ken-ichi Kameyama. Virtual clay modeling system. In *VRST '97: ACM Symposium on Virtual Reality Software and Technology*, pages 197–200, 1997.

[105] N. Karmakar, A. Biswas, P. Bhowmick, and B. B. Bhattacharya. Construction of 3d isothetic cover of a digital object. In *Proc. 14th International Workshop on Combinatorial Image Analysis: IWCIA11*, volume 6636 of *LNCS*, pages 70–83. Springer, 2011.

[106] A. Kaufman. An algorithm for 3d scan-conversion of polygons. In *Proc. Eurographics '87*, pages 197–208, 1987.

[107] A. Kaufman. Efficient algorithms for 3d scan-conversion of parametric curves, surfaces, and volumes. *Computer Graphics*, 21(4):171–179, 1987.

[108] A. Kaufman and E. Shimony. 3d scan-conversion algorithms for voxel-based graphics. In *Proc. ACM Workshop on Interactive 3D Graphics, Chapell Hill, NC*, pages 45–76, 1986.

[109] C. E. Kim. On cellular straight line segments. *Computer Vision Graphics and Image Processing*, 18:369–381, 1982.

[110] C. E. Kim. Three-dimensional digital planes. *IEEE Transactions on Pattern Analysis and Machine Intelligence*, 6:357–393, 1984.

[111] C. E. Kim and A. Rosenfeld. Digital straight line and convexity of digital regions. *IEEE Transactions on Pattern Analysis and Machine Intelligence*, 4:149–153, 1982.

[112] J.-H. Kim. Polygonal approximation method and apparatus for use in a contour encoding system. US Patent No. 5,978,512, 1999.

[113] R. Klette. Digital geometry: The birth of a new discipline. In L. S. Davis, editor, *Foundations of Image Understanding*, pages 33–71. Kluwer, Boston, Massachusetts, 2001.

[114] R. Klette and A. Rosenfeld. *Digital Geometry: Geometric Methods for Digital Picture Analysis.* Morgan Kaufmann Publishers, San Francisco, 2004.

[115] R. Klette and A. Rosenfeld. *Digital Geometry: Geometric Methods for Digital Picture Analysis.* Morgan Kaufmann, San Francisco, 2004.

[116] R. Klette and A. Rosenfeld. Digital straightness: A review. *Discrete Applied Mathematics*, 139(1–3):197–230, 2004.

[117] R. Klette and J. Žunić. Interactions between number theory and image analysis. In L. J. Latecki, D. M. Mount, and A. Y. Wu, editors, *Proc. SPIE vol. 4117, Vision Geometry IX*, pages 210–221. 2000.

[118] T. Y. Kong and A. Rosenfeld. Digital topology: Introduction and survey. *Computer Vision Graphics, and Image Processing*, 48:357–393, 1989.

[119] J. Koplowitz, M. Lindenbaum, and A. Bruckstein. The number of digital straight lines on an $n \times n$ grid. *IEEE Trans. Information Theory*, 36:192–197, 1990.

[120] Kazuyoshi Korida, Hiroaki Nishino, and Kouichi Utsumiya. An interactive 3D interface for a virtual ceramic art work environment. In *VSMM '97: International Conference on Virtual Systems and MultiMedia*, page 227, Washington, DC, USA, 1997. IEEE Computer Society.

[121] V. A. Kovalevsky. Finite topology as applied to image analysis. *Computer Vision, Graphics, and Image Processing*, 46:141–161, 1989.

[122] V. A. Kovalevsky. New definition and fast recognition of digital straight segments and arcs. In *Proc. 10th Intl. Conf. Pattern Recognition (ICPR)*, IEEE CS Press, pages 31–34, 1990.

[123] E. Krusinska. A valuation of state of object based on weighted Mahalanobis distance. *Pattern Recognition*, 20:413–418, 1987.

[124] Z. Kulpa. A note on "circle generator for display devices." *Computer Graphics and Image Processing*, 9:102–103, January 1979.

[125] G. Kumar, N. K. Sharma, and P. Bhowmick. Wheel-throwing in digital space using number-theoretic approach. *International Journal of Arts and Technology*, 4:196–215, 2011.

[126] Gautam Kumar, Naveen Kumar Sharma, and Partha Bhowmick. Creating wheel-thrown potteries in digital space. In *Proc. International Conference on Arts and Technology (ArtsIT2009)*, volume (to appear) of *Lecture Notes of ICST (LNICST) series*. Springer, 2009.

[127] M. A. Kumar. Medial circle and medial sphere representation of binary images. PhD thesis, Dept. of Electronics and Electrical Communication Engg., IIT Kharagpur, India, August 1995.

[128] M. A. Kumar, J. Mukherjee, B. N. Chatterji, and P. P. Das. A geometric approach to obtain best octagonal distances. In *9th Scandinavian Conf. Image Processing*, pages 491–498, 1995.

[129] M. A. Kumar, J. Mukherjee, B. N. Chatterji, and P. P. Das. Representation of 2D and 3D binary images using medial circles and spheres. *Intl. Jrnl. Pattern Recognition and Artificial Intelligence*, 10:365–387, 1996.

[130] L. Lam, S. W. Lee, and C. Y. Suen. Thinning methodologies: A comprehensive survey. *IEEE trans. on Pattern Recognition and Machine Intelligence*, 14(9):869–885, Sept. 1992.

[131] Y. Langsam, M. J. Augenstein, and A. M. Tenenbaum. *Data Structurs using C and C++*. Prentice-Hall of India, New Delhi, 2000.

[132] J. Lee, G. Han, and S. Choi. Haptic pottery modeling using circular sector element method. In *6th Intl. Conference on Haptics: Perception, Devices and Scenarios*, volume 5024 of *LNCS*, pages 668–674, 2008.

[133] F. Leymarie and M.D. Levine. Fast raster scan distance propagation on the discrete rectangular lattice. *CVGIP: Image Under.*, 55(1):84–94, January 1992.

[134] M. Lindenbaum and J. Koplowitz. A new parametrization of digital straight lines. *IEEE Transactions on Pattern Analysis and Machine Intelligence*, 13(8):847–852, 1991.

[135] Y. Livnat, H.-W. Shen, and C. Johnson. A near optimal isosurface extraction algorithm using span space. *IEEE Trans. Visualization and Computer Graphics*, 2:73–84, 1996.

[136] W. E. Lorensen and H. E. Cline. Marching cubes: A high-resolution 3D surface construction algorithm. *Computer Graphics*, 21:163–169, 1987.

[137] M. DeHaemer Jr. and M. J. Zyda. Simplification of objects rendered by polygonal approximations. *Computers and Graphics*, 15(2):175–184, 1991.

[138] C. Maurer, R. Qi, and V. Raghavan. A linear time algorithm for computing the Euclidean distance transform in arbitrary dimensions. *IEEE Trans. on Pattern Recognition and Machine Intelligence*, 25(2):265–270, Feb. 2003.

[139] M. D. McIlroy. A note on discrete representation of lines. *AT&T Technical Journal*, 64(2), 1985.

[140] M. D. McIlroy. Best approximate circles on integer grids. *ACM Trans. Graphics*, 2(4):237–263, 1983.

[141] R. A. Melter and I. Tomescu. Path generated digital metrics. *Pattern Recognition Letters*, 1:151–154, 1983.

[142] J. A. V. Mieghem, H. I. Avi-Itzhak, and R. D. Melen. Straight line extraction using iterative total least squares methods. *J. Visual Commun. and Image Representation*, 6:59–68, 1995.

[143] Filippo Mignosi. On the number of factors of Sturmian words. *Theoretical Computer Science*, 82(1):71–84, 1991.

[144] F. Mokhtarian and F. Mohanna. Content-based video database retrieval through robust corner tracking. In *Proc. IEEE Workshop on Multimedia Signal Processing*, pages 224–228, 2002.

[145] W. Mokrzycki. Algorithms for discretization of algebraic spatial curves on homogeneous cubical graphs. *Computers & Graphics*, 12(3/4):477–487, 1988.

[146] J. Mukherejee, P. P. Das, and B. N. Chatterji. Thinning of 3-D images using safe point thinning algorithm (SPTA). *Pattern Recognition Letters*, 10:167–173, 1989.

[147] J. Mukherejee, P. P. Das, and B. N. Chatterji. On connectivity issues of ESPTA. *Pattern Recognition Letters*, 11(9):643–648, 1990.

[148] J. Mukherejee, P. P. Das, and B. N. Chatterji. Segmentation of three-dimensional surfaces. *Pattern Recognition Letters*, 11(3):215–223, 1990.

[149] J. Mukherejee, P. P. Das, and B. N. Chatterji. Segmentation of range images. *Pattern Recognition*, 25(10):1141–1156, 1992.

[150] J. Mukherjee. On approximating Euclidean metrics by weighted t-cost distances in arbitrary dimension. *Pattern Recognition Letters*, 32:824–831, 2011.

[151] J. Mukherjee, M. A. Kumar, B. N. Chatterji, and P. P. Das. Discrete shading of three-dimensional objects from medial axis transform. *Pattern Recognition Letter*, 20:1533–1544, 1999.

[152] J. Mukherjee, M. A. Kumar, P. P. Das, and B. N. Chatterji. Fast computation of cross-sections of 3d objects from their medial axis transforms. *Pattern Recognition Letter*, 21:605–613, 2000.

[153] J. Mukherjee, M. A. Kumar, P. P. Das, and B. N. Chatterji. Use of medial axis transforms for computing normals at boundary points. *Pattern Recognition Letter*, 23:1649–1656, 2002.

[154] J. Mukherjee, P. P. Das, M. A. Kumar, and B. N. Chatterji. On approximating Euclidean metrics by digital distances in 2d and 3d. *Pattern Recognition Letter*, 21:573–582, 2000.

[155] J. Mylopoulos and T. Pavlidis. On the topological properties of quantized spaces II: The notion of dimension. *J. Assoc. Comput. Mach.*, 18:247–254, 1971.

[156] N. J. Naccache and R. Shingal. SPTA: A proposed algorithm for thinning binary patterns. *IEEE Trans. Systems Man Cybernet.*, 14:409–418, 1984.

[157] N. Nakamura and K. Aizawa. Digital circles. *Computer Vision Graphics and Image Processing*, 26:242–255, 1984.

[158] T. S. Newman and H. Yi. A survey of the Marching Cubes algorithm. *Computers & Graphics*, 30:854–879, 2006.

[159] K. J. O'Connell. Object-adaptive vertex-based shape coding method. *IEEE Trans. Circuits Syst. Video Technol.*, 7:251–255, 1997.

[160] T. Pavlidis. *Structural Pattern Recognition*. Springer, New York, 1977.

[161] T. Pavlidis. Algorithms for shape analysis and waveforms. *IEEE Trans. PAMI*, 2:301–312, 1980.

[162] T. Pavlidis. *Algorithms for graphics and image processing*. Computer Science Press, Rockville, MD, 1982.

[163] J. C. Perez and E. Vidal. Optimum polygonal approximation of digitized curves. *Pattern Recognition Letters*, 15:743–750, 1994.

[164] B. T. Phong. Illumination for computer generated pictures. *Communications of ACM*, 18(6):311–317, 1975.

[165] M. L. V. Pitteway. Integer circles, etc.: Some further thoughts. *Computer Graphics and Image Processing*, 3:262–265, 1974.

[166] I. Povazan and L. Uher. The structure of digital straight line segments and Euclid's algorithm. In *Proc. Spring Conf. Computer Graphics*, pages 205–209, 1998.

[167] P. P. Chakrabarti, P. P. Das, and B. N. Chatterji. The t-cost-m-neighbor distance in digital geometry. *Journal of Geometry*, 42:42–58, 1991.

[168] S. Pratihar and P. Bhowmick. A thinning-free algorithm for straight edge detection in a gray-scale image. In *Proc. 7th Intl. Conf. Advances in Pattern Recognition (ICAPR 2009)*, pages 341–344. IEEE CS Press, 2009.

[169] F. P. Preparata and M. I. Shamos. *Computational Geometry: An Introduction*. Spring-Verlag, New York, 1985.

[170] I. Ragnemalm. The Euclidean distance transformation in arbitrary dimensions. *Pattern Recognition Letters*, 14:883–888, 1993.

[171] M. Ren, J. Yang, and H. Sun. Tracing boundary contours in a binary image. *Image and Vision Computing*, 20:125–131, 2002.

[172] J. P. Reveilles. Geometrie discrete, calcul en nombres entiers et algorithmique. PhD thesis, Univ. Louis Pasteur, Strasbourg, 1991.

[173] Jairo Rocha and Rafael Bernardino. Singularities and regularities on line pictures via symmetrical trapezoids. *IEEE Trans. PAMI*, 20(4):391–395, 1998.

[174] C. Ronse. A simple proof of Rosenfeld's characterization of digital straight line segments. *Recognition Pattern Letters*, 3:323–326, 1985.

[175] A. Rosenfeld. Connectivity in digital pictures. *J. Assoc. Comput. Mach.*, 17:146–160, 1970.

[176] A. Rosenfeld. Digital straight line segments. *IEEE Transactions on Computers*, 23:1264–1269, 1974.

[177] A. Rosenfeld. Digital topology. *American Mathematical Monthly*, 86(8):621–630, Oct. 1979.

[178] A. Rosenfeld and A.V. Kak. *Digital Picture Processing, Vol. II*. Academic Press, 1982.

[179] A. Rosenfeld and C. E. Kim. How a computer can tell whether a line is straight. *American Mathematical Monthly*, 89:230–235, 1982.

[180] A. Rosenfeld and R. Klette. Digital straightness. *Electronic Notes in Theoretical Computer Sc.*, 46, 2001. http://www.elsevier.nl/locate/-entcs/volume46.html.

[181] A. Rosenfeld and J. L. Pfaltz. Sequential operations in digital picture processing. *Journal of ACM*, 13:471–494, 1966.

[182] A. Rosenfeld and J. L. Pfaltz. Distance functions in digital pictures. *Pattern Recognition*, 1:33–61, 1968.

[183] P. L. Rosin. Techniques for assessing polygonal approximation of curves. *IEEE Trans. PAMI*, 19(6):659–666, 1997.

[184] P. L. Rosin and G. A. W. West. Non-parametric segmentation of curves into various representations. *IEEE Trans. PAMI*, 17:1140–1153, 1995.

[185] T. Saito and J. Toriwaki. New algorithms for Euclidean distance transformations of an n-dimensional digitised picture with applications. *Pattern Recognition*, 27(11):1551–1565, 1994.

[186] H. Samet. *Design and Analysis of Spatial Data Structures*. Addison-Wesley Pub. Co. Inc, 1990.

[187] D. Sarkar. A simple algorithm for detection of significant vertices for polygonal approximation of chain-coded curves. *Pattern Recognition Letters*, 14:959–964, 1993.

[188] K. Schröder and P. Laurent. Efficient polygon approximations for shape signatures. In *Proc. Intl. Conf. Image Processing (ICIP)*, IEEE CS Press, pages 811–814, 1999.

[189] H. Schulz. Polyhedral surface approximation of non-convex voxel sets through the modification of convex hulls. In *Proc. 12th Intl. Workshop on Combinatorial Image Analysis (IWCIA'08)*, pages 38–50. Springer-Verlag, Berlin, Heidelberg, 2008.

[190] H. Schulz. Polyhedral approximation and practical convex hull algorithm for certain classes of voxel sets. *Discrete Appl. Math.*, 157:3485–3493, 2009.

[191] G. M. Schuster and A. K. Katsaggelos. An optimal polygonal boundary encoding scheme in the rate distortion sense. *IEEE Trans. Circuits and Systems for Video Technology*, 7:13–26, 1998.

[192] Daniel Shanks. *Solved and Unsolved Problems in Number Theory*. AMS Chelsea Publishing, New York, 1993.

[193] K. Shimizu. Algorithm for generating a digital circle on a triangular grid. *Computer Graphics and Image Processing*, 15(4):401–402, 1981.

[194] P. Shirley and A. A. Tuchman. A polygonal approximation to direct scalar volume rendering. *ACM SIGGRAPH Computer Graphics*, 24(5):63–70, 1990.

[195] A. W. M. Smeulders and L. Dorst. Decomposition of discrete curves into piecewise segments in linear time. *Contemporary Math.*, 119:169–195, 1991.

[196] I. Sobel. Neighborhood coding of binary images for fast contour following and general binary array processing. *Computer Graphics and Image Processing*, 8:127–135, 1978.

[197] R. Stefanelli and A. Rosenfeld. Some parallel thinning algorithms for digital pictures. *J. Assoc. Comput. Mach.*, 18:255–264, 1971.

[198] P. Stelldinger and L. J. Latecki. 3d object digitization: Topology preserving reconstruction. In *Proc. ICPR*, volume 3, pages 693–696, 2006.

[199] P. Stelldinger, L. J. Latecki, and M. Siqueira. Topological equivalence between a 3d object and the reconstruction of its digital image. *IEEE Trans. PAMI*, 29:126–140, 2007.

[200] P. Stelldinger and R. Strand. Topology preserving digitization with FCC and BCC grids. In *Proc. 12th Intl. Workshop on Combinatorial Image Analysis (IWCIA '06)*, pages 226–240, 2006.

[201] I. Stojmenovic and R. Tosic. Digitization schemes and the recognition of digital straight lines, hyperplanes and flats in arbitrary dimensions. *Vision Geometry: Contemporary Mathematics Series*, 119:196–212, 1991.

[202] R. Strand and G. Borgefors. Distance transforms for three-dimensional grids with non-cubic voxels. *Computer Vision and Image Understanding*, 100:294–311, 2005.

[203] Y. Suenaga, T. Kamae, and T. Kobayashi. A high speed algorithm for the generation of straight lines and circular arcs. *IEEE Trans. Comput.*, 28:728–736, 1979.

[204] S. Svensson, G. Borgefors, and I. Nystrom. On reversible skeletonization using anchor-points from distance transforms. *Journal of Visual Communication and Image Representation*, 10:379–397, 1999.

[205] C.-H. Teh and R. T. Chin. On the detection of dominant points on digital curves. *IEEE Trans. PAMI*, 2(8):859–872, 1989.

[206] D. M. Tsai and M. F. Chen. Curve fitting approach for tangent angle and curvature measurements. *Pattern Recognition*, 27(5):699–711, 1994.

[207] Etsuko Ueda, Yoshio Matsumoto, and Tsukasa Ogasawara. Virtual clay modeling system using multi-viewpoint images. In *3DIM '05: Fifth International Conference on 3-D Digital Imaging and Modeling*, pages 134–141, 2005.

[208] J. A. Ventura and J. M. Chen. Segmentation of two-dimensional curve contours. *Pattern Recognition*, 25:1129–1140, 1992.

[209] K. Voss. Coding of digital straight lines by continued fractions. *Comput. Artif. Intelligence*, 10:75–80, 1991.

[210] K. Wall and P.-E. Danielsson. A fast sequential method for polygonal approximation of digitized curves. *Computer Vision, Graphics, and Image Processing*, 28:220–227, 1984.

[211] G. Weber, O. Kreylos, T. Ligocki, J. Shalf, H. Hagen, and B. Hamann. Extraction of crack-free isosurfaces from adaptive mesh refinement data. In *Proc. VisSym '01, Ascona, Switzerland*, pages 25–34, 2001.

[212] J. Wilhelms and A. van Gelder. Topological considerations in isosurface generation. *Computer Graphics*, 24:79–86, 1990.

[213] Elsbeth S. Woody. *Pottery on the Wheel.* Allworth Press, New York, 2008. Original publisher: Farrar, Straus and Giroux (1975).

[214] W. E. Wright. Parallelization of Bresenham's line and circle algorithms. *IEEE Computer Graphics and Applications*, 10(5):60–67, 1990.

[215] L. D. Wu. On the Freeman's conjecture about the chaincode of a line. In *Proceedings of the 5th International Conference on Pattern Recognition*, pages 32–34, 1980.

[216] L. D. Wu. On the chaincode of a line. *IEEE Transactions on Pattern Analysis and Machine Intelligence*, 4:347–353, 1982.

[217] L.-D. Wu. A piecewise linear approximation based on a statistical model. *IEEE Trans. PAMI*, 6:41–45, 1984.

[218] X. Wu and J. G. Rokne. Double-step incremental generation of lines and circles. *Computer Vision, Graphics, and Image Processing*, 37(3):331–344, 1987.

[219] D. M. Wuescher and K. L. Boyer. Robust contour decomposition using a constant curvature criterion. *IEEE Trans. PAMI*, 13(1):41–51, 1991.

[220] Y. Xie and Q. Ji. Effective line detection with error propagation. In *Proc. Intl. Conf. Image Processing (ICIP)*, IEEE CS Press, pages 181–184, 2001.

[221] M. Yamashita and T. Ibaraki. Distances defined by neighborhood sequences. *Pattern Recognition*, 19:237–246, 1986.

[222] C. Yao and J. G. Rokne. Hybrid scan-conversion of circles. *IEEE Trans. Visualization and Computer Graphics*, 1(4):311–318, 1995.

[223] P. Y. Yin. A new method for polygonal approximation using genetic algorithms. *Pattern Recognition Letters*, 19(11):1017–1026, 1998.

[224] P. Y. Yin. Ant colony search algorithms for optimal polygonal approximation of plane curves. *Pattern Recognition*, 36:1783–1797, 2003.

[225] P. Y. Yin. A discrete particle swarm algorithm for optimal polygonal approximation of digital curves. *J. Visual Comm. Image Representation*, 15(2):241–260, 2004.

Index

T - #0382 - 071024 - C43 - 234/156/14 - PB - 9780367380212 - Gloss Lamination